Chemistry and Technology of Soft Drinks and Fruit Juices

Chemistry and Technology of Soft Drinks and Fruit Juices

EDITED BY

Philip R. Ashurst

THIRD EDITION

WILEY Blackwell

This edition first published 2016 © 2016 by John Wiley & Sons, Ltd.

Registered Office
John Wiley & Sons, Ltd, The Atrium, Southern Gate, Chichester, West Sussex, PO19 8SQ, UK

Editorial Offices
9600 Garsington Road, Oxford, OX4 2DQ, UK
The Atrium, Southern Gate, Chichester, West Sussex, PO19 8SQ, UK
111 River Street, Hoboken, NJ 07030-5774, USA

For details of our global editorial offices, for customer services and for information about how to apply for permission to reuse the copyright material in this book please see our website at www.wiley.com/wiley-blackwell.

The right of the author to be identified as the author of this work has been asserted in accordance with the UK Copyright, Designs and Patents Act 1988.

All rights reserved. No part of this publication may be reproduced, stored in a retrieval system, or transmitted, in any form or by any means, electronic, mechanical, photocopying, recording or otherwise, except as permitted by the UK Copyright, Designs and Patents Act 1988, without the prior permission of the publisher.

Designations used by companies to distinguish their products are often claimed as trademarks. All brand names and product names used in this book are trade names, service marks, trademarks or registered trademarks of their respective owners. The publisher is not associated with any product or vendor mentioned in this book.

Limit of Liability/Disclaimer of Warranty: While the publisher and author(s) have used their best efforts in preparing this book, they make no representations or warranties with respect to the accuracy or completeness of the contents of this book and specifically disclaim any implied warranties of merchantability or fitness for a particular purpose. It is sold on the understanding that the publisher is not engaged in rendering professional services and neither the publisher nor the author shall be liable for damages arising herefrom. If professional advice or other expert assistance is required, the services of a competent professional should be sought.

Library of Congress Cataloging-in-Publication data applied for

ISBN: 9781444333817

A catalogue record for this book is available from the British Library.

Wiley also publishes its books in a variety of electronic formats. Some content that appears in print may not be available in electronic books.

Cover image: © azureforest/DigitalVision Vectors

Set in 9.5/13pt Meridien by SPi Global, Pondicherry, India
Printed and bound in Malaysia by Vivar Printing Sdn Bhd

1   2016

Contents

Contributors, xv

Preface, xvi

1 Introduction, 1
P.R. Ashurst
  1.1 Overview, 1
  1.2 Soft drinks, 1
    1.2.1 Ready-to-drink products, 2
    1.2.2 Concentrated soft drinks, 2
    1.2.3 Legislation, 3
    1.2.4 Product types, 4
    1.2.5 Development trends, 6
    1.2.6 Nutrition, 7
    1.2.7 New product trends, 8
  1.3 Fruit juices, 8
    1.3.1 Processing technology, 9
    1.3.2 Adulteration, 10
    1.3.3 Other processes, 12
    1.3.4 Nutrition, 12
  1.4 Packaging, 13
  1.5 Summary, 14
  References and further reading, 14

2 Trends in beverage markets, 15
E.C. Renfrew
  2.1 Introduction, 15
  2.2 Definitions, 15
  2.3 Beverage consumption trends, 16
    2.3.1 Bottled water, 17
    2.3.2 Carbonated soft drinks, 17
    2.3.3 100% juices, nectars and fruit drinks, 19
    2.3.4 Energy drinks, 19
    2.3.5 Ready-to-drink (RTD) tea and ready-to-drink coffee, 20
    2.3.6 Coffee, 20
    2.3.7 Tea, 21
    2.3.8 Beer, 21
    2.3.9 Wine, 22
    2.3.10 Milk and flavoured milks, 22

- 2.4 Consumption charts, 23
- 2.5 Regions and markets, 25
- 2.6 Market share charts, 26
- 2.7 Main drivers in consumption, 28
  - 2.7.1 The search for 'natural', 28
  - 2.7.2 Adult soft drinks, 29
  - 2.7.3 Protein drinks, 29
- 2.8 Conclusion, 29

3 Fruit and juice processing, 31
B. Taylor
- 3.1 Introduction, 31
- 3.2 Fruit types, 32
  - 3.2.1 Botanical aspects and classification of fruit types, 32
  - 3.2.2 Harvesting considerations for berry, citrus, pome, stone and exotic fruits, 35
- 3.3 Fruit types for processing, 36
  - 3.3.1 Pome fruits, 36
  - 3.3.2 Citrus fruits, 38
- 3.4 General comments on fruit juice processing, 39
  - 3.4.1 Processing of 'fleshy' fruits, 40
  - 3.4.2 The use of enzymes in fruit juice processing, 43
  - 3.4.3 Extraction of citrus juices, 46
- 3.5 Juice processing following extraction, 'cleaning' and clarification, 48
  - 3.5.1 Juice concentration by evaporation, 49
  - 3.5.2 Freeze concentration, 50
  - 3.5.3 Hyper- and ultrafiltration, 50
- 3.6 Volatile components, 51
  - 3.6.1 Spinning cone column, 52
  - 3.6.2 Composition of fruit juice volatiles, 53
- 3.7 Legislative concerns, 54
  - 3.7.1 European fruit juice and nectars directive and associated regulations, 54
  - 3.7.2 AIJN Guidelines, 56
  - 3.7.3 Labelling regulations and authenticity, 57
  - 3.7.4 Juice in the diet – 'five-a-day', 58
- 3.8 Quality issues, 58
  - 3.8.1 Absolute requirements, 58
- 3.9 In conclusion, 62
- References and further reading, 64

4 Water and the soft drinks industry, 65
T. Griffiths
- 4.1 Usage of water in the industry, 65
- 4.2 Sources of water, 66

       4.2.1   Water cycle, 66
       4.2.2   Surface water, 67
       4.2.3   Ground water, 67
  4.3 Quality standards relating to water, 68
       4.3.1   UK legislative standards, 68
       4.3.2   Internal and customer standards, 68
  4.4 Processing water, 69
       4.4.1   Required quality, 69
       4.4.2   Starting quality, 72
       4.4.3   Processing options, 75
  4.5 Analytical and microbiological testing of water, 83
       4.5.1   Chemical tests, 83
       4.5.2   Microbiological tests, 84
  4.6 Effluents, 84
       4.6.1   Potential contaminants of water waste, 84
       4.6.2   Use of 'grey' water, 85
       4.6.3   Clean-up and reuse of effluents, 85
Further reading, 87
References, 87

5 Other beverage ingredients, 88
B. Taylor
  5.1 Introduction, 88
  5.2 Factors influencing development of the industry, 88
  5.3 The move towards standardisation, 91
  5.4 The constituents of a soft drink, 94
  5.5 Water, 94
       5.5.1   Requirements, 94
       5.5.2   Quality of fresh water, 96
       5.5.3   Water hardness, 96
       5.5.4   Water treatment, 96
       5.5.5   Water impurities and their effect, 97
  5.6 Acidulents, 98
       5.6.1   Citric acid, 98
       5.6.2   Tartaric acid, 99
       5.6.3   Phosphoric acid, 100
       5.6.4   Lactic acid, 101
       5.6.5   Acetic acid, 101
       5.6.6   Malic acid, 101
       5.6.7   Fumaric acid, 101
       5.6.8   Ascorbic acid, 102
  5.7 Flavourings, 102
       5.7.1   Flavourings and legislation, 104
       5.7.2   Flavourings in beverage application, 106

        5.7.3   Water-miscible flavourings, 106
        5.7.4   Water-dispersible flavourings, 107
   5.8 Colours, 112
   5.9 Preservatives, 115
        5.9.1   Microorganisms and beverages, 116
        5.9.2   Sulphur dioxide, 117
        5.9.3   Benzoic acid and benzoates, 119
        5.9.4   Sorbic acid and sorbates, 119
   5.10 Other functional ingredients, 120
        5.10.1 Stabilisers, 120
        5.10.2 Saponins, 120
        5.10.3 Antioxidants, 121
        5.10.4 Calcium disodium EDTA, 121
   5.11 Food safety, 122
   5.12 Future trends, 123
   Further reading and references, 125

6 **Non-carbonated beverages, 126**
   P.R. Ashurst
   6.1 Introduction, 126
   6.2 Dilutable beverages, 127
        6.2.1   Overview, 127
        6.2.2   Nomenclature, 127
        6.2.3   Ingredients, 128
        6.2.4   Manufacturing operations, 137
        6.2.5   Filling and packaging, 139
        6.2.6   Product range, 140
   6.3 Ready-to-drink non-carbonated products, 140
        6.3.1   Overview, 140
        6.3.2   Formulations, 140
        6.3.3   Special problems, 140
        6.3.4   Manufacturing and packing, 141
        6.3.5   Packaging types, 142
   6.4 Fruit juices and nectars, 142
        6.4.1   Processing, 142
        6.4.2   Packaging, 144
   Further reading, 145

7 **Carbonated beverages, 146**
   D. Steen
   7.1 Introduction, 146
   7.2 Carbon dioxide, 147
   7.3 Carbon dioxide production, 148
        7.3.1   Fermentation, 148
        7.3.2   Direct combustion, 148

		7.3.3	Quality standards, 149

       7.3.3   Quality standards, 149
       7.3.4   Delivery to the customer, 149
       7.3.5   Precautions, 150
  7.4 Carbonation, 152
       7.4.1   Basic considerations, 152
       7.4.2   Carbonation measurement, 154
  7.5 Syrup preparation, 156
  7.6 De-aeration, 157
  7.7 Carbonators, 158
  7.8 Filling principles, 160
       7.8.1   Gravity filler, 161
       7.8.2   Counter-pressure filler, 163
       7.8.3   Other filler types, 167
       7.8.4   Clean-in-place systems, 169
  7.9 Process control, 171
  7.10 Future trends, 172
  Further reading, 173

8 Processing and packaging, 174
  R.A.W. Lea
  8.1 Introduction, 174
  8.2 Juice extraction, 174
  8.3 Blending, 175
       8.3.1   Batch blending, 176
       8.3.2   Flip-flop blending, 176
       8.3.3   Continuous blending, 176
  8.4 Processing, 177
       8.4.1   Flash pasteurisation, 177
       8.4.2   Hot filling, 178
       8.4.3   In-pack pasteurisation, 179
       8.4.4   Aseptic filling, 179
       8.4.5   Chilled distribution, 181
       8.4.6   Summary, 181
  8.5 Control of process plant, 181
  8.6 Factory layout and operation, 182
  8.7 Hazard Analysis Critical Control Points, 186
  8.8 Good manufacturing practice, 186
  8.9 Cleaning in place, 187
  8.10 Packaging, 188
  8.11 Conclusion, 191

9 Packaging materials, 192
  D. Rose
  9.1 Introduction, 192
  9.2 Commercial and technical considerations, 193

         9.2.1   General considerations, 193
         9.2.2   Packaging materials, 195
  9.3 Processing, 197
         9.3.1   Cold-filling, 197
         9.3.2   In-pack pasteurising, 197
         9.3.3   Hot-filling, 198
         9.3.4   Aseptic filling of bottles, 198
         9.3.5   Liquid nitrogen injection, 202
  9.4 Bottles, 202
         9.4.1   Glass, 202
         9.4.2   Polyethylene terephthalate, 203
         9.4.3   High-density polyethylene, 207
         9.4.4   Polypropylene, 207
         9.4.5   Polyvinyl chloride, 207
         9.4.6   Plastic properties, 208
  9.5 Closures, 209
         9.5.1   Metal roll-on or roll-on pilfer-proof closures, 209
         9.5.2   Vacuum seal closures, 210
         9.5.3   Plastic closures, 211
         9.5.4   Crown corks, 213
  9.6 Cans, 213
         9.6.1   Metal bottles, 218
         9.6.2   Plastic cans, 218
  9.7 Cartons, 218
  9.8 Flexible pouches, 221
  9.9 Multipacks, 222
  9.10 Secondary packaging, 223
  9.11 Pack decoration, 224
  9.12 Environmental considerations, 225
  9.13 Conclusions, 228
  Acknowledgements, 230

10 Analysis of soft drinks and fruit juices, 231
  D.A. Hammond
  10.1 Introduction, 231
  10.2 Laboratory accreditation, 234
  10.3 Sensory evaluation, 236
  10.4 Water, 237
  10.5 Sweeteners, 239
         10.5.1 Analysis of natural sweeteners, 240
         10.5.2 Analysis of high-intensity sweeteners, 245
  10.6 Preservatives, 249
         10.6.1 Benzoic and sorbic acids, 249
         10.6.2 Sulphur dioxide, 251
         10.6.3 Dimethyldicarbonate, 252

- 10.7 Acidulants, 252
- 10.8 Carbonation, 256
- 10.9 Miscellaneous additives, 257
  - 10.9.1 Caffeine, 257
  - 10.9.2 Quinine, 258
  - 10.9.3 Other additives, 258
  - 10.9.4 Fibre analysis, 259
  - 10.9.5 Herbal drinks, 260
  - 10.9.6 Osmolality, 261
- 10.10 Analysis of colours used in soft drinks, 261
  - 10.10.1 Assessment of colour, 263
  - 10.10.2 Synthetic colours, 265
  - 10.10.3 Natural pigments, 267
- 10.11 Vitamin analysis in soft drinks systems, 272
  - 10.11.1 Fat-soluble vitamins, 274
  - 10.11.2 Vitamin B class, 274
  - 10.11.3 Vitamin C, 275
  - 10.11.4 Vitamin analysis using immunological procedures, 275
- 10.12 Methods used to detect juice adulteration, 276
- 10.13 Methods used to assess the juice or fruit content of soft drinks, 280
- 10.14 Conclusions, 282
- References, 283

11 Microbiology of soft drinks and fruit juices, 290
P. Wareing
- 11.1 Introduction, 290
- 11.2 Composition of soft drinks and fruit juices in relation to spoilage, 291
- 11.3 Background microbiology – spoilage, 293
  - 11.3.1 Sources, 293
  - 11.3.2 Yeasts, 294
  - 11.3.3 Bacteria, 295
  - 11.3.4 Moulds, 297
- 11.4 Microbiological safety problems, 299
  - 11.4.1 Escherichia coli, 299
  - 11.4.2 Salmonella, 299
- 11.5 Preservation and control measures, 299
- 11.6 Sampling for microbial problems, 301
- 11.7 Identification schemes and interpretation, 301
  - 11.7.1 Sample isolation, 301
  - 11.7.2 Non-molecular methods, 302
  - 11.7.3 Molecular identification, 302
- 11.8 Brief spoilage case studies, 303

11.9   Conclusions, 304
References, 306
Further reading, 309

12 Functional drinks containing herbal extracts, 310
E.F. Shaw and S. Charters
- 12.1   History, 310
- 12.2   The extraction process, 313
    - 12.2.1   Extraction heritage, 314
- 12.3   An extraction operation, 320
    - 12.3.1   Raw materials, 321
    - 12.3.2   Extraction, 323
    - 12.3.3   Organic extracts, 329
    - 12.3.4   Extract costs, 329
- 12.4   Extract characteristics and their problems, 331
    - 12.4.1   Specifications, 331
    - 12.4.2   Stability, 331
    - 12.4.3   Hazing, 332
    - 12.4.4   Availability, 333
- 12.5   Incorporation of extracts in beverages, 333
    - 12.5.1   Fruit juice-based and fruit-flavoured drinks, 333
    - 12.5.2   Mineral-water based and flavoured water drinks, 334
    - 12.5.3   Carbonated and dilutable drinks, 334
    - 12.5.4   Energy and sports drinks, 334
    - 12.5.5   Regulatory issues, 335
- 12.6   Some commonly used herbs, 337
References, 354

13 Miscellaneous topics, 356
P.R. Ashurst and Q. Palmer
- 13.1   Introduction, 356
- 13.2   Nutrition, 356
    - 13.2.1   Nutritional components, 357
    - 13.2.2   Calculation and declaration of nutrition information, 360
- 13.3   Sports drinks, 363
    - 13.3.1   Definition and purpose, 363
    - 13.3.2   Physiological needs, 363
    - 13.3.3   The absorption of drinks, 365
    - 13.3.4   Formulation, 366
- 13.4   Niche drinks, 369
    - 13.4.1   Alcoholic-type drinks, 369
    - 13.4.2   Energy drinks, 370
    - 13.4.3   Functional drinks or nutraceuticals, 371
    - 13.4.4   Powder drinks, 372

13.5   Dispensed soft drinks and juices, 372
        13.5.1   Introduction, 372
        13.5.2   Pre-mix and post-mix compared, 373
        13.5.3   Equipment, 373
        13.5.4   Outlets, 375
        13.5.5   Hygiene, 375
        13.5.6   Post-mix syrup formulation, 376
        13.5.7   Post-mix syrup packaging, 377
13.6   Ingredient specifications, 378
        13.6.1   Why have specifications?, 378
        13.6.2   What a specification should include, 378
        13.6.3   Preparation of a specification, 378
        13.6.4   Supplier performance, 379
13.7   Complaints and enquiries, 380
        13.7.1   Complaints, 380
        13.7.2   Enquiries, 382
13.8   Health issues, 383
        13.8.1   Soft drinks and dental damage, 383
        13.8.2   Effect of colourings and preservatives, 386
        13.8.3   Obesity, 387
13.9   Alternative processing methods, 388
        13.9.1   Microwave pasteurisation technology, 388
        13.9.2   High-pressure processing, 393
        13.9.3   Irradiation, 395
References, 396

Index, 398

Contributors

Philip R. Ashurst
Dr. P R Ashurst and Associates, Ludlow, UK.

Stuart Charters
Bwlch Garneddog, Gwynedd, UK.

Tony Griffiths
Quality Systems Consultant, Chelmsford, UK.

David A. Hammond
Fruit Juice and Authenticity Expert, Wolverhampton, UK.

Robert A.W. Lea
GlaxoSmithKline, Weybridge, UK.

Quentin Palmer
Schweppes Europe, Watford, UK.

Esther C. Renfrew
Market Intelligence Director, Zenith International Ltd, Bath, UK.

David Rose
A Pkg Prf. Packaging Development Manager, Britvic Soft Drinks, Hemel Hempstead, UK.

Ellen F. Shaw
Flavex International Limited, Kingstone, UK.

David Steen
Casa Davann, Murcia, Spain.

Barry Taylor
Firmenich (UK) Ltd., Wellingborough, UK.

Peter Wareing
Principal Food Safety Advisor, Leatherhead Food Research, Leatherhead, UK.

Preface

The first edition of this book was published in 1998, and the second in 2005. Now, some ten year later, this third edition is published. Its aim remains that of providing an overview of the science and technology of soft drinks and fruit juice products, and of the industries that manufacture and support them. The book is written for students and graduates in food science, chemistry and microbiology, those who are working in the beverage industry or its supply chain, or simply for any reader who wishes to know more about the subject.

There can be few, if any countries, in the world that do not have a plant manufacturing these products, as they have a universal appeal to most consumers of all ages. Soft drinks are available in virtually every city, town and village on the planet, as well as in aircraft and ships. The range of flavours and packaging formats is remarkable, as producers are always attempting to stimulate new interest.

There is no single definition for soft drinks but, in general terms, they are essentially non-alcoholic beverages, excluding dairy products, tea and coffee. Fruit juices are, by definition, self-explanatory, although the term is often misused to cover products that contain a proportion of juice.

Soft drinks and fruit juices have, from time to time, been the subject of criticism by various sections of the health community. In the past, this has been mainly because of the risk of tooth erosion. There is now an issue of obesity in almost all developed countries, as well as associated diseases, such as type 2 diabetes. Sugar consumption is cited as a significant contributory factor.

At the time of writing, there is considerable pressure on beverage manufacturers to reduce, or even remove, the sugar content of products. The issue is complicated by the wrong assumption that is often made, that the sugars occurring naturally in fruit juices are added to the products. Further confusion also arises because, in the United Kingdom and some other countries, public health bodies promote fruit juices as part of a healthy diet. Beverages, however, are far from the only source of sugar/carbohydrate intake, and it is important to recognise that they continue to play an important part in hydration and enjoyment and, particularly for fruit juices, are part of a balanced and healthy intake of nutrients.

The fundamental composition of soft drinks changes little, although manufacturers continually strive for new flavours and ingredients. The main changes that occur relate mostly to processing and packaging innovations and, in consequence, this book has a similar format to earlier editions. The opening introductory chapter sets the scene and is followed by a review of beverage markets. This second chapter includes references to beverage markets other than soft drinks and fruit juices, which helps put the various product categories into perspective.

Chapter 3 is an updated review of fruit processing and fruit juices, while chapter 4 is a new inclusion, dealing with the most important, but most easily overlooked, common ingredient – water. The following two chapters are updates from the second edition and deal, respectively, with a comprehensive review of ingredients and non-carbonated soft drinks.

Chapters 7 and 8 are inclusions from the second edition and deal, respectively, with the particular issues of carbonated drinks and typical processing and packaging operations. Although the second edition included a chapter on packaging materials, the corresponding contribution in this edition is a completely new inclusion.

Another area where there is almost constant development is that of analytical chemistry. Consumers, manufacturers and enforcement bodies are all concerned to ensure that products are of consistent quality, have correct label information, are free from harmful substances and, in some cases, ingredients that may relate to particular dietary needs. In the case of juices, authenticity is also of great significance. Chapter 10 is a very comprehensive review of this subject.

Although soft drinks and fruit juices are very rarely the source of any serious food health issues, product spoilage as a result of yeast or fungal contamination is probably the most common source of complaint and concern. The important inclusion on microbiology is also a completely new contribution dealing with this topic.

Earlier reference was made to the constant search by manufacturers for new ingredients. Botanical extracts such as infusions have been sources of flavour and herbal remedies over many centuries, and these are increasingly featured as inclusions in a wide range of beverages. Label claims on the remedial effects of such inclusions are, in most countries, highly regulated, in order to avoid confusion with medicines. Consequently, unless used as sources of flavour (e.g. ginger), the amounts of such botanical inclusions are usually very limited, and often rely on their listing on a product label for effect. The newly written chapter on such components is a valuable addition on this subject.

The final chapter is an attempt to incorporate a wide range of miscellaneous topics that are likely to be of significance to readers.

What has been excluded? One obvious topic is that of legislation relating to these products. In the United Kingdom and EU countries, there is now little compositional limitation, apart from constraints on some additives, with reliance being placed on label information to enable consumer choice. It is this area of labelling, including product claims, that is now perhaps of particular significance. Legislation changes frequently, and varies from country to country, although general approaches are becoming commonplace across the globe. For readers for whom this topic is important, there are many sources of up-to-date information which may be consulted. For obvious reasons, a book such as this cannot offer the range and timeliness of such information.

Other topics excluded are those of nutrition and environmental matters and, again, there are many comprehensive sources of this information available to enable readers to obtain the information they require.

Readers will find, in some instances, that there is overlap between chapter contents, but these have been left in the manuscript because of the significance of the topics within the respective chapters. There are also likely to be other topics that readers have not found covered and, for these and other shortcomings, the editor takes full responsibility.

Because of continuous commercial and other pressures, it is becoming extremely difficult to find authors with the knowledge, expertise and the ability to take time out of their day job to contribute to a work such as this. Essentially, this is the reason why this third edition has taken so many years to produce. The editor is greatly indebted to all the authors, who are all acknowledged experts in their respective fields, for their valuable contributions.

It is to be hoped that the reader will find this third edition to be a worthy and useful successor to earlier works, and that it meets the needs of those seeking to learn more about this important industry.

Philip R. Ashurst

CHAPTER 1

Introduction

Philip R. Ashurst

Dr. P R Ashurst and Associates, Ludlow, UK

1.1 Overview

Fruit juices and soft drinks are available in essentially the same form almost anywhere in the world. From polar bases to the tropics, and from the largest developed nations to small and less developed countries, soft drinks and fruit juices are available in bottles, cans, laminated paper packs, pouches, cups and almost every other form of packaging known.

This chapter outlines what soft drinks are, describes the various types of products available and sets the scene for later chapters, which deal with the more specialised aspects of the chemistry and technology of these products.

1.2 Soft drinks

What are soft drinks? There is no single definition available, but it is generally accepted that they are sweetened, water-based beverages, usually with a balancing acidity. They are flavoured by the use of natural or artificial materials, are frequently coloured, and often contain an amount of fruit juice, fruit pulp or other natural ingredients. The predominant ingredient is water – often ignored and frequently maligned – and it should be considered that the primary function of soft drinks is hydration. The sweetness and other characteristics enhance the enjoyment of consumption and make the products more appealing to consumers. They are, in some respects, secondary, and yet have importance in the provision of energy and some of the minor essential nutrients needed to meet daily requirements.

It is generally accepted that the description of soft drinks excludes tea, coffee, dairy-based beverages and, until recently, alcohol. However, in many countries, the production of 'soft' drinks containing alcohol is growing. Many see this as an undesirable trend because, traditionally, the taste of alcoholic beverages has

---

Chemistry and Technology of Soft Drinks and Fruit Juices, Third Edition. Edited by Philip R. Ashurst.
© 2016 John Wiley & Sons, Ltd. Published 2016 by John Wiley & Sons, Ltd.

been associated with adulthood. The blurring of the edges between the markets and tastes for alcoholic drinks and soft drinks appears to facilitate an easy transition for children and young people to the consumption of alcohol. It should be noted that, in many soft drinks, small amounts of alcohol (less than 0.5% alcohol by volume (ABV)) may be present as a consequence of alcohol being used as a solvent for many flavourings. Small amounts of alcohol may also be present in fruit juices.

There are two basic types of soft drinks: the so-called ready-to-drink (RTD) products that dominate the world market and the concentrated, or dilute-to-taste, products that are still important in some markets. These include syrups and so-called squashes and cordials.

Whether RTD or dilutable, soft drinks characteristically contain water, a sweetener (usually a carbohydrate, although artificial sweeteners are increasingly important), an acid (citric or malic and phosphoric in colas are the most common), flavouring, colouring and preservatives. There is a large range of additional ingredients that can be used for various effects.

1.2.1 Ready-to-drink products

This sector accounts for the largest volume of soft drinks production, and is divided into products that are carbonated – that is, they contain carbon dioxide – and those that are not. Carbonated RTD soft drinks dominate the world market, and detailed consumption trends are discussed in Chapter 2 of this volume.

The market for carbonated soft drinks is dominated by two giant brands of cola drinks that, together with their associated brand names, account for just over half the world's consumption of such products.

Non-carbonated RTD beverages have shown some considerable growth in recent years, mainly because of the availability of aseptic packaging forms. Non-carbonated drinks that rely on chemical preservation, or hot-pack/in-pack pasteurisation, often suffer from a number of potential problems, including rapid deterioration of flavour and colour.

1.2.2 Concentrated soft drinks

Concentrated soft drinks became very important during the Second World War, and in the early years following that conflict. Many were based on concentrated orange juice, which was widely available as a nutritional supplement in the United Kingdom, packed in flat-walled medicine bottles.

The main markets for concentrated soft drinks developed mainly in the United Kingdom and its former empire. The products became universally known as 'squashes' or 'cordials', and became enshrined as such in UK food legislation in the 1960s.

Another very important development was the production of citrus comminutes. These were produced by mixing together, in appropriate proportions, the juice, peel components and essential oils of citrus fruits, and comminuting the mixture in a suitable mill. The resulting product delivered a more intense flavour and cloud than could be obtained from juice alone, and allowed the creation of 'whole fruit drinks', which have dominated the concentrates market in the United Kingdom over the past 40–50 years.

1.2.3 Legislation

It is not the intention of this chapter to cover legislation affecting soft drinks in any detail – not least, because it varies from country to country, and there is often a continuous variation of legislation within countries.

Legislation is, however, important from an historical perspective. For example, in the United Kingdom, the Soft Drinks Regulations 1964 (as amended) codified the products according to the way in which the industry was then organised, and set into law definitions not only of 'soft drinks', but also of many different product types, such as crushes, squashes and cordials. These names subsequently became generic household names in the United Kingdom and in many parts of the English-speaking world.

The above regulations were probably among the most proscriptive compositional statutes that existed for any food products in the United Kingdom, and for beverages anywhere in the world. As well as defining soft drinks, they laid down the requirement for minimum levels of sugars in certain product types, the maximum levels of saccharin (the only artificial sweetener then permitted) and the minimum levels of comminuted fruit and fruit juices that defined the best-known product categories. These regulations were eventually revoked in 1995.

The current trend is to move away from compositional legislation, to a much freer approach in which carbohydrates and other nutritional components can be used at will, and additives are taken from 'positive' lists of functional components. Other ingredients are frequently controlled by negative usage (i.e. they must not be present, or must not exceed closely defined limits).

This move to remove controls on formulations is now backed by informed labelling that contains increasing amounts of information for the consumer. This approach is now used widely throughout the world, with only relatively minor variations from country to country.

At the time of writing, the relevant European Union regulations (EU 1169/2011) require that a food which includes fruit juices and soft drinks must be labelled, and that labels must contain the following information:
1 the name of the food;
2 the list of ingredients;
3 any ingredient or processing aid listed in Annex II (of the regulations), or derived from a substance or product listed in Annex II causing allergies or intolerances that is used in the manufacture or preparation of the food and is still present in the finished product, even if in an altered form;
4 the quantity of certain ingredients or categories of ingredients (see below);
5 the net quantity of the food;

6 the date of minimum durability or the use-by date;
7 any special storage conditions and or conditions of use;
8 the name or business name and address of the food business referred to in Article 8(1);
9 the country of origin or place of provenance, where provided for in Article 26;
10 instructions for use where it would be difficult to make use of the food without such information;
11 In beverages containing more than 1.2% of alcohol by volume of alcohol, the actual alcoholic strength by volume;
12 a nutritional declaration.

It will be apparent that not all of the above will apply to fruit juices and soft drinks, but the declaration of the quantity of key ingredients (fruit or fruit juice in soft drinks) became law through earlier quantitative ingredient declaration regulations in Europe. Where artificial sweeteners and carbohydrates are used together, an appropriate statement is necessary. A warning about the product being a source of phenylalanine must be incorporated when aspartame is used as a sweetener.

Other additional regulations may also apply, and the above information should only be considered as a general guide. There is a wealth of additional information on this topic available on the internet. Readers requiring more specific information should consult appropriate authorities.

Because, in most countries, legislation is a rapidly changing sphere, it is essential for those formulating, producing and marketing soft drinks to update themselves regularly as regards the legislation of consumer countries, and to ensure label compliance.

1.2.4 Product types
1.2.4.1 Ready-to-drink products

Historically, soft drinks were refreshing beverages that copied or extended fruit juices. Fruit juices typically have around 10–12% naturally occurring sugars, mostly with a pleasant balancing acidity that varies from about 1% down to 0.1%. It is, therefore, not surprising that soft drinks were typically formulated to contain around 10–11% sugar content, with about 0.3–0.5% of added acid (usually citric acid). The simplest form of beverage contained such a mix of these basic nutritional components in water, with flavouring, colouring and chemical preservatives added as necessary.

With the addition of carbon dioxide to render the product 'sparkling', 'effervescent' or 'fizzy', the manufacturer had a lemonade or similar product. With the addition of fruit juice to a level of 5–10%, a pleasing effect of both taste and appearance could be achieved. Such products were typically described as 'fruit juice drinks', 'fruit drinks' or 'crushes' (a reserved description in the old UK regulations). Various other additions could be made, including vitamins and minerals, clouding agents and foaming agents, and plant extracts.

RTD beverages are mostly carbonated (i.e. contain carbon dioxide). This, as well as giving sensory characteristics, provides some antimicrobial effect, especially against yeasts and moulds. Carbon dioxide is effective against yeasts, because it tends to suppress the production of more CO_2 as a by-product of the fermentation of sucrose to ethanol. In addition, it deprives moulds of the oxygen that most require for growth. Good hygiene standards are the norm in most soft drink-bottling operations today, and it is possible to produce carbonated drinks without chemical preservatives, by flash-pasteurising the syrup before it is mixed with carbonated water. The risk of microbiological spoilage is then low but, where large containers are used, the risk is increased because of the potential for subsequent contamination as product is removed and the container ullaged.

Carbon dioxide levels vary widely, and are usually expressed as 'volumes of CO_2 gas' (i.e. the volume of carbon dioxide contained in solution in one volume of product). Lightly carbonated products will contain around 2.0–3.0 volumes of the gas, while moderate carbonation usually refers to about 3.5–4.0 volumes, and high carbonation levels are around 4.5–5.0 volumes. Large bottles that are likely to become partially full will be relatively highly carbonated, while mixer drinks contain among the highest carbonation levels, because the resultant mixture (e.g. gin and tonic) needs to have a satisfactory residual level of dissolved carbon dioxide.

RTD beverages are also produced in non-carbonated forms. The most popular current form of these is distributed in aseptic card/foil laminate packs, such as Tetra Pak or Combibloc. These drinks are typically unpreserved, and come in volumes of 200–330 ml.

An alternative form of non-carbonated beverage comes in form-fill-seal plastic containers, which are typically square or round section cups, with a foil or plastic laminate lid. Such products are difficult to produce to a quality that will satisfactorily compete with the shelf life of aseptic foil/laminate packs. Form-fill-seal containers leave their contents vulnerable to oxidative degradation, are especially at risk of mould spoilage, and require the use of preservatives. The packs are now increasingly produced in aseptic conditions and are free from preservatives.

Some manufacturers produce RTD products at drinking strength, but this is wasteful of plant and requires large-volume production tanks. The usual approach is to manufacture a syrup or concentrated form of the beverage, which is then diluted with water and carbonated as required. Alternatively, the syrup, which can be flash-pasteurised, can be dosed into bottles which are then topped up with water. This is known as the 'post-mix' method. Where the alternative ('pre-mix') method is employed, syrup and water are mixed in the correct proportions in special equipment, prior to bottle filling.

1.2.4.2 Dilutables

As indicated in Section 1.2.2, some markets are substantial consumers of concentrated soft drinks. These products are purchased in concentrate form by the consumer, who then adds water (which can be carbonated if required) to achieve the desired taste.

In the United Kingdom and many parts of the English-speaking world, these products are referred to generically as 'squashes' or 'cordials'. Chapter 6 of this volume covers this topic in detail.

Most concentrated beverages contain fruit juice or 'whole fruit', a term that refers to a comminuted form of citrus that includes components of juice, essential oils, peel (flavedo) and pith (albedo). Concentrated soft drinks are usually flash-pasteurised and chemically preserved, where permitted. Their dilutable form means that they are often held in partially filled bottles for significant lengths of time (often many weeks, or even months). They are thus extremely vulnerable to spoilage by micro-organisms.

Some manufacturers do produce unpreserved concentrates, but such products are invariably pasteurised in the bottle, and carry a warning that the contents should be refrigerated after opening and consumed within a short time-span (typically two weeks).

Concentrates are normally produced at their packed strength, flash-pasteurised and transferred immediately to their final packaging.

1.2.5 Development trends

Probably the most significant trend in soft drinks manufacture in recent years has been towards the use of non-calorific artificial sweeteners. The best known of these – saccharin – was used in soft drinks during and after the Second World War, when sugar was in short supply. Saccharin, in its soluble form, is about 450 times sweeter than sugar and can be a significant cost-reducer, but its sweetness is marred by a bitter taste to which many consumers are sensitive.

In more recent years, other artificial sweeteners have been developed, and it is now possible to produce soft drinks with almost all the characteristics of the sugar taste. Such products are almost free of any energy (calorific) content, and lack much of the cariogenic property for which many have criticised sugar-containing beverages. Increasingly, however, the acidity of soft drinks is considered equally, if not more, damaging for dental health.

Almost all soft drinks are now available in 'low calorie', 'diet' or 'light' formulations. These products have a low energy content, and may be cheaper to manufacture than the corresponding sugar-containing products. The issue of sugar consumption and its potential contribution to obesity and diabetes is increasingly under scrutiny. Considerable effort is being made by pressure groups to encourage manufacturers to reduce the sugar content of all products – and particularly, soft drinks.

Another, and perhaps more obvious, development area is the constant search for new flavours and unusual ingredients. There is currently a great interest in the use of various botanical extracts, such as guarana and ginseng, because of their implied qualities, but it is noteworthy that one of the oldest, and certainly the most successful, flavours – cola – was originally formulated with, and still contains, a natural vegetable extract of cola nut.

The third major area for development is that of soft drinks containing ingredients that enable some special nutritional or physiological claim to be made for the product. This will usually be an energy claim because soft drinks are an ideal vehicle for delivering carbohydrates, some in specially formulated mixtures, in a readily and rapidly assimilable form. Of the other nutrients that can be included, fruit juice, vitamins and minerals are the most common, but some products contain significant levels of protein or even fibre (as non-metabolisable carbohydrate).

1.2.6 Nutrition

The nutritional value of soft drinks is currently under scrutiny in many countries, because of their sugar content and its possible impact on health. That said, the value of soft drinks must not be understated, because they are an important vehicle for hydration. Soft drinks, depending on their formulation, may be absorbed more readily than water (because of their osmolality), can replace lost salts and energy quickly, and are rapidly thirst-quenching. Their balance of sweetness and acidity, coupled with pleasant flavours, makes them attractive to all ages of consumers. Products are specially formulated to meet the tastes, nutritional needs and physiological constraints of the whole population, from babies to geriatrics.

The claims that are legally permitted for soft drinks vary from country to country but, for the most part, are limited to nutritional claims concerning energy, proteins, vitamins and/or minerals. Any form of medicinal claim (i.e. curative or symptomatic relief) will almost always be excluded by corresponding medicines legislation. There is, nevertheless, a growing trend to include natural extracts in many soft drinks (e.g. ginseng or ginkgo), and then rely on the general understanding and folklore that surrounds such ingredients to impart the special values that have been attributed to them.

There are three main areas of particular nutritional significance for soft drinks. The first is energy. Some soft drinks are formulated to deliver a rapidly assimilated energy boost to the consumer. All carbohydrates are important sources of energy, but soft drinks generally contain soluble sugars, which are easy to administer. However, because high levels of sugars are often intensely sweet and even sickly, and leave a cloying sensation in the mouth, energy drinks are often formulated around glucose syrup. For a given solid carbohydrate content, this raw material is much less sweet than sucrose. Selection of the method of hydrolysis used for the corn starch as the starting point allows glucose syrup to be tailored, to some extent, to include mixed carbohydrates – that is, mono-, di-, tri- and oligosaccharides. Such blends are the basis of some very effective products which are used by athletes and those recovering from illness.

The second area of nutritional significance is that of so-called isotonic drinks, which are of equivalent osmolality to body fluids. They promote extremely rapid uptake of body salts and water, and are thus very important products for participants in sports, as well as others requiring almost instant hydration.

Third, soft drinks have been widely formulated to low-calorie forms, and these are now available for those who wish to enjoy such beverages and yet minimise their calorific intake.

Other nutritional benefits that are claimed by some producers include the delivery of essential vitamins and minerals, especially to children.

On the negative side, soft drinks have acquired a reputation for being an agent in the development of dental caries, as well as contributing to obesity and diabetes. The issue of dental caries is claimed to arise when sugar residues remain in the mouth, or when (especially) young children have an acidic drink almost constantly in their mouths. It is perhaps now accepted that the dental caries problem is related more to the misuse, or even abuse, of soft drinks, than to the effects of normal consumption of such products as part of a balanced diet.

1.2.7 New product trends

New product development is a constant activity for most soft drinks producers. For the most part, there are few really new products. Alternative flavours and different forms of packaging are widespread, and no doubt will continue to be so.

The development of specialised energy drinks or isotonic beverages is, perhaps, an example of a truly new product area. As new raw materials become available (for example, soluble whey protein), whole ranges of products are likely to be spawned.

A recent trend has been to incorporate alcohol into soft drinks. Depending on the level of addition, such products can no longer be classified as soft drinks when their alcohol content exceeds 1.2% ABV.

As indicated above, another area of interest is the reintroduction of botanical extracts into soft drinks. It is sometimes overlooked that one of the earliest widely available soft drinks was based on an extract of cola nut.

Packaging developments are likely to offer some exciting new opportunities in the future, and soft drinks are likely to remain at the forefront of product innovation in many countries.

1.3 Fruit juices

What is a fruit juice? Various definitions have been suggested, but the one used in the UK Fruit Juice and Fruit Nectars Regulations of 2013 is helpful, as it provides various specifications. These regulations implement EU Directive 2012/12/EU,

amending directive 2001/112/EC. Specifications are provided in schedules for fruit juice within the United Kingdom as follows:
1   Fruit juice (schedule 2);
2   Fruit juice from concentrate (schedule 3);
3   Concentrated fruit juice (schedule 4);
4   Water extracted fruit juice;
5   Dehydrated and powdered fruit juices.

An additional description refers to fruit nectars. The Schedules referred to above give more precise definitions of fruit juice.

Working from the above definitions, there are thus two principal juice product types, and it is these that dominate today's markets.

The majority of fruit juice as supplied to the consumer is made by reconstituting concentrated juice with water to a composition similar to that of the original state. However, since records are not usually kept of the exact quality of the original juice, such reconstitution normally relates to an agreed trade standard. Reconstituted juices are often packed in aseptic long-life containers, such as Tetra Pak or Combibloc.

In many countries, there is a growing market for fresh 'single-strength' juice, made by squeezing fruit, subjecting it to limited processing, and packaging and selling it within a cold chain distribution system. Such juice is usually referred to as 'not from concentrate', or direct juice, and it will have a shelf life that varies from 1–2 weeks to 2–3 months.

1.3.1 Processing technology

It is not intended to give a detailed description here of juice processing. The subject is covered in Chapter 3 of this volume, and in other volumes such as *Fruit Processing* (Arthey and Ashurst, 2001) or *Production and Packaging of Non-Carbonated Fruit Juices and Fruit Beverages* (Ashurst, 1995), which may help the reader who wishes to obtain more detail.

In general terms, fruits are collected, sorted and washed, and then subjected to a type of mechanical compression appropriate for the fruit concerned. Although there are general fruit presses that can be used for more than one fruit type, fruits such as citrus, pineapple and stone fruits are usually processed in specially designed equipment.

Some fruit types (e.g. pome fruits, such as apples and pears) require mechanical treatment (milling), coupled with a biochemical process (involving enzymes) to break down the cellular structure and obtain the best yields. It is possible to achieve almost total liquefaction by means of an appropriate enzyme cocktail.

Additionally, a diffusion or water extraction process can be used to obtain best yields from certain fruits.

If juice is to be sold as 'not from concentrate', it is usually screened and pasteurised immediately after pressing – an operation with two main objectives.

The first is to control the growth of spoilage micro-organisms that live on the fruit surface (mainly yeasts and moulds). The second is to destroy the pectolytic enzymes that occur naturally in fruit and would otherwise break down the cloudy nature of the juice. If, however, a clear juice is required (e.g. apple or raspberry), enzymes can be added to accelerate this natural process.

Juice for concentration is normally subjected to screening to remove cellular debris, and then fed to a one- or multi-stage evaporation process to remove most of the water and other volatile material. Evaporators today are highly efficient processing units. Up to nine stages are used, sometimes with thermal recompression to obtain maximum efficiency. Increasingly, evaporators also recover the volatile aromatic substances that are partly responsible for giving fruit juices their sensory characteristics. The re-addition of such volatiles is widely practised at the point when concentrates are reconstituted into single-strength juices, and the issue of whether this should be obligatory has been clarified. European Council Directive 2001/112/EC (the Fruit Juices Directive) makes the addition of such volatiles at reconstitution obligatory. The UK 2013 Regulations (which are based on this Directive) state that reconstituted fruit juice is the product 'obtained by replacing, in concentrated fruit juice, water extracted from that juice by concentration, and by restoring the flavours'.

After concentration, juices are normally held in storage until they are reconstituted. Some concentrated juices, particularly orange, require freezing at below −10°C for effective preservation. Others, particularly apple, can be held at around 10–15°C without risk of deterioration. The degree of concentration plays an important part in determining storage conditions. In the above examples, orange juice is normally concentrated to about 65° Brix and apple to 70° Brix.

An alternative method of storage is to hold juices under aseptic conditions, in drums or other containers. No particular temperature constraints then apply for microbial stability, but there is a substantially increased risk of colour browning and taste deterioration if juices are held aseptically at temperatures above about 10°C.

Some juices are held in sulphited conditions (e.g. 1500–2000 ppm sulphur dioxide), but this is suitable only for juices destined for uses other than reconstitution as fruit juice.

1.3.2 Adulteration

The adulteration of fruit juices has been widespread at times. As with any commodity, juice manufacturers, blenders, and those using juices as ingredients, can secure considerable financial benefit from adulterating fruit juice. It should be emphasised that food safety issues are not normally an issue in fruit juice adulteration. The issue is simply the fact that traders and consumers are being defrauded; an adulterated fruit juice sold as pure fruit juice is not as it has been labelled.

Although adulteration is becoming increasingly sophisticated, it is normally seen as falling into one of three types:
1 over-dilution of juices with water;
2 use of cheaper solid ingredients (particularly sugars);
3 blending of cheaper with more expensive juices.

The issue of too much water being added to juices has largely been addressed through the application of a minimum solids content (measured in degrees Brix). European Union and many other countries now have in place a minimum Brix value for various juices. These minima are backed either by legal statute or industry code of practice. They normally apply to juices prepared by adding water to concentrate, rather than to 'not from concentrate' products.

The second category of adulteration is perhaps the most common. For example, apple juice will normally contain around 11% by weight of solids. At least 90% of these solids are carbohydrates – sucrose, dextrose and fructose predominating. Considerably cheaper sources of carbohydrates can be found, and the simple addition of a mixture of carbohydrates, in roughly the same proportions as those found naturally in apple juice, can be used to 'stretch' apple juice by a considerable proportion. In more sophisticated forms of adulteration, the added components can be made to carry a similar 'signature' to the juice.

In the third category, a cheaper juice can be used to adulterate a more expensive one. For example, elderberry juice can be used to extend strawberry or raspberry juice.

The detection of adulteration and its quantification have spawned some elegant scientific techniques – some borrowed from other fields, and some developed specifically for use in fruit juice work.

Detection of over-dilution and the presence of sugars of other origin, is now carried out largely by measuring key isotope ratios (such as carbon 13 : 12 ratios, deuterium/hydrogen ratios and oxygen 18 : 16 ratios) and comparing them with both those found naturally in fruit and agreed international standards. An important part of the fight against adulteration has been the development of databases that examine fruit of different origins and season.

Another elegant method of detecting sugar addition in particular has been the use of high performance liquid chromatography (HPLC), to determine the presence of oligosaccharides that are characteristic of the added sugars, but not the fruit. The use of enzymic methods for determining the presence of specific components (e.g. d-malic acid, which does not occur naturally) is also helpful.

The analytical detection and measurement of fruit juice adulterants is a rapidly developing field, and the interested reader is directed to works dealing specifically with the subject, such as *Food Authentication* (Ashurst and Dennis, 1996) and *The Analytical Methods of Food Authentication* (Ashurst and Dennis, 1997).

Finally, the addition of cheaper juices to more expensive ones can usually be detected and measured using techniques appropriate for the likely components. For example, the addition of elderberry to strawberry juice can readily be detected by examining the anthocyanins present, using HPLC, and comparing them with standards.

1.3.3 Other processes

A number of other processes have become commonplace in the manufacture of fruit juices. For example, if oranges of the varieties Navel or Navellina are processed, the juice becomes unpleasantly bitter, because of the biochemical development of a glycoside, limonin. This substance can be partially or totally removed by the use of appropriate ion-exchange resins to yield a juice of acceptable taste.

There has also been a range of developments leading to the removal of acidity, colour and minerals from clear juices such as apple. The product of such a combination of processes can be a clear, colourless carbohydrate syrup that can be used in a variety of food processes. There seems little doubt that the legal status of such a product is not fruit juice – yet it is often, optimistically, so called.

Enzyme and finishing treatments are widely used in the processing of fruit juices to obtain products of particular specification.

Another contentious issue is the further processing of fruit pulp, especially citrus pulp. The addition of water to such a pulp can give an extract containing around 5% solids, which can be concentrated to around 65% and used to dilute more expensive pure juice. These products are normally described as pulp-wash extracts. Note, however, that 'in-line' pulp-wash extract arises in normal citrus processing, and is becoming an acceptable component, at levels of addition not exceeding 5% of the juice.

By-products of the juice industry are important, but are not dealt with here. The interested reader is referred to *Fruit Processing* (Arthey and Ashurst, 2001).

1.3.4 Nutrition

Fruit juice is important in human nutrition, far beyond its use as a refreshing source of liquid. Many fruits contain a variety of minor ingredients – particularly vitamins and minerals – as well as carbohydrates, which are the predominant solid component. Although fruit contains small amounts of proteins and fats, these are not important ingredients of juices.

Nutrients frequently consumed in sub-optimal concentrations by humans are proteins, calcium, iron, vitamin A, thiamin (vitamin B1), riboflavin (vitamin B2) and ascorbic acid (vitamin C). Some of these nutrients occur in higher concentrations in fruit juices than in other foods. There have been claims that ascorbic acid of natural origin is nutritionally superior to that of synthetic origin.

It has been established that the above phenomenon is caused by the presence of certain flavonoid compounds in fruit juice that influence blood circulation, increasing the permeability and elasticity of capillaries. This action is known as vitamin P activity, but the flavonoids showing this property are not classified as vitamins, because there are several substances that demonstrate this activity, and no serious deficiency diseases occur if they are not consumed. There are indications that these flavonoids have a useful protective action – in particular against some respiratory diseases – but they are readily decomposed in the body, and it is impossible to maintain an effective concentration in the blood.

Apart from the more obvious benefits of fruit juice, such as being a source of potassium, it contains other substances that have or are claimed to have useful pharmacological activity. For example, limonin and other related limonoid substances present in citrus fruit are believed by some to have a role in inhibiting certain forms of cancer. Sorbitol, which occurs in many fruit juices, has a laxative effect.

Several components with antioxidant activity are found in fruit juices. These include ascorbic acid, tocopherols (vitamin E), beta-carotene and flavonoids. Beta-carotene has antioxidant activity, which can quench the singlet oxygen that can induce pre-cancerous cellular changes.

Whatever the nutritional interest, it should be noted that changes occur during storage, particularly to the minor components of juices, and particularly under adverse conditions (e.g. light, increasing temperature, time).

1.4 Packaging

Later chapters in this volume deal specifically with packaging. However, it is perhaps useful to look briefly at the trends in packaging that are important in the whole area of beverage development.

Traditionally, most beverages were packed in glass. Glass has many attractive features – not least that it is an excellent protective medium – but its overriding disadvantages are its weight and its brittleness. Despite this, high volumes of soft drinks and juices are still packaged in glass, some of it multi-trip packaging.

The development of the board-polymer-aluminium package used to form in-line boxes, which are packed aseptically, has been perhaps the outstanding packaging development for beverages. The pack provides an almost ideal combination of protection, minimal weight and economic size.

Another important packaging development area is plastic. Various plastics have been, and continue to be, used: for example, high- and low-density polyethylene (HDPE, LDPE), polyvinyl chloride (PVC), polystyrene (PS) and various barrier plastics. These can be formed into bottles of conventional shape, or fed into machines producing form-fill-seal packages, typically cups.

By far the most important plastic is polyethylene terephthalate (PET). Bottles of this material are formed in a two-stage process. So-called pre-forms are made by injection moulding and, in a second process, are then stretch-blow-moulded to produce a bottle. PET has properties surprisingly like those of glass, but it does not have the same disadvantages of weight and brittleness. Processing systems that pack small PET containers aseptically are now in regular use.

Developments are yet to feature in fruit juice packaging. PET can be laminated with other plastics, such as nylon and ethylene vinyl alcohol (EVOH), to give extremely good barrier properties, and polyethylene naphthalate (PEN) may enable production of a plastic bottle that can be pasteurised at high temperatures, although aseptic processing and packaging is often preferred.

1.5 Summary

Soft drinks and fruit juices are widely consumed in ever-increasing quantities, and are very important commodities in the trade of most countries. This volume sets out to introduce the reader with a good general science background to the more detailed aspects of these products, and it is hoped by this means to provide a useful reference work that will be widely used by those wishing to learn more about these products.

References and further reading

Arthey, D. and Ashurst, P.R. (eds, 2001). *Fruit Processing*, 2nd edition. Aspen Publishers Inc., Gaithersburg, MD.

Ashurst, P.R. (ed, 1995). *Production and Packaging of Non-carbonated Fruit Juices and Fruit Beverages*, 2nd edition. Blackie Academic & Professional, Chapman & Hall, London.

Ashurst, P.R. and Dennis, M.J. (eds, 1996). *Food Authentication*. Blackie Academic & Professional, Chapman & Hall, London.

Ashurst, P.R. and Dennis, M.J. (eds, 1997). *The Analytical Methods of Food Authentication*. Blackie Academic & Professional, Chapman & Hall, London.

European Communities Council Directive 2001/112/EC.

EU Regulation 1169/2011.

UK Fruit Juices and Fruit Nectars Regulations 2013, Statutory Instrument 2013 No. 2775, HMSO, London.

UK Soft Drinks Regulations 1964 (as amended), Statutory Instrument No. 760, HMSO, London.

CHAPTER 2
Trends in beverage markets

Esther C. Renfrew

Market Intelligence Director, Zenith International Ltd, Bath, UK

2.1 Introduction

This chapter describes the context of the global beverage industry. It begins with definitions of the beverage categories, and then considers the latest consumption trends by sector. An analysis of the world's most influential beverage markets is followed by a discussion of the fastest-growing regions and countries. This chapter examines the soft drinks sector across the world and, particularly, those segments that are leading the way for growth. Consumer trends are then explored, followed by a summary of the future growth outlook, and those market developments that should transform the beverage industry.

2.2 Definitions

There are four primary sectors of the global beverage market, as shown in Figure 2.1. The first is hot drinks, such as tea, coffee and hot, malt-based products. The second is milk drinks, including both white drinking milk and flavoured milk beverages. The third is soft drinks, which are subdivided into bottled water; carbonated soft drinks; dilutables (also known as squash, and including powders, cordials and syrups); 100% fruit juice and nectars (with 25–99% juice content); still drinks including ready-to-drink (RTD) teas; sports drinks; and other non-carbonated products with less than 25% fruit juice. The fourth category is alcoholic drinks, including beer, wine, spirits, cider, sake, and flavoured alcoholic beverages, sometimes referred to as pre-mixed spirits or alcopops.

Chemistry and Technology of Soft Drinks and Fruit Juices, Third Edition. Edited by Philip R. Ashurst.
© 2016 John Wiley & Sons, Ltd. Published 2016 by John Wiley & Sons, Ltd.

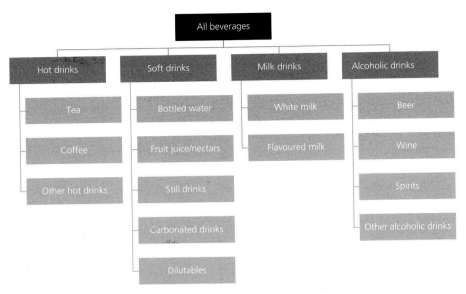

Figure 2.1 Beverage sectors and segments. *Source*: Zenith International.

2.3 Beverage consumption trends

Despite the recent worldwide economic downturn, overall consumption increased by an average annual rate of 5% between 2009 and 2013. This equates to an additional 210 billion litres consumed, rising from 1,476.8 billion litres in 2009 to 1,686.9 billion litres in 2013.

Per capita consumption stood at 325 litres in 2013, up from 294 litres in 2009, meaning that each person, on average, consumed 31 litres more in 2013, compared with 2009.

Of the main beverage sectors, energy drinks, flavoured water and plain bottled water were the three fastest-growing amongst all other types, with double-digit growth. Of the three, bottled water added the most increase in absolute volume terms, with an additional 70 billion litres consumed between 2009 and 2013.

Tea commands the highest percentage volume share, with 23% in 2013. This has not changed much since 2009 and is, in fact, expected to decrease slightly during the forecast period. The sector that has gained most is water. In 2009, its share stood at 14% and, in 2013, this grew to 17%. By 2018, its share is expected to rise by another two percentage points to 19%, equivalent to another 80 billion litres of water consumed between 2014 and 2018.

In *per capita* terms, tea is the most consumed beverage, standing at 75 litres per person in 2013. This is expected to increase to 82 litres per person by the end of 2018. In second place is bottled water, rising from 43 litres *per capita* in 2009 to 71 litres per person in 2018. Carbonates, which were ahead in volume and *per capita* terms, compared with bottled water, until 2006, are expected to remain around the 38–40 litres *per capita* over the period forecasted to 2018.

2.3.1 Bottled water

Bottled water is now the clear leader in the global drinks industry, both in volume and in the rate of growth, with 282.7 billion litres accounting for 17% of the total soft drinks volume in 2013. The booming global demand for bottled water is expected to drive an average annual growth of 6.1% until 2018, with the category delivering over 34% of the projected volume growth in soft drinks. However, profitability in this category is a real challenge. The global multinationals that operate in bottled water, such as Nestlé, Danone, The Coca-Cola Company and PepsiCo, are facing growing competition from smaller regional and private-label competitors that operate on lower margins.

In the high-volume growth markets of China, India and Brazil, the bottled water markets are each dominated by local brands, with the multinationals still struggling to gain share. Conversely, in Indonesia, Danone dominates the landscape with its Aqua brand (predominately in HOD), while FEMSA and Nestlé dominate the market in Mexico.

In the mature markets of West Europe and North America, bottled water volume grew by an average annual rate of 1.5% and 4.9% respectively between 2009 and 2013.

Enhanced bottled water, such as flavoured and flavoured-functional waters, should see continued growth, particularly in Europe where, predominately in Germany and Italy, sparkling water remains a more popular option than still water, and is favoured as a healthier alternative to carbonated soft drinks. Flavoured waters will continue to grow in South and Central America, especially in countries like Mexico, as more consumers are choosing this product rather than soft drinks with high-calorie content. Functional waters are expected to grow moderately in North America and Asia Pacific.

The popularity of bottled water can be explained by the simple concept of 'convenient hydration'. It is handy and portable and, while there has been a backlash in some countries relating to price and packaging, it is a necessity in countries where tap water quality is poor and/or inconsistent. Bottled water overtook carbonated soft drinks volumes in 2006, has been increasing year on year, and is expected to continue on this upward trajectory.

2.3.2 Carbonated soft drinks

In 2013, sales of carbonated soft drinks reached an annual volume of 196 billion litres and represented 12% of the global drinks volume. Despite the issue of sugar content surrounding carbonated soft drinks – in particular linked with the obesity epidemic – carbonated soft drinks have still managed to achieve an average annual growth rate of 2.6%. The USA was the biggest market in volume, with just under 51 billion litres consumed in 2013, though volumes have been declining each year over the review period. Conversely, consumption in Mexico and China, the next two biggest consumers of carbonated soft drinks, has increased by 2.5% and 8.1%, respectively, per year since 2009. These growth rates will be less impressive going forward, but volumes in China and Mexico are approximately 40–45% those of the USA in 2013.

With the 2014 FIFA World Cup and the 2016 Summer Olympics, The Coca-Cola Company has made a significant investment in the immediate consumption channel in Brazil, which will help to sustain growth. Like Mexico, *per capita* consumption of carbonated soft drinks in Brazil is high, though growth is limited somewhat as the market reaches maturity and consumer concerns over sugar in soft drinks increase.

Emerging markets, such as Nigeria, India and Iraq, are expected to be among the fastest-growing markets, with an annual CAGR through to 2018 in the range of 9–12% a year. Success in these dynamic markets requires significant capital investments and a long-term commitment to growing *per capita* consumption. In November 2013, PepsiCo announced a US$5.5 billion investment in its India operations through until 2020 – a commitment that slightly exceeded a similar promise made a year earlier in 2012 by the Chairman of Coca-Cola, Muhtar Kent, to invest US$5 billion in India through until 2020. He went on to say that Indian consumers, at the time, consumed only 12 eight-ounce (236 ml) bottles of Coca-Cola per year, compared with 230 bottles in Brazil and 92 bottles globally.

In Mexico, in one of the world's largest soft drinks market, the federal government recently passed a tax on sugared soft drinks equal to one peso (£0.05) per litre. Increasingly, local state and federal governments in developed markets are becoming more active in restricting or taxing the consumption of sugared soft drinks. This measure is likely to result in a decline in carbonated soft drinks volume.

In China, the negative health image of carbonated soft drinks has been rising among consumers, leading to muted growth in this category. The market is expected to grow less than 5% a year until 2018. Market volume stood at about 22 billion litres in 2013, accounting for 5.6% of consumption in China. Competition between Coca-Cola and Tingyi-PepsiCo is intensifying, especially in the light of Coca-Cola's announcement that it will invest US$4 billion in new bottling and distribution facilities between 2015 and 2017. Together, Coca-Cola and Tingyi-PepsiCo account for over 90% of the market in China.

In the mature markets of Western Europe and North America, both Coca-Cola and PepsiCo are struggling to find any growth in carbonated soft drinks, as consumers shift to healthier alternatives, such as bottled water and ready-to-drink teas. In the USA, carbonated soft drinks are expected to decline by an average of 0.3% a year through to 2018, and across Western Europe, carbonated soft drinks are projected to grow by an average rate of 1.4% during the same period.

2.3.3 100% juices, nectars and fruit drinks

These three categories registered average annual growth rates of 2.2%, 4.1% and 8.8% during 2009–13 and, together, accounted for 4% of the global beverages market. The volume consumption in 2013 for these categories stood at 23 billion litres for 100% fruit juices, 14 billion litres for nectars and 31 billion litres for fruit drinks. The majority of the consumption is from China (19 billion litres), followed by the USA (12 billion litres).

The largest producers of fruit are China, Brazil and India. China has a significant production of apples and, meanwhile, Brazil is focused on oranges and India is a major growing area for mangoes. The freshly pressed (also known as 'not from concentrate') juice market has always been an attractive alternative to processed (reconstituted from concentrate) juice. However, during the economic downturn, consumers traded from 100% juice and nectars to juice drinks, which commanded a lower price. As urbanisation and disposable income rises in emerging markets, the trend is that the juice category will also grow.

In China, juice drinks are by far the largest sub-group in the juice category, with 14 billion litres consumed in 2013, as they offer more acceptable tastes and varieties to the local consumer. China's *per capita* consumption of fruit drinks reached 10.4 litres in 2013 – a 3% growth over 2012. It is expected that growth will slow down, given the already high *per capita* levels. With slower growth in soft drinks in this country, major producers are actively developing innovative ideas and flavours in this category, which is expected to continue in the years to come. While Coca-Cola continues to lead the market share of juice drinks, Tingyi-PepsiCo is gaining momentum, with Tingyi continuously launching new products of fruit mix drinks featuring traditional Chinese health-preserving concepts. These have strong appeal among Chinese consumers.

In markets such as Egypt, Indonesia, Mexico, Iran, Nigeria, the Philippines, South Korea and Vietnam, fruit juice consumption is growing faster than in the BRIC (Brazil, Russia, India and China) countries, albeit from a lower base.

2.3.4 Energy drinks

Energy drinks is a segment of the market that cannot be ignored. It has experienced high growth, which is forecast to continue. Added to this, there has been a significant amount of innovation in this segment – so much so, that energy drinks are continuing to dominate the functional drinks category. The energy drinks market globally was over 5.6 billion litres in 2013, with an estimated value of US$40 billion and double-digit growth in both volume and value. This is a category which is evolving after an initial explosion on the beverages market, and which is continuing to grow and to innovate. One of the key emerging themes in the category is natural energy, and increased use of plant extracts such those from ginseng, guarana, tea and kola nut. Caffeine-free energy formulations are also increasingly being sought after by consumers in this category. Finally, on the theme of segmental blurring, there are a number of 'plus energy' segments emerging – for example, water plus energy and juices plus energy, among others.

2.3.5 Ready-to-drink (RTD) tea and ready-to-drink coffee

This is a dynamic category, and one which is continuing to innovate and strengthen on a global level, with ongoing and sustained focus on both the RTD tea and RTD coffee markets.

Zenith estimates that the combined market for RTD tea and coffee was approximately 27 billion litres globally in 2013, although this is not a market that is seeing huge year-on-year growth. The segment is starting to experiment with the use of a wide range of flavours and the use of herbs and spices, for example including hibiscus blossom, cherry, sweet vanilla and white teas. There has also been a surge in products available on the western markets that use green tea, which is high in anti-oxidants.

The consumer driver of convenience is also key to this segment, and the fact that this is being combined with functional properties, such as the anti-oxidant content of green teas, is important to growth. These products are now often seen as alternatives to carbonated soft drinks and bottled waters.

Zenith believes that the market for fortified and functional ready-to-drink tea products will increase in the future, and that the segment has potential for increased innovation.

2.3.6 Coffee

In 2013, over US$10.5 billion of new capital investment occurred in the coffee market across the globe. The high level of investment interest in coffee comes in the wake of a couple of years of lower coffee prices. Global coffee output, according to Rabobank, was expected to reach 140 million 60 kg bags in 2013/2014 and, with a bumper crop of 60 million bags expected from Brazil, the current surplus was forecast to last well into 2014.

In developed markets, innovation in new single-cup offerings, from the likes of Mondelez, DE Master Blenders and Green Mountain Coffee Roasters, is expected to drive demand in North America by 5%, and in Western Europe by just below 1%.

There is an intense level of competition among the coffee players in the developed markets, with innovation helping to expand and modify their overall offerings and impacting channels. In 2014, there was increased promotional activity in the coffee aisle in North America was expected and, with the outlook for lower prices, demand for coffee is expected to remain strong in the years ahead.

European coffee players are focusing on expanding into new markets such as Brazil. Instant coffee is also becoming 'premiumised' by combining traditional instant with higher quality micro-ground coffee, to create a 'hybrid' instant and fresh coffee that does not compromise taste and quality.

2.3.7 Tea

The global tea market was 387 billion litres in 2013, with nearly half of the volume being black tea, and about a quarter green tea. The marketplace for tea is highly fragmented and is relatively static, in comparison to global coffee or soft drinks markets. Within this category, there has been limited merger and acquisition activity, and industry consolidation has been slow. The top ten global brands represent less than a third of the total tea market, compared to coffee, in which the top ten brands represent over half of the market. The low level of globalisation in the tea category suggests that there is still ample room for large tea companies to expand across international markets.

Currently, the global market leader is Unilever, with its brand Lipton, followed by Tata Global Beverages, with its brands Tata Tea and Tetley, and American British Foods, with Twining's. These three companies are the only truly global competitors in tea and, while they remain strong in mature markets across Western Europe and North America, their challenge is to continue growing organically and to penetrate the important tea markets of China, Russia and Japan. China is forecasted to account for half of the global growth through to 2018, and Unilever's share of the tea market in this country is less than 2%, with the other two players having little presence there. Success for global tea companies in China will require investment through acquisitions, and joint ventures with local players or new product developments, to satisfy Chinese demand for local products.

2.3.8 Beer

In recent years, the global beer market has shown steady growth rates, with volumes reaching almost 180 billion litres in 2013. Volumes in mature markets have declined, but the decline has been more than offset by growth in emerging markets – in particular, in Brazil, India and China. These countries accounted for 89% of the absolute volume growth in litres over the past five years, with China being by far the largest contributor to beer volume growth.

For Russia, beer consumption has been affected by the government's measures since 2007 to reduce alcohol consumption. In the longer term, Russia and Brazil are fairly mature markets, with *per capita* consumption resembling levels in North America. Brazil's volumes were helped by the FIFA World Cup in 2014. Levels in China are much lower, but also above the global average.

The greatest potential for volume increases in the long run can be found in the Indian market, despite the existing asymmetrical taxation structure, which shifts the alcoholic mix towards spirits rather than beer. While India will maintain one of the highest growth rates, frequent tax increases, changes in bottle recycling policy in Maharashtra and early rains have had an impact on current trading, and the current growth has fallen to single digits. Growth was expected to recover slightly in 2014, helped by increased spending during elections, and improved prospects in rural India, due to a predicted bumper harvest following a timely monsoon.

Growth of beer consumption in countries such as Nigeria, the Philippines, Turkey, Mexico and Vietnam is expected, although not without risks, such as political unrest, inflation and government regulations impacting the beer sector. Nigeria is expected to be one of the major contributors to growth, although high inflation is hampering prospects somewhat. Other countries, such as Iran, Indonesia, Pakistan and Bangladesh, have limited forecast growth rates, mainly due to religious factors.

In mature markets, beer volumes have grown, though its share of the total beverage market has declined due to the increase in bottled water consumption (due to health concerns) and other alcoholic beverages (due to more indulgent propositions). As the economic crisis in 2008 took hold, beer volumes declined. With recovery in consumer confidence, beer volumes have improved and, in particular, craft beers have shown to be successful not only in the USA, but also in Europe, where beer is highly priced and commands higher margins.

2.3.9 Wine

While wine consumption, in more mature markets, struggled during the economic downturn, consumption in developing markets grew. In China, premium and high-end quality wines boomed, in particular linked to wine produced in 2008. The boom was due to the Olympics being held in China in that year, and the number 'eight', which is seen as lucky due to sounding very similar to the Chinese word for 'prosper' or 'health'. However, more recently, with the slowdown in the Chinese economy, paired with the political leadership transition in 2012 and the corresponding crackdown on lavish and conspicuous consumption, the result has been a significant reduction in the growth rate in the Chinese wine market. Competition among suppliers has also intensified, as companies looking to enter the market have found themselves competing for fewer opportunities.

2.3.10 Milk and flavoured milks

Around 44% of the world's population drinks milk every day. It is perceived as a healthy, nutritious and affordable food staple. Global *per capita* consumption of milk is approaching 40 litres, and is growing at around 1.6% a year. Asia is the most important consuming region, with 45% share of the total world consumption, followed by Europe (22%) and North America (14%). That said, Europe and Oceania have the highest liquid milk consumption per person in the world, and Asia and Africa the lowest. On average, developed economies *per capita* consumption is around 71 litres, and is growing by as much as 0.4% annually. In developing economies, *per capita* consumption stands at around 31 litres, up nearly 3.5%.

China has one of the lowest levels of *per capita* milk consumption worldwide. In recent years, the strongest growth has occurred in Asia Pacific, Africa and South America – respectively, 28%, 12% and 7% since 2009. Milk and drinking milk products are key drivers of dairy growth in emerging markets, especially in countries like China. Product quality and ongoing product innovations are expected to be the main means by which leading manufacturers gain share.

For consumers unwilling to compromise on taste, health or convenience, flavoured milk is proving to be an ever more popular alternative to other beverages. Although consumption of flavoured milk globally remains low, relative to white milk, carbonated soft drinks and juices, demand for flavoured milk is rising. It is expected to enjoy compound annual growth rates of at least double that of white milk, and more than triple that of carbonated soft drinks, over the coming years. While flavoured milk has traditionally been consumed by children, there are opportunities going forwards to significantly broaden and deepen its appeal. Whether it is a breakfast replacement, a post-workout drink or an indulgent drink, flavoured milk is enhancing the consumer experience of liquid dairy products. It is offering variety, versatility and adding value.

White milk consumption is expected to grow from around 191 billion litres in 2013, to just over 215 billion litres by 2018, with flavoured milk expected to grow by 26% worldwide between the same period, from 12.4 billion litres to 15.7 billion litres, driven by developing countries – China, India, Indonesia and Brazil.

2.4 Consumption charts

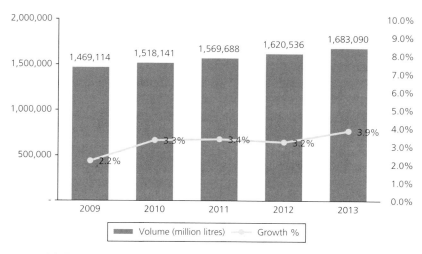

Figure 2.2 Global beverage consumption, 2009–2013.

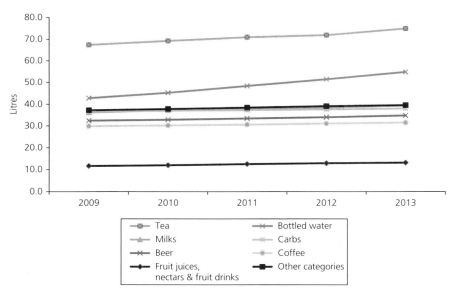

Figure 2.3 Global beverage consumption per person, 2009–2013.

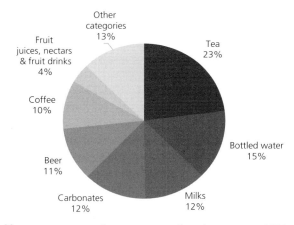

Figure 2.4 Global beverage consumption percentage share by category, 2009.

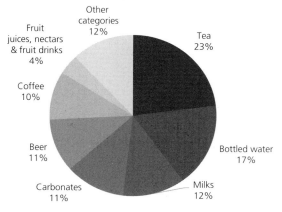

Figure 2.5 Global beverage consumption percentage share by category, 2013.

2.5 Regions and markets

In terms of global volume consumption, the top six categories are tea, bottled water, milk, carbonates (excluding sparkling water, energy drinks and flavoured/functional water), beer and coffee. Except for carbonates and coffee, the Asia Pacific region accounts for 40% or more volume consumption in all top six categories, with the region's share in tea at 74% in 2013. China is the leading consumer of three of the top six categories, while India leads in milk and milk beverages consumption, and the USA leads in both carbonates and coffee consumption. The tables below show the top six categories and top six countries by consumption in each category in 2013.

As volume has moved to lower priced emerging markets and lower priced bottled water, volume growth is expected to exceed value growth.

Table 2.1 The top six categories in terms of global volume consumption, and the top six countries by consumption in each category in 2013.

Tea	Volume (m litres)	Share%	Bottled water	Volume (m litres)	Share%
China	145,079	37.5%	China	63,651	22.5%
India	111,686	28.9%	USA	35,431	12.5%
Japan	16,351	4.2%	Indonesia	21,911	7.7%
Russia	16,215	4.2%	Mexico	21,220	7.5%
Turkey	15,444	4.0%	Brazil	15,159	5.4%
USA	11,019	2.8%	Germany	13,997	5.0%
Other countries	70,908	18.3%	Other countries	111,365	39.4%
Total	386,703	100%	Total	282,734	100%
Milk	**Volume (m litres)**	**Share%**	**Carbonates**	**Volume (m litres)**	**Share%**
India	53,829	26.5%	USA	50,982	26.0%
China	27,810	13.7%	China	21,975	11.2%
USA	26,183	12.9%	Mexico	20,138	10.3%
Brazil	12,221	6.0%	Brazil	15,869	8.1%
UK	6,836	3.4%	Germany	8,179	4.2%
Russia	6,230	3.1%	UK	6,332	3.2%
Other countries	70,198	34.5%	Other countries	72,583	37.0%
Total	203,308	100%	Total	196,058	100%
Beer	**Volume (m litres)**	**Share%**	**Coffee**	**Volume (m litres)**	**Share%**
China	57,108	31.8%	USA	30,700	18.9%
USA	27,202	15.1%	Brazil	27,190	16.7%
Brazil	11,096	6.2%	Germany	12,551	7.7%
Germany	9,085	5.1%	Japan	9,677	5.9%
Russia	7,875	4.4%	Italy	8,112	5.0%
Mexico	7,108	4.0%	France	7,848	4.8%
Other countries	60,151	33.5%	Other countries	66,663	41.0%
Total	179,625	100%	Total	162,741	100%

Growth in global beverages has long been driven by demand from the BRIC markets, but it is apparent that this growth is moderating, leaving global beverage companies with the task of finding new market opportunities and growth strategies for the future. Without doubt, beverage companies will need to adapt to this new landscape by looking for the next growth opportunities, and by applying strategies for mature markets where consumer demand is slowly recovering. Nevertheless, the global beverage industry remains highly competitive.

For Brazil, the challenge revolves around weaker growth than first envisaged, and high inflation. The market projects that GDP growth is forecast to be lower than the 2.5% expansion in 2013. Limited real income gains will continue to have an impact on the consumption of beverages, with salary increases just about keeping pace with inflation. Consumer price inflation is expected to remain high (at just under 6%), and higher interest rates will contribute to moderate economic activity and consumption. Brazil's worsening fiscal position meant that the government had limited room to boost spending in 2014. Added to which, the Brazilian Real is likely to continue weakening, and a higher exchange rate will impact on key beverage sectors such as beer and soft drinks, that rely on key imported materials and ingredients. Meanwhile, the beverage industries remain highly taxed and any price hikes will negatively impact on sales, as was the case in 2013. On a more positive note, the 2014 FIFA World Cup provided a boost to some of the beverage categories.

In Russia, alcoholic beverages, in particular, remain under intense pressure due to the government regulations aimed at curbing alcohol consumption. The current government programme runs until 2015. Despite the slowdown in economic growth, Zenith expects the trend to continue towards premium beverages, such as craft beer, organic juice and natural carbonated soft drinks, especially in Moscow and Saint Petersburg.

2.6 Market share charts

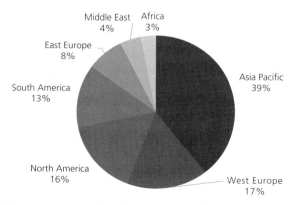

Figure 2.6 Global beverage consumption share percentage by region, 2009.

Trends in beverage markets 27

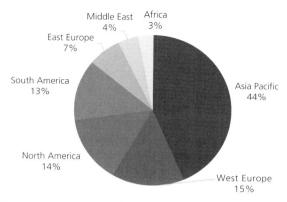

Figure 2.7 Global beverage consumption share percentage by region, 2013.

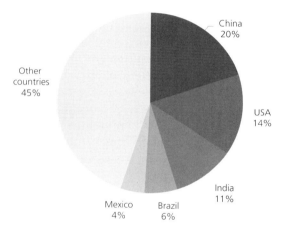

Figure 2.8 Global beverage consumption – top five countries, 2009.

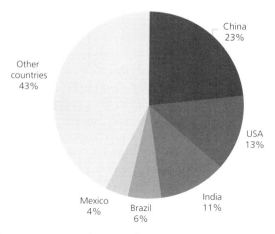

Figure 2.9 Global beverage consumption – top five countries, 2013.

2.7 Main drivers in consumption

Innovation is significant in the beverage industry. It is increasingly necessary to put the consumer first and to offer something of personal relevance in an increasingly cluttered global marketplace. It is essential that products resonate with consumers and, due to the tough economic conditions since 2008, consumers are increasingly looking for more from the products they purchase. They want products that hydrate, and which also offer an added functional benefit. Functional properties are seen as offering added value, and are more likely to appeal.

The major trends driven by consumers include: health and well-being; age concerns and ageing populations; the time/energy conundrum; convenience; and the search for natural products. These macro consumer trends underpin the emerging beverage segments.

An important ongoing stimulus, but with significant variation by market, is the focus on calories and health. Reduced and no-calorie offerings are in continued demand and are also driving wider trends, such as the use of alternative sweeteners, such as Stevia. Products that combine functional ingredients with a natural, healthy and, preferably, low-calorie offering are set to continue on their upward trajectory. Despite considerable legislative constraints, there has been an increase in the number of products with a health claim attached to them, such as: weight loss and 'fat blocking'; immune support (leveraging a growing consumer focus on preventative health solutions); beauty drinks; anti-ageing drinks containing collagen protein; body skin care (e.g. Aloe vera); and alertness and relaxation drinks.

2.7.1 The search for 'natural'

A key emerging segment within the beverages market is 'naturalness', and this is starting to become a serious prerequisite of new product formation, particularly within the developed global beverages markets. Due to the rising worldwide obesity levels, with concerns about weight and diabetes, consumers are becoming increasingly likely to seek low-calorie produce, but 'reduced sugar' is increasingly being associated with 'artificial'. Natural solutions to replace synthetic/artificial ingredients are being sought, and key to this is the increased use of the natural intense sweetener, Stevia. The European Commission approved the use of this late in 2011, but it is also being used globally.

Many new products have been launched using Stevia, including versions of Sprite, Fanta and Drench. Zenith's report on the global market for Stevia shows rapidly increasing use, with current growth rates estimated to be well over 20% per year, as defined by volume. The sweetener is being used in a wide range of beverages – not only soft drinks, but also waters, ready-to-drink teas and coffees and dairy products. Stevia is opening the door for innovation in this area, and it is likely that 'low-calorie' or 'diet' will not always be associated with the carbonated drinks industry.

In addition to Stevia, there has been a growth in the use of natural ingredients, such as extracts of guarana, ginseng and moringa. So-called 'superfruits', such as pomegranate and blueberries, which are high in anti-oxidants, are being more widely used, in addition to products using green teas. Zenith believes that health and well-being products with a mainstream consumer appeal have the potential to do well, and they see this as a growth area.

2.7.2 Adult soft drinks

Zenith has observed a surge in the number of adult-focused drinks which are emerging in the market. Adults have become somewhat jaded with traditional soft drinks offerings, which are not innovating to any great degree. The potential for this non-alcoholic market is considerable especially as, in some countries, there has been a reduction in the proportion of the adult population who consume alcohol, mainly due to government-sponsored health campaigns and taxation. Growth has been noted generally in the more premium end of the market, where products are borrowing from alcohol in positioning and presentation – for example, with carbonated fruit/botanic drinks aimed at high-end bars and restaurants, and premium fruit juices with flavours such as pineapple mojito and spicy tomato, and posh mixers.

2.7.3 Protein drinks

Protein drinks have the potential to grow in importance. While this is still a niche market, it is an area that is relevant to some consumers. The market is moving beyond its traditional 'sports and bodybuilders' image, and is moving towards becoming a more mainstream interest. This accounts for the success of such beverages as coconut water, which is a strong proposition within the market. These products are becoming readily available and more affordable. With growth in population and urbanisation, worldwide demand for protein is ever increasing, while traditional sources such as meat and milk are failing to meet global demand. Zenith anticipates further activity in this category which, even in developed markets, is still small and in its infancy. There is also potential for product development that utilises vegetable and plant proteins, such as nuts, grains and legumes.

2.8 Conclusion

In conclusion, consumers are moving away from artificial ingredients and towards more natural products. There is increased demand for lower calorie products, but ones that use more natural sweeteners. There is also a focus on products claiming functional benefits for health and well-being, for example weight loss or anti-ageing, which adds value in a tough climate. There is growth and increased innovation within the ready-to-drink tea and ready-to-drink coffee segments. There is also an increase in non-alcoholic products catering for the adult market. Finally, there are opportunities for innovation within both the protein and energy drinks sectors, to ensure appeal to a wider audience.

CHAPTER 3
Fruit and juice processing

Barry Taylor

Firmenich (UK) Ltd., Wellingborough, UK

3.1 Introduction

Naturally sourced products have been, and will always be, featured as major components of the human diet. With the passage of time, increasing attention is being paid by the consumer to the provenance of what is now available for consumption.

During recent decades, at least, the term 'processed', as applied to food products, has been increasingly viewed with scepticism and suspicion by the consumer, leading inevitably to tighter control and legislation by the authorities. Statutory controls are widely applied in the European Economic Community (EEC), and in the United States of America by the Federal Food and Drink Authority (FDA). Most other countries also operate some form of legislative control on food and drink.

The challenge facing the relevant manufacturing industries to provide a continuous and adequate supply, from the field to the table, of good-quality foods of high nutritional value is immense. The fruit juice industries are no exception. The raw materials in this sector require careful husbandry to ensure that they will be harvested at optimum maturity in good yield, to qualify suitability for processing.

In the past, freshly squeezed orange juice or apple juice has been suitable for immediate consumption, but to expect it to maintain its quality for even a day or two has been tempting providence. The use of preservatives, sometimes with pasteurisation, has been widespread. Today, with the benefit of aseptic packaging techniques and systems, it is possible to produce and store pressed juices for extended periods, with very little deterioration in quality and without the use of chemical preservatives.

Advances in instrumental analytical techniques have now made it possible to identify those chemicals in natural extracts (whether from fruit or botanical origin) which provide the characteristic flavour profile of most beverages.

Chemistry and Technology of Soft Drinks and Fruit Juices, Third Edition. Edited by Philip R. Ashurst.
© 2016 John Wiley & Sons, Ltd. Published 2016 by John Wiley & Sons, Ltd.

This analytical knowledge, apart from leading the way to 'designer drinks', has served also to maintain and standardise the quality of a wide range of beverage types, which still base their success upon traditional fruit juice systems. The majority of the active flavour components of most fruit types have been mainly identified (TNO-CIVO), and provide the beverage technologist with a basis for the adjustment of certain characteristics in the development of a new product.

It is the emphasis placed upon specific flavour characteristics which can provide the drink with its identity in the marketplace. This can be achieved by the use of both natural and synthetic flavourings, artificial or nature-identical, to create the desired top-note effects.

A fruit-based drink will always declare, under its ingredients listing, reference to the juice or juices (if mixed) used, at natural or 'single-strength' status. Usually appearing fairly high in order of concentration, the fruit juice will provide the generic character of the drink, with its specific identity being given by the flavouring materials chosen for a particular formulation and, preferably, in synergy with the flavour contribution of the juice ingredient.

The beverage technologist has a wide range of fruit types to choose from, and this chapter will investigate some of the procedures associated with the processing of these in order to produce fruit juices commercially.

In addition to the formulated beverage, there now exists in today's market an increasingly diverse range of natural strength juice products, aseptically packed, that provide nutritional benefits for the consumer as an intake of natural fruit equivalent. The impact that this has had upon the concept of healthy food consumption will also be referred to in the following pages.

3.2 Fruit types

3.2.1 Botanical aspects and classification of fruit types

The term 'fruit' is applied to a critical stage in the reproduction of botanical species throughout the plant kingdom, describes the structure that encloses, protects, or contains the seeds until they are ripe, and often assists in their dispersal.

Broadly speaking, fruits can be classified into two groups, according to their physical condition when ripe.

The widest diversity in the manner of seed dispersal is exhibited by the 'dry fruits'. These are typified by windborne types, such as dandelion 'parachutes' or sycamore 'keys'. Mechanical scattering is exhibited by many legumes, whose seed pods, when fully ripened and dried out, can split with explosive force to scatter their contents in readiness for a 'follow-on' crop. Another type is made up of those fruits, such as 'cleavers' or 'dock-burrs', that possess small hooks, whereby the fruit can be caught up in the fur of animals for transportation.

However, it is the second group that will be the focus of this chapter – the succulent or fleshy fruits, where the seeds are ripened or supported within a soft or fleshy mass, containing food materials which may attract animals to eat the fruit.

After being eaten by animals, the fruit is digested, while the seeds, protected by a hard, shell-like coating, will progress through the alimentary canal of the animal, to be passed out in the faeces. This basic method of propagation has been Nature's way, almost since the dawn of time, having the advantage of a 'built-in' seedbed growing system that is rich in nitrogenous compounds and often-essential trace elements. *Prunus aves*, or bird cherry, gained its name in acknowledgement of this manner of propagation.

Many of the commercial fruit varieties that are popular in the western world have been developed from specimens whose origins can be traced back to regions east of the Mediterranean, where stone fruits such as peach, apricot, cherry and pome fruit, apple, pear, and so on, grew in fertile surroundings and became part of the staple diet of the inhabitants of the region.

Figures 3.1–3.3 show the structures of the various succulent or fleshy fruits.

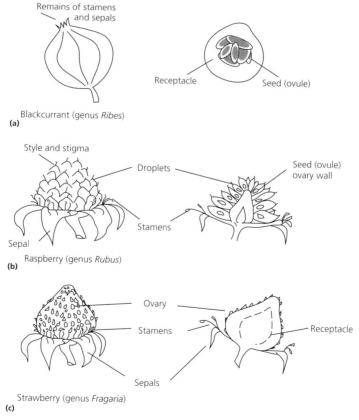

Figure 3.1 The structure of soft fruits: (a) currants, e.g. blackcurrant (*Ribes*); (b) raspberry (*Rubus*); (c) achenes, e.g. strawberry (*Fragaria*).

34 Chapter 3

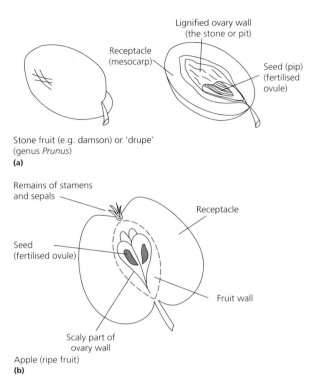

Figure 3.2 (a) Typical stone fruit of drupe, e.g. the damson (*Prunus*); (b) structure of a ripe apple (*Malus*).

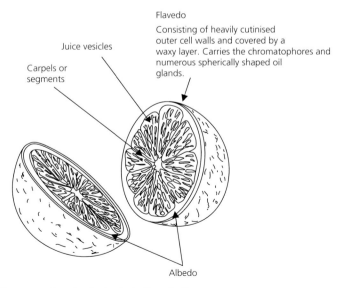

Figure 3.3 Component parts of a typical citrus fruit.

3.2.1.1 The basics of plant reproduction and fruit formation

The flower is the reproductive centre of a plant, and the series of changes resulting in the formation of a fruit start here, with the process of pollination, whereby pollen is transferred from anthers to stigma by a series of mechanisms that are usually dependent upon plant species. Essentially, these mechanisms may be initiated by an insect visiting a flower and becoming dusted with pollen from the ripe stamens (carrying the pollen bearing anther), then visiting another flower, where the pollen on its body will adhere to the stigma of the second flower. At this stage, the pollen grain, containing the male nucleus, fuses with the stigma, absorbing nutrient and sending out a growing tube that eventually reaches an ovule contained in the plant ovary. The male nucleus has been passing down the tube during this period, and is ideally positioned to fuse with the female nucleus contained in the ovule. From here on, rapid changes occur, resulting in the development of fruit, receptacle and seed formation.

3.2.1.2 Respiration climacteric

During their ripening or maturation stages, certain, but not all, fruit varieties undergo a phase of upsurge in metabolic activity known as the climacteric, a term coined by Kidd and West (1922) to describe the increase in respiration rate and heat evolution as the fruit softens and develops flavour and aroma. Fruits such as apples, pears, bananas, and most stone fruit, have stored reserves of starch which, during the climacteric, are converted to sugars by starch-degrading enzymes. The carboxylic acid content of the fruit will also take part in the conversion, with acidity levels consequently reducing as the maturation takes place.

3.2.2 Harvesting considerations for berry, citrus, pome, stone and exotic fruits

In order to achieve a year-round supply of sound, good quality fruit, a highly efficient international quality management system is required to operate throughout the whole procedure of growing, harvesting, ripening, storing and handling of the fruit. Apples and pears can be stored under prescribed temperature and environmental conditions in order to ripen gradually and, after grading to meet the requirements for direct sale, those fruits falling outside size limits may be sent for juicing purposes. Careful handling of the fruit during harvesting is an essential requirement as, also, is post-harvest management. If the fruit is already destined for juicing, then it is more usual for mechanically harvested fruit to be transferred directly from orchard to processing plant to meet the demands of a tight production programme during the season. In Europe, the season usually runs from the end of June to mid-December.

The determination of correct maturity for fruit to be processed is not easy. The soluble solids content of the fruit juice is a reliable indicator in the case of non-climacteric fruit, whose composition will show little change following harvest and storage under suitable conditions. Citrus fruit and grapes, both non-climacteric fruits, can be assessed successfully as to their suitability for harvesting, and the ratio of soluble solids to titratable acidity is frequently used in establishing maturity criteria.

Fruits that undergo climacteric change are more difficult to deal with, as the full potential of the fruit will not be known until they are fully ripened, yet the important commercial decisions on harvesting cannot be left until then.

Citrus fruits are harvested directly for commercial sale or juice processing, undergoing washing, grading and, thereafter, packaging or juicing. Various types, such as Valencia, Jaffa and Shamouti, follow on in a sequence of harvesting seasons. Around 70% of the world's citrus production is grown in the northern hemisphere; in particular, this includes countries around the Mediterranean, China and the United States, although Brazil is also one of the largest citrus producers.

About one third of the world's citrus production goes for processing, and more than 80% of this is for orange juice production. The current annual world citrus production has been estimated at over 105 million tonnes, with over 50% of this being oranges (UNCTAD – United Nations Conference for Trade Development).

Unlike the pome and citrus fruits, soft fruits are subject to rapid deterioration if even the slightest 'bruising' takes place and, unless harvested for direct pressing, they are best subjected to rapid freezing and held in the region of −18°C to −26°C. Two forms of grading are employed for soft fruit. Selected top-quality berries will be individually quick-frozen (IQF), to be later used as whole fruit pieces in jams, yoghurts and culinary preparations and so on, whereas fruit for conversion and use as fruit pulp may be cleaned by washing them free from leaves, twigs and other detritus, and 'block-frozen'. Berry fruit intended for juice production will be 'block-frozen' as harvested, complete with any incidental stray foliage. In the case of blackcurrants, redcurrants and so on, this comprises the stalks (strigs) to which the individual berries are attached, and serves a useful function as a natural 'filter-aid' during the pressing stage to release juice from the fruit.

While the effect of freezing will disrupt the cell structure of soft fruit and render it 'pulpy' upon thawing out, any adverse changes to juice quality will be minimal, and flavour and colour can be easily preserved by this treatment.

3.3 Fruit types for processing

3.3.1 Pome fruits

These include the apple, pear, medlar and quince. The two latter fruits are commercially of little importance, but are occasionally seen in speciality outlets. Medlar fruits are brown-skinned, apple-shaped and are best eaten after 'bletting', a process where, upon storage, the flesh softens and sweetens. This can take up to two to three months from harvest and explains why, in Europe in the Middle Ages, the medlar was a useful fruit to store for winter consumption.

Likewise, the quince, similar to the pear in shape and at one time a very popular fruit, is now something of a speciality. It is ripe when the fruit turns a bright yellow. High in natural pectin, it finds its main use in jams and jellies.

3.3.1.1 Apple (genus *Malus*) and pear (genus *Pyrus*)

Although these fruit types display a large number of distinct species within each genus, the domestic fruit is selectively farmed from an even greater number of cultivars. The ancestors of *Malus domestica* are generally thought to be *Malus sieversii* and *Malus silvestris* which are found growing wild in the mountains of Central Asia. Likewise, the cultivated European pear (*Pyrus communis* subsp. *communis*) originated from one or two wild subspecies (*Pyrus communis* subsp. *pyraster* and *P.communis* subsp. *caucasica*).

The apple and pear are of major commercial importance, and are grown in most of the temperate regions of the world. Argentina, Australia, Bulgaria, Canada, China, France, Germany, Hungary, Italy, Japan, the Netherlands, New Zealand, Poland, South Africa, Spain, the United Kingdom, and the United States are among the foremost of those countries growing pome fruit on a considerable scale for both home and export use. Although local varieties are found to have followings in their own regions, the world markets are dominated by perhaps no more than 20 dessert and culinary varieties, which have been selectively bred to display such characteristics as disease resistance, winter hardiness, appearance (colour and shape) and texture, together with high average yield. Among the apple varieties will be found Bramley's Seedling, Brayburn, Cox's Orange Pippin, Red Delicious, Golden Delicious, Discovery, Granny Smith, Jonathon and Newtown Pippin.

The main varieties of pears encountered in the market are the Bartlett or Williams, Bon Chrétien, the Comice (Doyenné du Comice) and Conference.

The World Apple and Pear Association (WAPA) is a trade association, established in 2001, whose membership comprises major apple and pear growing countries globally. Each member country is represented by its nationally recognised industry organisation, through which are channelled the interests of the respective major national apple and pear producers. WAPA provides a forum for the discussion of issues affecting the international marketing of apples and pears. World production forecasts are released in August for the northern hemisphere, and in February for the southern hemisphere.

The Secretariat operates from Brussels.

3.3.2 Citrus fruits

Citrus fruit varieties are grown for commercial use in many parts of the world. Originating in the southern and eastern regions of Asia, China and Cochin and the Malay Archipelago, the citron (*citrus medica*) is said to have first arrived in Europe during the third century BC, when Alexander the Great conquered Western Asia. Later, the orange and lemon were introduced to the Mediterranean regions in the days of the Roman Empire, when trade routes from the Red Sea to India became established. Cultivation of citrus fruits has since spread worldwide to all regions where the climate is not too severe during the winter months, and where suitable soil conditions are available. In the United States, the notable growing areas are in Florida and California. In South America, Brazil has taken over the largest share of the world market for oranges and orange juice products. Morocco, South Africa and parts of Australia have shown increased output during recent years although, within the latter two areas, yields are frequently affected by variable weather conditions. China is the largest producer after the United States and Brazil, but 90% plus of its output is destined for the home market.

In the area of citrus juice production, Brazil, California and Florida are major players, with Spain and Israel being notable producers of specialised concentrates. Israel enjoyed a thriving citrus industry in the 1980s, but has since lost much ground, due to strong competition from South America with the emergence of the Brazilian market. In recent years, a spate of drought conditions, labour shortages and political instability has done nothing to improve the situation.

The main citrus varieties for juice processing are the orange, lemon, lime and grapefruit.

3.3.2.1 Orange

The most important of all citrus fruit is the sweet orange (*C. sinensis*), which is widely grown in those world regions adapted to citrus. Each region usually has its own characteristic varieties. Common examples to be found growing in various parts include Navel, Valencia, Shamouti, Hamlin and Parson Brown. The Mandarin orange (*C. reticulate*) is representative of the 'soft citrus' loose-skinned oranges, or 'easy peelers', hitherto of primary importance to the Far East and now popular in other parts, including the United States and Europe. The group includes Satsumas, an important crop in Japan, and the Clementine, an important cultivar to be found in the Mediterranean areas. Other cultivars of note are the tangor, a hybrid of mandarin and orange, and the tangelo, a hybrid of mandarin and grapefruit.

A third distinctive variety of orange is the bitter orange (*C. aurantium*) chiefly represented by the Seville orange, which is commercially grown in southern Europe, mainly for products such as marmalade. Compared with other citrus crops its yield is small and of little use in the juice market.

3.3.2.2 The lemon (*c. limon*)

An important crop in Italy and some other Mediterranean countries, the lemon is also grown commercially in the United States and Argentina. The characteristic oval shaped, yellow fruits, apart from their culinary usage, are an important source of juice and flavouring for the soft drinks industry.

3.3.2.3 The grapefruit (*c. paradisi*)

A large round citrus fruit with a thick yellow skin and a somewhat bitter pulp, grapefruit is generally accepted to be a hybrid between the pomelo and the orange. The pomelo, also known as the pummelo (*c. grandis*) originated in Asia and is grown in many eastern countries, including China, Japan, India, Fiji and Malaysia. It was introduced to the West Indies during the 17th century by Captain Shaddock, and thus in that region it is sometimes referred to as a 'Shaddock'. Today, the commercially important grapefruit is grown in many parts of the world. Notable producing countries are Argentina, Cuba, Cyprus, the Dominican Republic, Egypt, Honduras, Israel, Mexico, Mozambique, Pakistan, South Africa, Spain, Turkey and the United States.

The most predominant cultivar to be seen in the market is the 'Marsh Seedless', followed by a red, pigmented version known as the 'Star Ruby'.

3.3.2.4 The lime (*c. aurantifolia*)

Limes require warm and humid weather conditions in order to thrive on a commercial scale. India, Egypt, Africa, Mexico, Peru and the West Indies are, therefore, prime growing areas. Mexico, Peru and the West Indies together produce a large percentage of the world's lime crop. Relative to the other citrus fruits, limes are a small, round fruit, green or greenish-yellow in colour, and not more than 8–10 cm in diameter, with a sharp, fresh and characteristic flavour.

3.4 General comments on fruit juice processing

Because of their nature, shape, size, harvesting characteristics and so on, the various types of fruit may require specialised treatment during processing. In all instances, however, the operation is seen to involve a number of stages. The fruit supply must be obtained in a correct state of maturity and the juice expressed in the most efficient manner possible. Then, if required, the juice is treated with enzymes (e.g. pectolases, cellulases, etc.) as a prerequisite to clarification, to be followed by a suitable filtration stage before concentration and eventual packaging or storage.

In citrus fruits, where the outer skin or epicarp is a composite structure containing citrus oils that are useful as flavouring substances, it would be detrimental to juice quality were the fruit to be subjected to direct pressure. This is also the case with the 'fleshy fruits' (i.e. soft fruits, pome fruits and 'stone fruits').

It should be noted that stone fruits, before being processed for juice separation, must firstly be separated from their stones, or pits, in order to facilitate the ease of handling and the avoidance of unwanted notes in the finished product. The pits can be further processed, to yield both fixed oils for application in the cosmetics industry and glycosides, from which may be sourced other natural flavouring ingredients (such as benzaldehyde, a characteristic of marzipan, almond flavourings, etc).

3.4.1 Processing of 'fleshy' fruits

In the separation of juice from its fruit, the traditional method has been to apply pressure to the mashed, or pulped, fruit in order to force the liquid portion through a cloth or some form of screen. There are several styles of 'separator' available for both batch and continuous production. A few of these are referred in the following.

3.4.1.1 Pack press

This is based on the traditional 'rustic' or 'cheese' press widely used in the cider industry. The design comprises a set of frames for containing the fruit. These are loaded, in stages, by placing a loose-weave cloth over each rectangular frame and adding an appropriate quantity of the fruit mash from a hopper feed above the assembly. The mash is then smoothed, or trowelled, over the frame by the operator, who then folds the cloth across to cover over, creating a form of sandwich. Another frame is then placed on top, and the process repeated until several filled layers are formed. The stack is built up inside a rectangular tray, or bed, which serves both as a collecting device and a platform. This is then raised by the action of a vertical hydraulic ram, to bring the top of the stack in contact with a fixed frame, where pressure is exerted to express the juice, which runs down into the tray for collection.

Early versions were made of hardwood, as are some current ones. Most modern pack presses, however, are usually constructed from stainless steel and are frequently designed to accommodate two stacks for improved efficiency. These are assembled in sequence in their respective collecting trays, before moving across the hydraulic ram so that, while one pressing operation is under way, the next stack is being prepared and charged in readiness to follow on. Thus an almost continuous, albeit labour-intensive, pressing operation can be carried out.

The pack press is ideally suited for a relatively small production output for specialised fruit varieties but, for large-scale production, fully mechanised systems are necessary.

3.4.1.2 The horizontal rotary press

Perhaps the most successful of the mechanised systems to date, has been the horizontal rotary press, designed and developed some 45 years ago by a Swiss company, Bucher Guyer AG of Niederwenigen (now Bucher Unipektin AG).

Fruit and juice processing 41

Figure 3.4 Horizontal rotary press: Universal Fruit Press HP5000 (*Source*: Bucher-Unipektin AG).

This design (See Figure 3.4) carries a horizontal hydraulic piston (HP), operating within the cylindrical hollow body of the press. Between the specially designed endplate and the piston faceplate run a large number of flexible rubber lines of solid cross-section, with well-defined ribbing along their lengths to act as juice drains. Each line is covered with a coarsely woven nylon sleeve throughout its length.

The press can be programmed to operate, self-optimising, in strict sequence with pressing periods appropriate to the fruit in process. Mash is pumped into the press to partly fill the cylinder space. The piston then moves forward, under hydraulic pressure, to express the juice by consequently forcing it through the filter sleeves and along the channels in the rubber lines, to be fed through outlets in the specially designed plates for collection in a juice tank.

The piston then withdraws repeatedly, while further mash charges are received and the process repeats. During the pressing operation, the endplate of the unit rotates, with consequent meshing of the filter-drains so that, at the end of a pressing cycle, the drains are loosely meshed together within the press-cake, or fruit pomace. As the piston moves back in conjunction with incoming mash, reverse rotation occurs, disentangling the lines and redistributing the pomace amongst the fresh charge of mash. Hence, there is no build-up of press-cake, and maximum efficiency in terms of juice removal.

At the end of a complete pressing sequence, the pomace residue is discharged by rotating the whole press body, with the piston fully retracted, and moving the cylinder flanges away from the endplate to create an opening through which the dry pomace is released for its collection (usually along a belt conveyer, with a suitable storage hopper).

The standard Bucher Unipektin press, the HPX 6007, is designed to comfortably accept a 10–12 tonne loading although, from this author's experience, with suitable selection and pre-treatment of the fruit, quantities well in excess of this can be handled quite easily.

Good quality soft fruit (e.g. apples) will yield 85–95% by weight of expressed juice. It is quite feasible, therefore, for a press of this size to handle up to ten tonnes of fruit mash, in stages, leaving around one tonne of pomace to discharge.

3.4.1.3 The use of centrifuges in processing

While direct pressure has previously been an obvious choice, there has been a move in comparatively recent times towards employing centrifugal separation of solids from juice from a continuous fruit mash stream. The modern decanter centrifuge may be used, in conjunction with a pressing system as a preliminary step, to increase throughput efficiency, or as a complete separation system, where two units are involved providing a coarse primary, followed by a final clarification stage.

The decanter is a horizontal scroll centrifuge, with a cylindrical-conical solid wall bowl for the continuous separation of solids out of suspensions (see Figure 3.5).

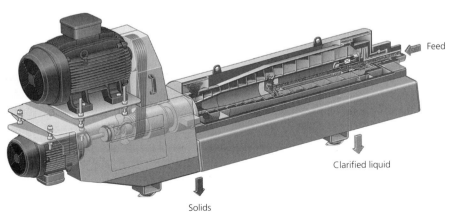

Figure 3.5 Clarifying decanter (horizontal scroll centrifuge). *Source*: Courtesy of GEA Westfalia Separator Group.

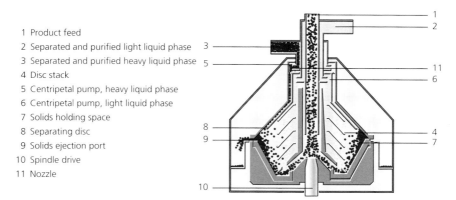

1 Product feed
2 Separated and purified light liquid phase
3 Separated and purified heavy liquid phase
4 Disc stack
5 Centripetal pump, heavy liquid phase
6 Centripetal pump, light liquid phase
7 Solids holding space
8 Separating disc
9 Solids ejection port
10 Spindle drive
11 Nozzle

Figure 3.6 Self-cleaning clarifier (disc stack centrifuge). *Source*: Courtesy of GEA Westfalia Separator Group.

Centrifugation is of particular advantage when producing single-strength cloudy juices for direct consumption, as a better definition in terms of particle size distribution can be attained. Typically, with decanter juice, 60% of the particles in suspension are smaller than 1 micron (1 µm), although this figure reduces to around 20% for pressed juice. Hence, there is greater likelihood in the latter for instability and sedimentation. It should be noted that the major factor in the production of 'naturally cloudy' juices is the rate of processing. To ensure stability, the juicing stage should be followed immediately by pasteurisation, in order to deactivate the naturally occurring enzymes present in the fruit.

Decanters are also of use in the production of fruit purée, where the aim is to remove only the undesired particles, such as pips, stalk fragments, skin fragments and coarse tissue material, leaving the crushed fruit flesh evenly distributed throughout the juice. By setting the machine parameters accordingly, the undesired components can be selectively removed from the liquid stream output of purée.

Decanters are frequently used in conjunction with disc stack-type centrifuges in the pre-preparation of clear juices and juice concentrates, where the initial decanter treatment results in a partially clarified juice with a low level of suspended solids. This is followed by a clarification stage using a disc stack, whereby the solids are thrown outwards from the through-flow juice stream into a 'solids holding' space and automatically discharged from there as, and when, an optimum level of solids is reached (see Figure 3.6.).

3.4.2 The use of enzymes in fruit juice processing

Pectin is an essential structural component of fruits where, in combination with hemi-celluloses, it binds single cells to form the fruit tissues. Pectins are chains formed almost exclusively of galacturonic acid (α-D-galacturopyranoside acid) units, partially esterified with methanol. These chains are often referred to as 'polygalacturonic acid', or its synonym 'homogalacturonane'.

In the immature fruit, pectins are mainly insoluble but, as the fruit ripens, there is a gradual breakdown of some of the pectic substances in the skin and flesh cell walls, resulting in the formation of polysaccharide component materials.

The general term 'pectic substances' covers not just pectins, but just about everything resulting from the degradation processes involving pectin that take place as the fruit maturates – soluble forms included.

As the fruit becomes softer, less acidic, sweeter, and heads towards its optimum state of maturity, then such changes need to be taken into account by the juice processor. Apples, in particular, are best processed prior to their fully ripened state, as solubilised pectin and softened fruit tissues will seriously affect the efficiency of separation of the juice, and low yields will result. Other fruits, such as the berry fruits, need to be fully ripened, in order to optimise flavour and so on.

Broadly speaking, therefore, if the resulting juice is to be clarified, then enzyme treatment is required at some stage, in order to break down the pectin and to enable precipitation or sedimentation of the resulting pectic substances. It is particularly important, when the final product is to be a clarified juice concentrate, that no pectin should be left available. As a general rule, it is the juice of the apple and other soft fruits, rather than the fruit itself, that is to be treated with pectolytic enzymes, (i.e. after pressing) to produce a clear juice.

However, to improve juice yields and to facilitate easier pressing operations, apples and other soft fruit varieties are often 'de-pectinised' at the mash stage before pressing. Naturally occurring enzymes are found in the proximity of the fruit, borne within traces of surface moulds where present, and the pre-process washing stage is designed partly to reduce the effect of these. Milling of fruit is violently disruptive and, apart from accessing juice for ease of separation, the process will initiate a host of interactions as the internal enzyme systems natural to the fruit are released and are introduced to suitable substrate materials. Thus, the presence of natural pectinesterases and, where necessary, added enzymes will improve yields and processing operations. Enzymes are often added during the milling stage, prior to any pasteurisation or heat treatment. The result is de-methoxylation of the pectin chain, liberating methyl alcohol, which will appear later as a trace contaminant in the aroma volatiles fraction.

In some processes, for example the production of cloudy juices, it is necessary to denature naturally occurring rogue enzymes by a pasteurisation stage prior to juice separation. This is achieved by passing milled fruit pulp through a tubular heat exchanger. Where the juice is destined for concentration, it is essential for the pectin to be destroyed.

Pectinase or poly-(1,4-alpha-D-galacturonide) glycanohydrolase, in its commercially available form, is produced from fungal sources (i.e. *Aspergillus sp., Rhizopus sp.*). It possesses a wide variety of component activities and can operate comfortably between pH 2.5–6.0. Subject to supplier type, it can function well at specified temperatures between 30–60°C.

Activities include:

1 *Esterase* (polymethylgalacturonase esterase); where the action is to de-esterify pectins, with the removal of methoxy groups, to form pectic acid.
2 *Depolymerases* (polymethylgalacturonases with either endo- or exo- activity); Several different mechanisms take place, in which the polymer chain is completely disrupted into fragments. The term 'endo-' refers to those polygalacturonases which act at random within the chain, and 'exo-' to those where the attack is sequential along its length, starting at one end.
3 Another form of depolymerase activity is given by *pectinlyases*, which operate at glycosidic linkages, either side of which carries an esterified or methoxylated group.
4 *Amylases*; in the case of pome fruits, other enzyme activities are sometimes required. Where fruit has been picked before maturity and then ripened under controlled atmosphere conditions in a cool store, there is a likelihood of starch retention originating from the unripe fruit. This can become gelatinised during juice processing, and can give rise to precipitation and haze effects in the final product. Amylases are used here to break down any residual starch and overcome the problem.
5 *Cellulases* have been used to facilitate the rapid removal of colour during fruit processing. Such enzymes have also been employed to good effect in recent years in the 'total' liquefaction of plant tissues during processing, obviating the use of a press, yet increasing yields.

The usual way of employing enzymes in the case of soft fruits is to dose at the recommended level (e.g. 0.1%) into the pre-warmed mash, which is well mixed and left to stand for a recommended period and constant temperature.

For a typical pectinase, the standing period for blackcurrants and other soft fruits will be around 1.5 hours. The exact time needs to be determined by taking a series of samples of the mash during the enzymation stage, pressing or filtering off some of the juice, and treating this with an excess of alcohol (e.g. 40 ml single strength juices + 60 ml alcohol, 100 ml measuring cylinder). Dissolved pectin, if present, will be thrown out of solution as an insoluble gel whereas, if fully degraded, it will form a flocculent precipitate of pectic substances that settles to the bottom of the measuring cylinder.

This is a purely empirical test but, with experience, it will give the operator correct information on the required process parameters.

3.4.3 Extraction of citrus juices

As discussed, 'fleshy fruits' yield juice upon pressing. A pre-treatment is necessary but, effectively, the whole fruit (or de-pitted fruit in the case of drupe fruit) is used for the resulting extraction. Citrus fruits, however, because of their structure, are handled in an entirely different manner. The epicarp, or outer peel of the citrus fruit, contains a rich source of the essential oil in oblate spherically shaped oil glands situated in that part of the flavedo, just below the waxy surface layer. Citrus oils are of great importance in the flavour industries, being widely used, as might be expected, in beverage formulations, and they command a strong place in the market.

Many different processes have been employed worldwide across the range of citrus types for separation of both the oil and juice. Typically, the fruit will be passed over a rasping device (e.g. an abrasive roller), to pierce and disrupt the oil glands within the flavedo layers, thus releasing the oil to be washed away for collection by a water spray. From here, it may be recovered by centrifugation and dried. The rasped fruit is moved into an extractor, where the juice is expressed and recovered, to leave the albedo (pith) and flavedo (outer peel). The expressed juice is subjected to screening before being further processed.

A rotary brush sieve, followed by a centrifugal separator, may be used to bring the suspended particles into the stability range of below 1 μm diameter, where it can be pasteurised and aseptically packed for direct consumption, as single strength juice or concentrated, as required.

Lemons and limes are processed in a similar manner to the above, by direct pressure on the washed fruit.

3.4.3.1 The FMC extractor

Perhaps the most frequently encountered processing sequence is that provided by an extractor manufactured by the Fruit Machinery Corporation (FMC), which is generally employed for orange types. The extraction unit is designed to process individual fruit in rapid succession. In the processing hall, these units are usually set up in banks of eight to, to accommodate a continuous stream of washed and clean fruit, which has been separated into size bands, along appropriate feed channels to the correct size of extractor.

Each extractor, constructed in stainless steel, comprises two cups, one inverted above the other. During the operation, fruit is received into the lower cup and the upper cup descends to press down upon it. Simultaneously, a perforated stainless steel tube forces up through a channel in the lower cup, cutting a plug in the bottom part of the fruit. As pressure continues to be applied, the juice is forced out through the perforations in the wall of the tube, which also acts as a screen to retain the plug, seeds and any pulp debris. The solids are later ejected at the end of the process cycle, in readiness for the next pressing operation. As the two cups move together to enclose the fruit, the oil vesicles in the outer peel are ruptured to release the oil. While the juice is being expressed from inside the fruit, the oil is removed from the outside surface of the fruit, together with some skin debris, by water spray via a conveyer, to a cold-pressed oil recovery system.

The expressed juice flows into a manifold attached to the line of extractors, and thence to the so-called 'finishing stages', wherein the juice is progressively screened to remove excess pulp and to bring it into the range set by plant quality standards, effectively minimising the level of insoluble solid material and rendering the juice suitable for further processing (e.g. pasteurised single strength or for concentrate).

The pulp recovered during screening may be transferred to a pulp-wash operation to yield further soluble solids by counter-current extraction with water.

The washed pulp may be held for further processing or included with the bulk of ejected peel material from the extractors. This is milled, treated with lime (calcium hydroxide or calcium oxide) to break down pectin and reduce water retention, pressed, dried to about 10–12% moisture, and finally converted to pellets. These, being high in carbohydrates, are used as 'filler' in livestock feed blends.

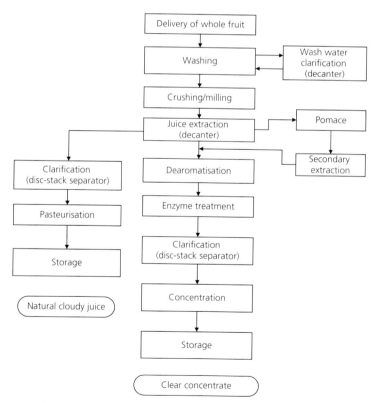

Figure 3.7 Flow diagram of the fruit juice concentration process.

3.5 Juice processing following extraction, 'cleaning' and clarification

Depending upon the juice type, the requirements for further processing will alter to some extent. In the case of cloudy juices, heat treatment via pasteurisation is used to denature any residual enzymes released from the fruit during processing, and also to eliminate spoilage organisms, yeasts and moulds, which may also arise from the natural fruit source. The stability of a cloudy juice will very much depend upon the 'cleaning' stages, after pressing designed to rapidly remove pulpy and sedimentary material. Brush screens, decanter and disc-stack centrifuges may be used 'on-line' to remove potentially 'unstable' suspended solid material, to provide a juice of uniform consistency and cloud, in readiness for pasteurisation.

Pasteurisation of citrus juice is carried out at temperatures above 95°C to eliminate the undesirable effects of pectolytic enzymes, particularly pectinesterase, whose action in the demethoxylation of pectin will give rise to cross-linking between resulting polygalacturonic acid molecules, gelation effects and loss of cloud stability.

Natural strength juices for direct consumption are pasteurised and processed under aseptic conditions, where the product is then filled without contamination into sterile containers and hermetically sealed.

There are a number of well-established systems for the aseptic packaging of liquids. Notable among these are those packs constructed in box form, *in situ* on the filling line, from a cardboard, aluminium or plastic laminate sheet, such as Tetra Pak or Combibloc. In the Tetra Pak system, the packing material enters the filling machine from a feed roll, and the sheet contact surface is sterilised with warm hydrogen peroxide solution and is formed into a tube. Its lower end is then heat-sealed across the width and the tube is filled, sealed at the upper end, cut and, finally, is folded into a box shape. This enables the production of a continuous output of filled cartons, with the resulting premium utilisation of bulk storage capacity.

A second, and highly efficient, aseptic packaging system uses a plastic laminated bag with a novel filling and sterile sealing facility. There are several designs available but, in general, the bags are supplied in a sterilised condition (having been γ-irradiated for that purpose), and the neck of the bag is sealed with a plastic membrane which, during the filling action, is automatically steam-sterilised and then ruptured upon the introduction of the pasteurised juice stream. Upon completion of the filling operation, a non-return valve enables removal of the inlet feed tube without loss of liquid, and the neck is capped and sealed. The bags are available in a wide range of sizes and are designed for filling into boxes from 5–25 litre capacity, 200 litre drums, or rigid outer containers of up to 1000 litre capacity.

Clarified juices, where pectin has been actively removed by enzyme treatment as an integral part of the process, are filtered bright and pasteurised, or may be rendered sterile by the use of membrane filters in order to eliminate yeasts and moulds directly. Ultra-filtration techniques are used, in which the juice feed flows transversely, under pressure, across a membrane support tube, to avoid 'blinding' of the filter surfaces. For sterile filtrations, such membranes provide a porosity value of, typically, 0.02 µm, ensuring the removal of spoilage organisms.

Clarified, sterile, juice passes from within the tube for storage under sterile conditions. Periodically, the circulating liquors on the feed side of the filter are run off and replaced with fresh juice input.

3.5.1 Juice concentration by evaporation

Where the final juice product is a concentrate, the filtered 'clear' or 'cleaned' cloudy juices are automatically subjected to heat treatment during the course of the concentration process.

Heat treatment of juices is an area where the design of process requires careful consideration, in order to avoid any detrimental effects on flavour of appearance of the product. Early evaporators had shown that high vacuum, low temperature processing produced concentrate of good flavour quality, but it was soon discovered there was a drawback, in that the heat treatment did not necessarily deactivate pectinmethylesterase, which gave rise to gelation in the final product. The effect was not immediately apparent, but was seen to occur after a few weeks' storage when, for example, the contents of a filled drum of, say, orange concentrate was found to have gone 'solid'.

Short-term high-temperature pasteurisation was introduced into the concentration process in order to offset this effect. The process brings the juice temperature to around 95°C, with sufficient 'holding' time to eliminate microorganisms and to denature the enzyme. The early evaporators operated by recycling the juice feed until the desired level of concentration was reached, which also increased the heating effect on the juice. The industry has since progressed to evaporators of a multi-effect, multi-stage, single pass design which, although being highly efficient in heat utilisation, has the added advantage of being easier to control when there are Brix variations on the incoming single-strength juice feed. At the present time, thermally-accelerated-short-time evaporators (TASTE) are in worldwide use across the processing industry.

While a combination of product quality and cost considerations will dictate the methods used for bulk processing of fruit juices, there are instances where the flavour components present in the juice are vulnerable to any form of heating during concentration. Strawberry juice is perhaps the best example of this, being one of the most heat-sensitive of fruits, and it works well with alternative processes for concentration, such as freeze-concentration and hyperfiltration.

3.5.2 Freeze concentration

In general, the advantage of freeze concentration is that there is no loss of volatile flavour components, as in the evaporation procedures. Freeze concentration is carried out by crystallisation of water from the juice. Depectinised juice is cooled to low temperature within a scraped surface heat exchanger, to form a slurry of ripening water crystals as the heat of crystallisation is withdrawn from a well-mixed suspension of crystals. The crystals are removed by screening or centrifugation. On an industrial scale, freeze concentration is generally carried out as a multi-stage operation, where the overall crystallisation rate, which strongly decreases with increasing concentration, is higher than in a single stage operation.

Viscosity is a determining value in the degree of concentration attainable. An upper limit of 55% RS is claimed, below that of standard evaporative processes. In this process, a large majority of the volatile flavour components are retained, although some traces (<10 ppm in ice removed) are lost during the removal of the ice crystals.

3.5.3 Hyper- and ultrafiltration

By the use of selective membranes, water may be removed by filtration from the juice, in order to effect its concentration. Depending upon the molecular size of the compounds and the cut-off value of the membrane used, there is likely to be some loss of flavour components. These may be recovered from its permeate by distillation, and returned to the juice concentrate. Concentration by these methods is less effective in terms of folding than other methods, but it can provide advantages in specific cases, for example capital costs associated with hyperfiltration are somewhat less than ≈ 10–30% for evaporative systems with aroma recovery equipment.

Table 3.1 Range of product concentrations technically attainable from depectinised juices.

Equipment type	Juice concentrate (% m/m)
Scraped film evaporators	75–85
Plate evaporators (recycling effect)	65–75
Falling film evaporators	65–75
Freeze concentration	45–55
Hyperfiltration	15–25

3.6 Volatile components

The characterisation of fruit type or variety will be reflected in the flavour profile of its volatile components. Although analytical techniques can produce an accurate peak profile using gas chromatography, in simpler terms, the sensory receptors of most individuals can quickly differentiate between fruit varieties.

We have four basic taste categories, sometimes described as, sweet, sour, acid and bitter, and these are identified by taste receptors situated mainly on the tongue, the key area of flavour differentiation. So-called 'top-notes' and such are not so much by taste as by aroma, detectable in the nasal cavity. Thus, during the process of consuming food, eating and drinking, the release of aroma volatiles can be identified, and an assessment of their value can be arrived at.

It is not always necessary to return aroma volatiles to the juice concentrate, as this is dependent upon intended application of the latter but, should the juice concentrate be reconstituted for use in a soft drink formulation, then the addition of flavouring top-notes will almost certainly be a necessity.

Aroma volatiles recovered during evaporative concentration of juices provide a prime source of natural flavouring. However, as referred to previously, from the moment the fruit is milled in readiness for processing, there will be changes occurring due to the release of natural enzymes and the initiation of biosynthetic pathways. This may result in the formation of some atypical flavour components.

Therefore, to create a more representative extraction of flavour, it is preferable to remove volatiles at the earliest possible stage in the process. The term 'pulp-stripping' applies to the technique of removal, in-line, of volatiles from the pulped fruit. This is carried out prior to depectinisation, in readiness for the pressing stage. It may be effected by passing the pulped fruit through a tubular or scraped surface pre-heater/pasteuriser into a suitably sized evaporator, to remove something in the region of a 5% strip, carrying the larger part of the volatiles fraction, before cooling and returning back to the process. The volatiles may be further concentrated by distillation, or may be stored frozen at the strength produced (Figure 3.8).

The condition of incoming fruit is always of the greatest importance, and there will inevitably have been a certain amount of fermentation, due to natural yeasts present on the fruit. Rapid processing will have minimised the effect, but almost certainly there will be a certain amount of ethanol present in the 'stripped' volatiles fraction. Levels of ethanol are not uncommon up to values of 5% or 6%, and are often deemed acceptable but, above this level, the balance of flavour may be seriously affected. Design of plant is critical, and rate of heat input must be carefully controlled in order to enable a steady state and optimum level of volatiles strip for a particular fruit system. Separator designs vary from the

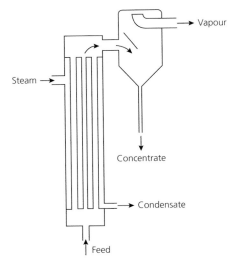

Figure 3.8 Rising film evaporator.

simple cylindrical chamber, with side entry and steady input of pre-heated pulp, to high-speed tangential entry units, which create a thin film effect and larger surface area for more efficient removal of volatiles. The spinning cone column is a prime example of the more technically efficient counter current system producing multi-stage effects.

3.6.1 Spinning cone column

Although the conventional separators already referred are successful in removing volatiles from juice streams, the resulting strip will usually require further rectification and concentration, in order to render it a flavour component in its own right.

The additional heat treatment inevitably involves losses in efficiency, as there are likely to be subtle changes in quality and yields, due to heat degradation of some of the components. The spinning cone column is designed to selectively isolate a volatiles concentrate in one operation. It is, in effect, a 'counter-current extractor', incorporating a central, vertically placed, rotating shaft, along the length of which are a series of upturned cone-shaped cups. In operation, the cups rotate smoothly within a similar series of cone-shaped stators, attached to the inner wall of the column (see Figure 3.9).

A juice feed is introduced at the top of the column onto a distributor disc, and runs down the column in a thinly layered stream between the stators and spinning cone assembly. Counter to its flow, and moving up the column, is the gas phase, which may be non-intrusive, such as nitrogen, or more commonly a 'live' steam feed. The multiple stage effect produces a highly efficient separation of the component volatiles from the juice stream and an enriched fraction is produced.

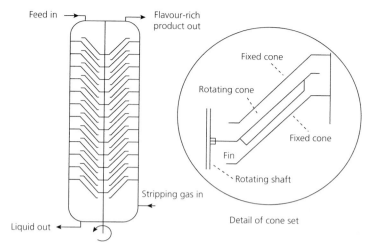

Figure 3.9 Spinning cone column.

Table 3.2 Volatile components identified in a 5% pulp-stripped fraction from raspberry fruit.

Compound	ppm m/m
Isobutyl alcohol	38.9
Isoamyl acetate	52.8
Isoamyl alcohol	141.1
Ethyl hexanoate	10.8
Octanal	0.2
Hexanol	2.2
Cis-3-hexanol	3.7
Nonanal	0.3
Ethyl octanoate	5.5
Decanal	0.3
Linalool	5.2
Alpha-terpineol	0.8
Benzyl acetate	0.8

3.6.2 Composition of fruit juice volatiles

As already mentioned, it is mainly the volatile constituents that serve to identify fruit type and variety. Broadly speaking, qualitative analysis will identify the principal substances present in the volatiles fraction as representative of a particular fruit type, but it is the relative proportions of those substances that will reflect the variety. Alcohols, volatile acids, esters, carbonyl compounds and low-boiling hydrocarbons are the principal groups represented. Analysis by GC/MS (gas chromatography coupled with mass spectroscopy) may be used to provide quantification and identification of the various constituents.

Table 3.2 illustrates a typical sample of volatiles obtained by pulp-stripping, in this case from raspberry fruit. The recovered volatile fraction (5%) contains the flavour constituents identified in ppm m/m, while the low concentrations emphasise the high potency of these natural flavouring components.

3.7 Legislative concerns

Fruit juices, whether of natural strength or concentrated, are materials of commerce, to be sold direct, or for use in a variety of food and drink applications. It is essential that they conform in the final product to legislative requirements for authenticity and purity, whether for labelling purposes (in avoidance of misleading statements), nutritional standards or food safety.

On a global scale, there is good correlation between quality standards for fruit and fruit juice processed in the different regions. In Europe, legislative controls are set up or modified by the central European Council, usually following discussion with, and between, trade organisations from the ECC member countries. The new directive, once agreed and approved, is translated into the statutory laws of the member country or countries concerned.

3.7.1 European fruit juice and nectars directive and associated regulations

In Europe, Council Directive (2001/112/EC) relating to fruit juices and certain other products intended for human consumption was published in the *European Official Journal* on 12th January 2002, thereby becoming law. This directive revoked several earlier regulations and their amendments dating from 1977.

This directive, in UK Law, was effectively an update on a range of previous regulations, invoked various provisions from the Food Safety Act 1990, Council Directive 95/2/EC (Miscellaneous additives), Council Directive 2001/111/EC (Sugars).

Definitions and descriptions are given (Schedule 1) for fruit juice and concentrated fruit juice, and also for fruit juice from concentrate. It is this latter category that has involved some difficulty in the interpretation of the regulations and, following progress in the harmonisation of international legislation, the Directive was amended in August 2009 by Commission Directive 2009/106/EC.

The revised form took into account other international standards, including the Codex Standard for fruit juices and nectars (Codex Stan 247-2005) and Code of Practice for the evaluation and quality and authenticity of juices, drawn up within AJIN guidelines (3.7.2).

As is the practice, the new directive has now been integrated with the relevant legislation of the European Community member countries, a process often taking about two to three years from the first publication of the document in the Official Journal of the EEC. Within the United Kingdom, this necessitated new regulations being approved by the parliamentary authorities of the member regions. In England, Statutory Instrument 2011 No.1135, known as 'The Fruit Juices and Fruit Nectars (England) (Amendment) Regulations 2011' was laid before Parliament on 20th April, and was scheduled to come into force on 16th May 2011.

As always, legislative and regulatory adjustments are designed to clarify where there has been confusion, to maintain product standards, and to protect the consumer from being misled by descriptive ambiguity.

The amended regulations include, in 'Schedule 6', a listing of fruit types with minimum Brix levels, to be complied with when reconstituting fruit juices and fruit nectars from their concentrates. This list is reproduced in Table 3.3, and it can be seen from the footnotes therein that what might be considered as straightforward is often anything but!

Table 3.3 Minimum Brix levels for Fruit Juices from concentrate, taken from recent European legislation (Schedule 6 of Statutory Instrument 2011 No.1135).

Fruit's common name	Botanical name	Minimum degree Brix values for reconstituted fruit juice and reconstituted fruit purée
Apple (*)	*Malus domesticus* Borkh.	11.2
Apricot (**)	*Prunus armeniaca* L.	11.2
Banana (**)	*Musa sp.*	21.0
Blackcurrant (*)	*Ribes nigrum* L.	11.6
Grape (*)	*Vitis vinifera* L. or hybrids thereof *Vitis labrusca* L. or hybrids thereof	15.9
Grapefruit (*)	*Citrus x paradise* Macfad.	10.0
Guava (**)	*Psidium guajava* L.	9.5
Lemon (*)	*Citrus limon* (L.) Burm.f.	8.0
Mandarin (*)	*Citrus reticulate* Blanco.	11.2
Mango (**)	*Mangifera indica* L.	15.0
Orange (*)	*Citrus sinensis* (L.) Osbeck	11.2
Passion Fruit (*)	*Passiflora edulis* Sims	13.5
Peach (**)	*Prunus persica* (L.) Batsch	10.0
Pear (**)	var.*Persica*	11.9
Pineapple (*)	*Pyrus communis* L.	12.8
Raspberry (*)	*Ananas comosus* (L.) Merr.	7.0
Sour Cherry (*)	*Rubus idaeus* L.	13.5
Strawberry (*)	*Prunus cerasus* L. *Fragaria x ananassa* Duch.	11.2

1. If a juice from concentrate is manufactured from fruit not mentioned in the above list, the minimum Brix level of the reconstituted juice shall be the Brix level of the juice as extracted from the fruit used to make the concentrate.
2. For those products marked with an asterisk (*), which are produced as a juice, a minimum relative density is determined as such in relation to water at 20/20°C.
3. For those products with two asterisks (**), which are produced as a purée, only a minimum uncorrected Brix reading (without correction of acid) is determined.
4. In respect of blackcurrant, guava, mango and passion fruit, the minimum Brix levels only apply to reconstituted fruit juice and reconstituted fruit purée produced in the EU.

The commercial importance of juice concentration lies, of course, in the area of transportation. The removal of a large volatile fraction (including both water and flavour top-note components) from the natural strength juice will reduce both weight and volume of product, resulting in worthwhile cost-savings on transport. At the same time, there are benefits over natural strength juice, in that the higher concentration of dissolved solids (usually 66–72% m/m RS) will exert an osmotic effect upon microorganisms, if present, and inhibit their growth. Hence, there will be greater stability and an extended shelf-life under cool or frozen storage conditions for the so-called 'commercially sterile' juice concentrates.

Some difficulty is likely to occur, however, when the concentrated juice is reconstituted. In an ideal situation, both actual water and flavour components removed during concentration should be returned when reproducing the diluted form. This approach would seem to be commercially untenable, yet is included in part of the legislation. Guidelines (Food Standards Authority) provide an extension upon the description of 'Fruit juice from concentrate', enabling a more practical approach to be made, in which the following designation is given:

'The product obtained by replacing, in concentrated fruit juice, water extracted from that juice during concentration, and by restoring the flavours and, if appropriate, pulp and cells lost from the juice but recovered during the process of producing the fruit juice in question or fruit juice of the same kind; in which the water added must display such chemical, microbiological, organoleptic and, if appropriate, other characteristics as will guarantee the essential qualities of the juice; and the product must display organoleptic and analytical characteristics at least equivalent to those of an average type of fruit obtained from fruit or fruits of the same kind.'

3.7.2 AIJN Guidelines

Central to processing of fruits and juice products within member counties of the European Union is the AIJN – European Fruit Juice Association. This body, through its technical committee, has issued guidelines detailing standards for the range of juice products manufactured in the EU, and also on GMP matters.

The 2001/112/EC Directive referred to above, during its draft stage, led to much discussion, as might be expected, on the subject of restoration of flavour to fruit juice reconstituted from concentrate. The AIJN guidelines have been published to clarify this part of the directive. Add-backs, restoration aromas or flavours are focused upon, in an effort to clarify any controversy over their use.

Section 5 of the guidelines is perhaps the most important, as it refers to permitted solvents to be used in the preparation of the restoration aromas;

'Additives & Solvents permitted in the manufacture of Natural Restoration Aromas: – The natural status of these aromas should not be jeopardised by the use of materials not from the named fruit. Water, food grade CO_2, and ethanol from non-GMO foodstuffs may be used as solvents, and also in extraction. Other solvents and additives, even those allowed by the EU Flavouring Directive (88/388/EU) and subsequent amendments are not permitted.'

3.7.3 Labelling regulations and authenticity

'All things should be laid bare, so that the buyer may not be in any way ignorant of anything the seller knows.' (Cicero)

This has always been an objective of fair trade, but the temptation to enhance profitability too frequently intervenes, and it is necessary for the buyer at times to carry out testing to ensure authenticity of a purchased raw material, so that the declaration subsequently given can meet requirements of labelling. Over the years, fruit juices have often been subjected to adulteration. As analytical techniques have become more objective and accurate in discovering fraud, so the methods used by the fraudsters have also become more technically devious.

The most frequent method of deception is to dilute the concentrated juice product with sugar syrup, and to adjust flavour levels and colour levels as necessary. Thirty years ago, the 'stretching' of juices would be carried out in a fairly basic fashion, using sugars that did not match the fruit, booster flavourings likewise, and sometimes colours that might have been more at home on an artist's palette.

Nowadays, such subterfuges would be easily recognised. Periodically, there have been attempts at using analytical profiling of certain of the juice components. Amino acid profiles were, at one time, thought to be an ideal route for the proof of authenticity – particularly in the case of citrus juice concentrate – until it was discovered that a carefully selected blend of amino acids could be used to restore balance, at least on an analytical scale. Therefore, such testing went out of vogue. However, the testing of a wide range of parameters was, for many years, still the only real method of arriving at a sensible assessment of authenticity.

Perhaps the most detailed system was that developed in the early 1970s, within the auspices of the Association of the German Fruit Juice Industry. A special working group of experts from research, industry and food control was set up, whose objective was to devise a method of evaluation of fruit juices. The initial results were published in 1977, under the title '**R**ichwerte und **S**chwankungbreiten bestimmter **K**ennzahlen für Apfel-, Trauben- und Orangen-säfte' (Guide values and ranges of specific reference numbers for apple juice, grape juice and orange juice). The system of RSK-values was officially recognised within the Federal Republic of Germany in 1980, and RSK-values for other fruit juices were published in the years following. The appraisal relies upon the degree of correlation within a full set of analytical values, and the process can be quite time-consuming, as something of the order of thirty values can be included in the assessment.

Although AIJN and RSK-value guidelines are of great use in the evaluation of fruit juices, there is always the risk of fraud.

3.7.4 Juice in the diet – 'five-a-day'

The term 'five-a-day' refers to an initiative operating in various countries worldwide, taken to encourage the consumption of at least five portions of fruit and vegetables each day. This follows recommendations given by the World Health Organisation to consume a minimum of 400 g of fruit and vegetables daily.

To receive maximum benefit to health, a variety of fruit and vegetable types should be taken, in order to access a wider spectrum of minerals, fibre content and other nutrients. Fresh natural strength fruit or vegetable juices qualify well in this respect, and 200 ml of 100% fruit or vegetable juice counts as a portion. A beverage 'smoothie', consisting primarily of mixtures of puréed fruit and vegetables, will also fit nicely into the five-a-day concept, and this type of product has become a major source of readily available fruit intake for the health conscious consumer.

In general, fruit juices are, of course, all very low in sodium, saturated fats and cholesterol. They are good sources of Vitamin C and a measurable quantity of other vitamins, as well as minerals, trace elements and amino acids, and therefore their inclusion in the diet can only be beneficial.

3.8 Quality issues

Quality can be regarded as a measure of suitability of the fruit juice, fruit juice concentrate or fruit juice extract for an intended application. In general, whatever the application, it will be the consistency in performance of the product, from batch to batch, from season to season, that is the prime concern. In order to meet quality targets, therefore, it becomes critical that processing is carried out in the correct manner, using fruit of an optimum level of maturity, and that the product is stored under suitable conditions in order to limit any effects of degradation during a required shelf-life.

3.8.1 Absolute requirements

As previously indicated, the evaluation of fruit juice will, as a rule, require an expert appraisal of the entire analysis. In practice, there will be key parameters to be noted and adhered to during the manufacture of fruit juices and related products. Of these, the soluble solids content and titratable acidity are the major indicators to be taken into account when identifying the status and suitability of a juice product for use in application.

3.8.1.1 Soluble solids

The soluble solids content will relate directly to both the sugars and fruit acids, as these are the main contributors. Pectins, glycosidic materials, and metals such as sodium, potassium, magnesium and calcium and so on, when present, will also register a small but insignificant influence on the solids figure.

Table 3.4 Brix table.

Degree Brix or % w/w sucrose	Apparent SG at 20/20°C	Grams of sucrose per 1,000 ml (in air)	Degree Brix or % w/w sucrose	Apparent SG at 20/20°C	Grams of sucrose per 1,000 ml (in air)	Degree Brix or % w/w sucrose	Apparent SG at 20/20°C	Grams of sucrose per 1,000 ml (in air)
1.0	1.003	10.8	31.0	1.134	340.5	61.0	1.295	787.7
2.0	1.008	21.7	32.0	1.139	363.4	62.0	1.301	804.3
3.0	1.012	32.8	33.0	1.144	376.4	63.0	1.307	821.1
4.0	1.016	44.0	34.0	1.149	389.4	64.0	1.313	838.0
5.0	1.020	50.8	35.0	1.154	402.6	65.0	1.319	855.0
6.0	1.024	61.3	36.0	1.159	415.9	66.0	1.325	872.1
7.0	1.028	71.7	37.0	1.164	429.3	67.0	1.331	889.5
8.0	1.032	82.3	38.0	1.169	442.8	68.0	1.338	907.0
10.0	1.040	103.7	40.0	1.179	470.2	70.0	1.350	942.3
11.0	1.044	114.5	41.0	1.184	484.0	71.0	1.356	960.3
12.0	1.048	125.5	42.0	1.189	498.0	72.0	1.363	978.3
13.0	1.053	136.5	43.0	1.194	512.1	73.0	1.369	996.5
14.0	1.057	147.7	44.0	1.120	526.3	74.0	1.375	1014.9
15.0	1.061	158.7	45.0	1.205	540.7	75.0	1.382	1033.5
16.0	1.065	170.0	46.0	1.210	555.2	76.0	1.388	1052.2
17.0	1.070	181.4	47.0	1.216	569.8	77.0	1.395	1071.1
18.0	1.074	192.8	48.0	1.221	584.5	78.0	1.401	1090.0
19.0	1.079	204.3	49.0	1.227	599.3	79.0	1.408	1109.3
20.0	1.083	216.0	50.0	1.232	614.3	80.0	1.415	1128.6
21.0	1.088	227.7	51.0	1.238	629.4	81.0	1.421	1148.7
22.0	1.092	239.6	52.0	1.243	644.6	82.0	1.428	1168.5
23.0	1.097	251.5	53.0	1.249	660.0	83.0	1.435	1189.6
24.0	1.101	263.5	54.0	1.254	675.5	84.0	1.441	1208.2
25.0	1.106	275.6	55.0	1.260	691.1	85.0	1.448	1228.3
26.0	1.110	287.9	56.0	1.266	706.9			
27.0	1.115	300.2	57.0	1.272	722.8			
28.0	1.120	312.6	58.0	1.277	738.8			
29.0	1.124	325.2	59.0	1.283	755.0			
30.0	1.129	337.8	60.0	1.289	771.3			

While the soluble solids can be determined gravimetrically from the juice sample (filtered clear, if necessary, from any suspended solids), it is usual to refer soluble solids to the more accessible determination of Brix value. As it is 'dissolved solids' that influence the measurement, there is direct relationship between Brix values and specific gravity of solution. Although accurately determined by a BRIX hydrometer, which reads directly in 'percentage' sucrose, with the higher viscosities of concentrated fruit juices, it is more convenient to use an optical refractometer, Brix calibrated, thereby providing a direct reading of % w/w sucrose.

Although the term '% w/w RS' (i.e. refractometric solids) would be more appropriate for fruit juices when measured in this way than 'degrees Brix', it is the latter that is in general use in the fruit juice industry to indicate the degree of concentration or 'folding'.

Refractometer readings may be affected by the presence of other dissolved solids. The presence of fruit acids can, as might be expected, influence the refractometric Brix reading, and should strictly be taken into account when calculating the degree of juice concentration but, in most cases, it does not have a significant effect. However, where there are appreciable levels of acid in the juice (e.g. lemon and lime juices), where the amount of acids usually exceeds the level of sugars, there will be a need to apply the correction.

The table of Stevens and Baier (1939) gives a correction for obtaining Brix from refractometer readings from juices or other acid-containing sugar solutions. Based on the citric acid content of juices, the corresponding correction is to be added to the refractometer reading.

3.8.1.2 Titratable acidity

The acid character of a juice contributes to its flavour type, and is taken into consideration when assessing the value of the juice for inclusion into new beverage product formulations.

Acid content (% w/w) is determined by direct titration against a standardised alkali solution (e.g. 0.1 Molar sodium hydroxide), using a pH meter, to an endpoint at pH 8.1. Where the juice is naturally clear, or has been clarified, and is of low colour intensity, the end-point may be accurately found using phenolphthalein as indicator.

Although there are other acids present in fruit juices (e.g. oxalic, iso-citric, tartaric, etc.), it is usual to record acidity in terms of citric acid, both for citrus fruit juices and for the majority of soft fruit juices. Where apple and other pome fruit juices are concerned, the major organic acids are malic and citric, and malic usually predominates. In some varieties of pears, the two acids can occur in approximately equal proportions. Nevertheless, it is general practice to quote titratable acidity for pome fruit juices as % w/w malic acid. The adjustments required, if the juice is to be used in the production of apple based drinks, will be effected by the addition of malic acid. Both acids are usually referred to by % w/w in their anhydrous form, although it is sometimes convenient to determine titratable acidity for citric acid in terms of its monohydrate, as this form of the acid may be used in formulation for certain beverages.

As a general rule, the acidity of juices will decrease with increasing maturity of the fruit source, or with increasing levels of sugars in the resulting juice. Thus, the ratio of soluble solids (e.g. Brix values) to acidity is an important value in the assessment of juice quality. The Brix/acid ratio is frequently used to establish standard sensory, or taste, qualities for incoming juice supplies, and to minimise the effect of seasonal variation. The higher the Brix in relation to the acid content of the juice, the higher the ratio and the 'sweeter' the taste.

3.8.1.3 Other quality considerations

In order to create a quality standard for a particular juice, natural strength or concentrate, there are a number of fairly routine analytical test parameters to be considered. These include:

- **Specific gravity** (20°C/20°C),
- **Acidity** (% w/w) as citric acid **anhydrous** (or malic acid, as appropriate)
- **Colour measurements**. For **clarified** juices, strawberry, blackcurrant and so on, this will be a spectrophotometric determination, recording absorption of monochromatic light at a specified wavelength on passing through a 1 cm pathway. Cloudy juices may be assessed directly by visual comparison, using a colorimeter against a recognised standard,
- **Pulp content** (cloudy juices only) – screened and suspended pulp determinations. 'Screened' pulp is a term applied in the citrus juice industry, where the natural juice cells, or sacs, present in citrus juice are described as screened or floating pulp. Such pulp is normally removed during processing, and can be added back, as desired, to meet a required specification. 'Suspended' pulp, is centrifugable to a degree, and may be determined as such, or by certain specialised tests, where a juice concentrate is diluted into a measured volume of water and allowed to stand and settle over a specified time. Measurements are generally empirical, but serve to provide comparisons within agreed parameters – for example, the Imhoff cone test, in which the sedimentable pulp is allowed to settle to the base of a calibrated (inverted) conical vessel, where its apparent volume may be read off.
- **Oil content.** Citrus juices are unique amongst other fruit juices, where normally no oil is present, in containing residual oil after processing, or oil that has been added back to its concentrates. It is essential to standardise oil levels, as much of the flavour character is established. The oil content is determined by using either the 'Clevenger' or 'Cocking-Middleton' procedures. These methods make use of an oil trap apparatus in distilling off, and collecting, the volatile oil fraction of the juice. Both apparatus types include a calibrated vertical glass tube into which the volatile oil, upon steam distillation, collects, and it may be easily measured and quantified with respect to the weight of juice taken.
- **Ascorbic acid**. Many juices contain ascorbic acid or vitamin C. Quantitatively, this is the most important vitamin in soft fruits, ranging from a negligible level in some whortleberries, to around 200 mg/100 gm in blackcurrants. Ascorbic acid performs a valuable function as an antioxidant in minimising the degradation of certain flavour principles, and it is often important for it to be included in processed juice or in a soft drink formulation. Levels in the region of 200–400 mg/kg are typical. It should be noted that ascorbic acid is to be added to the natural strength juice, if this is intended for direct use, or to the juice concentrate. Addition to natural strength juice, before its concentration, will result in its own degradation during the heating process, and ultimately cause spoilage of the product, when an intense browning reaction takes place.
- **Preservatives**. Modern aseptic processing techniques and packing of juices will largely obviate the use of preservatives, which are not permitted for use in

pure fruit juices. However, there are instances where they are required, so they will need to be specified. Preservatives are strictly controlled by legislation, with upper limits given for sorbic and benzoic acids used singly in soft drinks, as 300 mg/l and 150 mg/l, respectively. When used at higher levels for juice concentrates, it is important to recognise the statutory levels for these in application. Sulphur dioxide has been severely limited in recent years, and in soft drink formulations it is now permitted, under European legislation, at just 10 ppm 'carry-over' from the use and addition of juice concentrates. HPLC techniques for the analysis of sorbic and benzoic acids have largely replaced the earlier well-used method, where the acid type was isolated from its sample by steam distillation and the determination carried out spectrophotometrically on the resulting distillate.

- **Yeasts, moulds and bacteria**. Because of their low pH, fruit juices will present less than ideal conditions for pathogenic bacteria species. Therefore, these are generally of no major concern for juice producers who operate under good manufacturing practice. There are acid-tolerant bacteria, however, whose presence can give rise to off flavours, and this effect may be encountered with citrus juices, where it is thought that both diacetyl and acetylmethylcarbinol may be metabolic products of the growth of acid-tolerant bacteria. Diacetyl will introduce a mild cheesy or buttery note to the juice, while acetylmethylcarbinol produces no off-flavour (although its presence may indicate bacterial growth). Where appropriate, it is customary to determine the diacetyl value, and this is effected by a colorimetric method involving the reaction of diacetyl, isolated from a sample, with alpha-naphthol, and comparison of the optical density of resulting colour formation against a standard scale (diacetyl v. O.D.). The standard plate count (selected agar media, 29–31°C, three days) giving the total viable count (TVC), and the pour plate yeast and mould count method (OGY agar, 20–24°C, 3–5 days) are generally used for microbiological evaluation to ensure results will be within acceptable limits. Complications may arise with certain spoilage types, where specialised control procedures will need to be applied – in particular, for osmophilic yeasts, if present in concentrates, and also the presence of spore-forming yeast or mould species, which, in their inactive state, can withstand the effects of pasteurisation.

3.9 In conclusion

Fruit juices are of great commercial importance in their own right, as well as for direct use as ingredients in food and beverage products. As we have seen, they form the basis of a worldwide industry. While production of natural strength juices can be sustained locally for direct consumption, it is the concentrates that offer advantages in worldwide trade, in view of savings in terms of bulk transportation. Modern methods of processing are aimed at optimising all quality factors by use of highly efficient, short-time processing, followed by pasteurisation and aseptic filling. The advances in techniques for this latter procedure, coupled with improved sanitisation of plant facilities, have now reached a very high standard.

Improved conditions of storage, incorporating refrigeration, are used more and more to offset colour degradation effects and to maximise product shelf life.

This chapter has touched upon some of the mechanical techniques used to express juice from the fruit source, but it should be noted that there will be process variations on those listed that are subject to exclusivity within parts of the industry, and details of which are not publicly available.

Good improvements in the concentration of fruit juices have been achieved in recent years. As it is the case that the majority of concentrates are to be reconstituted in application, the quality achievable for the natural strength remake will be an important issue.

With world population figures scheduled to reach in excess of nine billion at the halfway mark of the current century, it is evident that, as with the food industry in general, the challenge facing juice producers will be one of major interest.

The forecast of changes in climatic conditions due to global climate change is cause for concern amongst fruit growers and fruit processors, and the industry will be forced to adapt accordingly. While the world production total will be expected to increase at perhaps the existing annual average of around 2%, the composition of that total will have altered considerably as the annual production output of individual countries is seen to fluctuate due to climatic changes. For instance, the increases not necessarily due to climate shown by China over the past decade, averaging around 7.5% for Pome fruits, have countered the decreases experienced by other country producers to maintain the worldwide annual increase in the region of 2%.

Other concerns which have potential to affect the food industry in general, and which are of particular importance to the fruit growers and fruit juice producers, include a threat that has appeared relatively recently, involving those essential contributors to a good fruit harvest –the insect pollinators. Other, and perhaps more immediate, problems include the effects of plant diseases, such as citrus sudden death.

Insect pollinators, primarily of the bee family, who work tirelessly amongst the plant flora, gathering nectar and ensuring the onset of the fruit growth, are now mysteriously in decline in many parts of the world. The issue is one that, as yet, needs to be fully understood.

Commercial honey producers have recently been faced with the loss of sometimes 40–50%, or more of their hives, due to a colony collapse disorder, where one of the chief suspect causes is the use in agriculture of neonicotinoids – now the world's most widely used form of pesticide. The matter has such serious implications that the European Commission announced a ban, from December 2013, of the use of three major nicotinoids. This is a controversial decision, as other factors have yet to be fully explored. These may include fluctuating temperatures conditions and long, extended winters, particularly in Europe.

It has long been known that the varroa mite, a parasitic insect whose life cycle is linked with the honey bee as its host, carries a destructive virus that can cause hive collapse. However, other pollinators in the bumblebee family are also disappearing in large numbers, where the varroa mite would not be causal.

A great deal of work is being carried out worldwide to supply answers to the global decline in bee populations, and one cannot underestimate the seriousness of the situation.

Commercial fruit juice represents the end of a carefully orchestrated chain of events, starting with the selection and cultivation of certain fruit-bearing botanical species. Harvesting yields, seasonal changes and maturation, among other factors, have to be taken into account before the processing and juice production takes place, so it is not surprising that we will encounter subtle variations in the taste profile of the final product. We can anticipate that, in future, more focus and research effort will be directed to the recovery and treatment of the aroma volatiles and other natural flavour ingredients, and the manner in which their reintroduction to the diluted concentrate is carried out. It is here, in the area of taste, that the consumer finally puts product quality to the test, and a resulting guarantee of satisfaction will be essential to future growth and commercial viability throughout the industry.

References and further reading

AIJN (1997). *Code of Practice for Evaluation of Fruit and Vegetable Juices*. AIJN, Brussels, Belgium.

Ashurst, P.R. and Taylor, R.B. (1995). In: Ashurst, P.R. (ed). *Food Flavourings*, Chapter 4. Blackie Academic & Professional, Glasgow.

Downes, J.W. (1990) In: Hicks, D. (ed). *Production and Packaging of Non-carbonated Fruit Juices and Fruit Beverages*, Chapter 6. Blackie & Son Ltd, Glasgow.

European Council Directive 93/77/EEC of 21 September 1993.

European Council Directive 2001/112/EEC of 20 December 2001.

European Council Directive 2009/106/EEC of August 2009.

Hulme, A.C. and Rhodes, M.J.C. (1974). In: Hulme, A.C. (ed). *The Biochemistry of Fruits and their Products*, Chapter 10. Academic Press, London.

Janda, W. (1983). In: Godfrey, T. and Reichelt, J. (eds). *Industrial Enzymology*, Chapter 4.10. Macmillan Publishers Ltd., New York.

Kidd, F. and West, C. (1922). Gt.Brit. Report of the Food Investment Board for 1921, p.17.

Redd, J.B. Hendrix, C.M. and Hendrix, D.L. (1986). *Quality Control Manual for Citrus Processing Plants*. Intercit Inc., Florida.

Rutledge, P. (1996). In: Arthy, D. and Ashurst, P.R. (eds). *Fruit Processing*. Blackie Academic & Professional, Glasgow.

Stevens, J.W. and Baier, W.E. (1939). *Refractive Determination of Soluble Solids on Citrus Juices, Industrial and Engineering Chemistry, Analytical Edition* **11**, 447–9.

TNO Nutrition and Food Research Institute (1996). *Volatile Compounds in Food: Qualitative Data*. 7th edition. TNO, The Netherlands.

World Apple and Pear Association (WAPA): www.wapa-association.org. Rue de Treves 49–51 bte 8.1040 Brussels, Belgium.

CHAPTER 4

Water and the soft drinks industry

Tony Griffiths
Quality Systems Consultant, Chelmsford, UK

4.1 Usage of water in the industry

There is surprisingly little up-to-date information on the exact usage of water in industry, but the UK Food and Drink Industry is a major user of water, and it is estimated that this industry uses nearly 700 megalitres (700 000 m^3) of water daily. This is mainly used for processing, cleaning, diluting, cooling, product conveyance and steam generation, as well as for the water incorporated in the products themselves. Water as a resource, and the discharge of trade effluent, contribute significantly to the operating costs of food and drink companies.

The ratio of water used to its content in a drink sold can be very high; an upper ratio of 10 : 1 has been quoted, although this is not readily appreciated by most consumers and possibly some producers. The UK Food and Drink Federation (FDF) has introduced initiatives to reduce water usage in the industry.

This could mean that a soft drinks factory producing one million litres of drink could use ten million litres of water in its production. Of those ten million litres, nine million are thrown away down the drain. With effluent costs often higher than the cost of incoming potable water, this should ensure that the effective use of water as a raw material has a high profile. The actual ratio of use to product is dependent upon the different sectors within the drinks industry, as well as the size and efficiency of each operation. Thus, a company producing fruit-based soft drinks in returnable glass may have a ratio of 10 : 1, because of the need to wash and clean the returned bottles, while a company bottling spring water into non-returnable PET might have a ratio as low as 1.5 : 1. These ratios are also affected by how effective a company is at controlling and reutilising water within its factories.

Chemistry and Technology of Soft Drinks and Fruit Juices, Third Edition. Edited by Philip R. Ashurst.
© 2016 John Wiley & Sons, Ltd. Published 2016 by John Wiley & Sons, Ltd.

4.2 Sources of water

4.2.1 Water cycle

Water is a natural raw material, derived almost entirely from rain. The rain forms part of the hydrological cycle (see Figure 4.1) where, starting with the oceans that cover about 70% of the worlds surface, heat from the sun causes evaporation into the atmosphere. When conditions are right, water vapour concentration exceeds the saturation limit and rain falls. Rain falls on both land and sea in very varying amounts (the average over the whole world is about 0.5 m/yr.) However, this can range from virtually zero, in some desert regions, to about 2.5 m/year in places like the Sierra Nevada mountains in the USA, which are fed by the moist winds from the Pacific Ocean. The extreme of the rainfall was recorded in Cherrapundi in India where, in 1860, in excess of 1000 inches (25.6m) fell in a year. Of the rain that falls on land (about 20% of the total), about two-thirds returns to the atmosphere by evaporation and transpiration. Of the remainder, most of this water eventually returns, in one form or another, to the hydrological cycle.

In theory, water is an inexhaustible raw material, with a large excess of availability over demand. However, UK statistics of recent years have shown that, where the rainfall has been below average, increase in demand has given rise to water shortages. Other years have seen significant and repeated flooding in some areas. Changes in seasonal rainfall and temperature patterns will also have an impact on water quality and availability, which need to be considered.

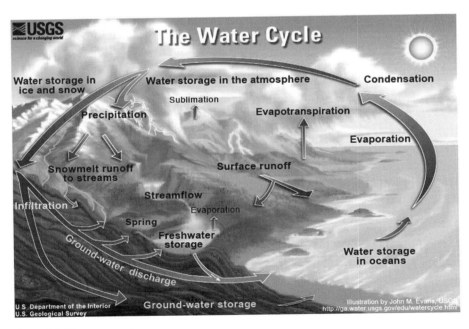

Figure 4.1 The water cycle.

4.2.2 Surface water

Surface waters will contain dissolved atmospheric gases, as well as those derived from industrial activities. Water can also pick up particulates of many kinds from the atmosphere. Once collected on the surface, water will pick up many materials, such as decaying vegetation, bacteria, mineral particles, organics from biodegradation, pesticides and herbicides. Surface water may also contain raw and partially decomposed sewage, which will include bacteria associated with faecal matter, as well as other pathogens and viruses. Industrial effluent will contribute detergents and a wide range of organics, as well as other chemical residues and metal ions, which will increase the biological oxygen demand (BOD) of the water. Agricultural run-offs will add fertilisers, herbicides and pesticides, as well as bacteria and viruses associated with livestock and their excreta.

4.2.3 Ground water

Materials found in ground waters are derived from the atmosphere, as they are for surface waters. Because the water permeates through various rock types, it also picks up (depending on the geology of the area) various dissolved anions, such as bicarbonates, sulphates, nitrates and chlorides. It will dissolve cations such as calcium, magnesium, sodium and potassium, as well as trace levels of iron and manganese. Though not usually collecting the detritus that can contaminate surface waters, many soluble organic and inorganic materials can leach through to the ground waters. Microbiological contamination can also occur.

4.2.3.1 Connection between surface and groundwater sources

The connection between surface water and groundwater is shown in Figure 4.2 below. Groundwater discharges to surface water at various rates, varying from

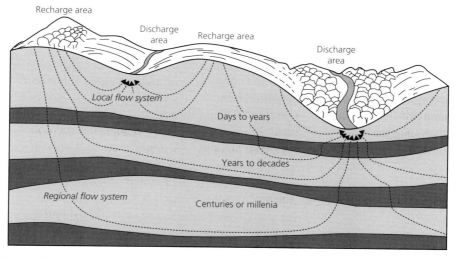

Figure 4.2 Connection between surface and groundwater sources.

the local flow cycle, which operates over a few days to years, to regional flow systems, where the transfer rate is of the order of centuries. The groundwater flow sustains surface flows during periods when little or no rain falls.

4.3 Quality standards relating to water

4.3.1 UK legislative standards

The UK Food Safety Act, 1990 (as amended) requires that any item of food or drink supplied for sale must be safe. Where drinks manufacturers are starting with potable water, the emphasis is on maintaining its quality and not contaminating it with undesirable materials, or encouraging microbiological growth. This is usually a hygiene issue, and is covered by the concept of HACCP (Hazard Analysis of Critical Control Points) to identify and control risks. EU Legislation (98/83/EC) specifies requirements for parametric values for most components and potential contaminants which might get into the water supplies. These are monitored and controlled by the water supply companies. Where companies process private potable or non-potable water sources, such as boreholes, they are responsible for ensuring that their processed water complies with the relevant EU/UK regulations. (UK Private Water Supplies Regulations 2009, SI 2009 No 3101).

The quality standards for water are derived mainly from World Health Organisation (WHO) findings, which are based on scientific and medical knowledge. The EU regulations were first published in 1980, with the UK regulations first appearing in 1989, plus further amendments over the years. As a consequence of the second edition, the EU Directives were reviewed and a third edition issued in 1998.WHO has since revised its advice again, issuing the 'Guidelines for Drinking Water Quality – 4th Edition' (WHO Press) It is likely that, in due course, revised EU regulations will be issued, based on the WHO fourth edition.

4.3.2 Internal and customer standards

Not only do manufacturers have to comply with EU, UK or other international regulations, but also with the requirements of their customers, franchisors and consumers, as well as their own internal standards and policies.

If a company acts as a franchisee, then the franchise holder will often impose constraints on the quality of water used in the product. In some cases, this impact may be minimal while, in others, such as if dealing with a worldwide franchise, the impact can be considerable. For example, major international beverage companies try to ensure that their products have an identical taste around the world. To this end, many companies have a standard water quality for beverage makeup, and expect their franchisees to install appropriate equipment to produce water of that quality.

Customers could be considered to include franchisors, but here, the consideration is major retailers and dispense vending operations. The major retailers may well impose their own constraints on the quality of water used in the manufacture of products that will be retailed under their brand name. Dispense operations, such as those in pubs, clubs, cinemas and restaurants, can have a major impact on the quality of products dispensed, due to variable water quality. Here, a company's concentrated product is added to water supplied by another company. While the water must be potable, its constituents and impact on flavour can vary greatly across the country.

Although consumers do not have any legal claim to set quality standards, they can have an impact in terms of preference determined either by sales figures or market research. A variable water supply may influence the taste of the product, and consumer perception of the products, when dispensed or made up locally. With regard to purity, the consumer is strongly influenced by the press, and can quickly turn away from any product which is believed (rightly or wrongly) to be potentially harmful. Manufacturers thus need to ensure that, while removing undesirable components, their water treatment processes do not add anything unnecessary to the water.

Finally, there are internal requirements, as manufacturers set their own standards which meet all the needs of all their customers, as well as meeting the relevant statutory obligations and industry codes of practice. A company must to respond to any changes by evaluating the impact on their business, be it economic, ethical or environmental. Such changes may require methods of processing water to be changed.

Current EU standards for potable water are shown in Table 4.1, and the requirements for spring and natural mineral water, both of which are widely used for manufacturing, as well as direct consumption, are shown in Table 4.2.

4.4 Processing water

4.4.1 Required quality

The specification for water used within the drinks industries will vary from company to company. In almost all cases, the source of water is the potable supply from the water utility main, or an equivalent borehole producing water that is already of a potable standard, or becomes so with suitable treatment. From an industry standpoint, what is ideally required is a water supply that meets all the varied needs at the turn of a tap, and which gives total consistency in quality (including physical, organoleptic, chemical and microbiological constituents), pressure of supply, flow rate and, in some cases, temperature.

In practice, most companies start from the base of available potable water, then adjust this to suit their needs – for example, reducing alkalinity or hardness, as appropriate, or removing chlorine and other unwanted taints or microorganisms.

There are several parameters of interest to the drinks industry, and some of these are given below, with typical quality values.

- **pH** – water with a low pH (<4.0) is acidic, and may be aggressive, while alkaline water with a high pH (> 8.5) may form scale, will have a soapy feel, and will adversely effect soft drinks. Potable water is required to have a

Table 4.1 Current EU water standards for potable water.

EU parameters and parametric values
(for indication only)

Note: for a number of these parameters, such as lead, the values allowed are being decreased over time and it is essential to read the directives and subsequent amendments.

PART A Microbiological parameters		PART B Chemical parameters	
Parameter (number/100 ml)	Parametric value	Parameter	Parametric value
Escherichia coli (E. coli)	0	Acrylamide	0.10 µg/l
Enterococci	0	Antimony	5.0 µg/l
		Arsenic	10 µg/l
The following applies to water offered for sale in bottles or containers:		Benzene	1.0 µg/l
		Benzo(a)pyrene	0.010 µg/l
		Boron	1.0 mg/l
Escherichia coli (E. coli)	0/250 ml	Bromate	10 µg/l
Enterococci	0/250 ml	Cadmium	5.0 µg/l
Pseudomonas aeruginosa	0/250 ml	Chromium	50 µg/l
Colony count 22°C	100/ml	Copper	2.0 mg/l
Colony count 37°C	20/ml	Cyanide	50 µg/l
		1,2-dichloroethane	3.0 µg/l
		Epichlorohydrin	0.10 µg/l
		Fluoride	1.5 mg/l
		Lead	10 µg/l
		Mercury	1.0 µg/l
		Nickel	20 µg/l
		Nitrate	50 mg/l
		Nitrite	0.50 mg/l
		Pesticides	0.10 µg/l
		Pesticides	Total 0.50 µg/l
		Polycyclic aromatic hydrocarbons	0.10 µg/l
		Selenium	10 µg/l
		Tetrachloroethene and Trichloroethene	10 µg/l
		Trihalomethanes	Total 100 µg/l
		Vinyl chloride	0.50 µg/l

Table 4.1 (*Continued*)

PART C Indicator parameters		Radioactivity	
Parameter	**Parametric value**	**Parameter**	**Parametric value**
Aluminium	200 µg/l	Tritium	100 Bq/l
Ammonium	0.50 mg/l	Total indicative dose	0.10 mSv/year
Chloride	250 mg/l		
Clostridium perfringens (including spores)	0/100 ml		
Colour	Acceptable		
Conductivity	2500 µS cm^{-1} at 20°C		
Hydrogen ion concentration	> 6.5 and < 9.5 pH units		
Iron	200 µg/l		
Manganese	50 µg/l		
Odour	Acceptable		
Oxidisability	5.0 mg/l O_2		
Sulphate	250 mg/l		
Sodium	200 mg/l		
Taste	Acceptable		
Colony count 22°	No abnormal change		
Coliform bacteria	0/100 ml		
Total organic carbon (TOC)	No abnormal change		
Turbidity	Acceptable		

Table 4.2 Inorganic constituents of water.

Major constituent	Minor constituent	Trace component
Calcium	Potassium	Cadmium
Magnesium	Iron	Lead
sodium	Manganese	Mercury
Sulphate	Copper	Rare earths
Chloride	Carbonate	Bromide
silica	Boron	
bicarbonate	Fluoride	
nitrate	Aluminium	
	Zinc	

range of 6.5–9.5, but for soft drinks it is best if it is 6.5 or below, except for specialist products.
- **Colour and turbidity** – critical in making of soft drinks where a clear, colourless product is required.

- **Suspended solids** – a nil value is ideal for the drinks industry.
- **Dissolved solids** – in contrast to suspended solids, the drinks industry can tolerate fairly high values (up to about 500 mg/l), but lower values (150–300 mg/l will usually be preferred
- **Calcium and magnesium** – responsible for hardness, and a major contributor to dissolved solids content. Although not particularly critical for most soft drinks, the presence of calcium may have an adverse effect if poorly processed fruit juice is used as an ingredient, when it can cause serious visual defects.
- **Iron and manganese** – less than 1 mg/l is desirable, because of their influence on taste and, possibly, colour.
- **Nitrate** – a lower limit than the legal maximum (50 mg/l) is desirable, with values of 10 or 20 mg/l being quoted. In the past, the level of nitrate was critical in those companies producing canned beverages, because it could lead to corrosion. This was also accelerated by high levels of sulphates and chlorides. With the extensive use of lacquered cans, this had become less critical but, with the continued light weighting of cans, physical damage to the can and, hence, the lacquer, is now more common, so this parameter (and sulphate and chloride) need continued control.
- **Silica** – can be critical, as excess silica (> 15 mg/l) can lead to increase in the risk of precipitating proteins.
- **Alkalinity** – needs to be below 50 mg/l, expressed as calcium carbonate, to reduce use of acidulants.
- **Chlorides and sulphates** – as with nitrates, these can lead to problems in corrosion of cans, though modern lacquered cans are less susceptible. Can also lead to off-tastes.

4.4.2 Starting quality
4.4.2.1 Main sources of potable water
Potable water supplies are generally derived either from surface sources, or from groundwaters such as boreholes. It is quite common for abstractions to be made from both ground and surface sources, then blended in a reservoir before treatment at a treatment works.

The potable water used as the starting point in most factories is derived from utility mains-treated water or private boreholes. The chemical and microbiological quality and character of this water is directly related to the soil structure, the geology of the surrounding catchment area, and the human and industrial impact on wastewater entering the treatment plants. Rain falling on the ground will pick up and dissolve inorganic salts, naturally occurring organic materials and bacteria, as well as man-made pollutants. For example, rain falling on chalk will produce water that is high in dissolved solids, and will probably have high alkalinity and total hardness, while that falling on granite will be low in dissolved solids, hardness and alkalinity. Water from peaty areas is usually pale yellow in colour, and contains appreciable amounts of organic matter.

Surface waters

These will have a component of dissolved gases, both natural to the atmosphere and those derived from industrial activities. Water can also pick up particulates of many kinds from atmospheric sources. Once collected on the surface, the water will pick up many materials, such as decaying vegetation, bacteria, mineral particles, organics from biodegradation, pesticides and herbicides, as well as undecomposed and partially decomposed sewage, which will include bacteria associated with faecal mater and other pathogens and viruses. Industrial effluent may contribute detergents and a wide range of organics, as well as other chemical residues and metal ions, all of which will increase the biological oxygen demand of the water. Agricultural run-offs are likely to add fertilisers, herbicides and pesticides, as well as bacteria and viruses associated with livestock and their excreta.

Groundwaters

Groundwater permeates through various rock types, from which it picks up (depending on the geology of the area) various dissolved anions, such as bicarbonates, sulphates, nitrates and chlorides. In addition, it will dissolve cations such as calcium, magnesium, sodium and potassium, as well as trace levels of iron and manganese. Though not picking up the detritus that can contaminate surface waters, many soluble organic and inorganic materials can leach through to the ground waters. Bacteriological contamination can also occur. Some components, such as tritium (^3H), found in groundwaters may be derived from the atmosphere. Groundwaters may be effectively filtered through the rock of the area (e.g. sandstone or chalk) from which they are drawn or, in some areas such as karstic limestone, may receive almost no filtration of surface or atmospheric components.

4.4.2.2 Gross chemical description of waters

A common way of characterising water is by describing its hardness, which is determined by the content of calcium and magnesium and is expressed as calcium carbonate. Water may be hard or soft, depending on its geological source, with groundwaters generally being harder than surface waters. Other waters may be described as brackish or humic, as indicated below.

Hard water

This contains calcium and magnesium salts in solution, frequently in the form of bicarbonates, when the water is drawn from a limestone or chalky source or, sometimes, in the form of chlorides or sulphates, when the source is sandstone or calcite. Hard waters have a fully mineralised taste and are thought to be most palatable and beneficial to consume.

Soft water
This is frequently obtained from surface sources, consisting of lakes and rivers flowing through rocky terrain. Soft waters are also obtained from certain underground sources, where the aquifer is igneous rock, gravel, or laterite, and these usually contain sodium and potassium salts in the form of bicarbonates, sulphates, chlorides, fluorides or nitrates. The taste of soft water tends to be slightly soapy.

Brackish water
This contains a high proportion of sodium or potassium chlorides in solution, and occurs frequently in the neighbourhood of underground salt deposits, or in the close proximity of the sea coast. The taste of brackish water is sometimes slightly salty.

Humic water
Soft waters that are derived from moorland peat areas are sometimes called humic waters. These arise from surface sources in areas where peat bogs are prevalent. They may also occur in borehole water abstracted from the marshland plain area of a river valley. The water is coloured slightly yellow by the humic and fulvic acids present, and may also have an unpleasant odour and, frequently, a slightly bitter taste.

Inorganic constituents of water
With appropriate analytical equipment, it is possible to detect almost all inorganic elements in water, as well as a very wide range of organic materials and many different micro-organisms. To put this into perspective, the loadings of these materials in water can be described in terms of major, minor and trace constituents. A major constituent is typically found at levels above 10 mg/l (possible up to several thousand mg/l), while a minor constituent might be in the range 0.01–10 mg/l, and a trace component below 0.01 mg/l (see Table 4.2).

Organic constituents of water
A wide range of different organic molecules, both of natural and man-made origin, can often be found in many water sources. As with inorganic substances, modern analytical techniques are capable of detecting organic materials at the ppt (10^{-12}) level. Because of this complexity, regulating bodies tend to concentrate on those chemicals which are known to have toxic or carcinogenic properties. Thus, there are limits for pesticides, poly aromatic hydrocarbons (PAHs), trihalomethanes (THMs) and disinfection by-products, as well as many other chlorinated or brominated compounds.

Microbiological constituents of water
Water is part of the natural world and, as described previously, it will pick any number of microbiological and biological constituents along the way, from the

smallest virus up to small crustaceans. From a quality point of view, it is the content of pathogenic and spoilage organisms that are of particular concern, which is why so much effort is spent in disinfecting water before use. There are a very large number of microorganisms, so a few key organisms are selected as indicative of the presence of both harmful and harmless species. For pathogens, *E. coli* is regarded as an indicator of faecal contamination of the water, while others tested for may include sulphite-reducing clostridia, thermo-tolerant bacteria in general, and pseudomonas. Environmental organisms such as *Lactobacilli*, *Alicyclobacillus*, yeasts and moulds are also of concern for beverage manufacturers. Parasitic organisms, such as *Cryptosporidium* and *Giardia*, must also be eliminated.

4.4.2.3 Environmental impacts on quality
Cyclical changes in water quality
The quality of the water entering water treatment plants is not necessarily constant. There are many factors that can affect the incoming water, some of which are cyclical. For example, run-offs in spring from snow meltwater would be classified as soft water, and would be likely to have a high loading of mud and a high bacterial count. Run-off after a period of drought would, in contrast, give a hard water with a high mineral content. Water sources tend to increase in salt content through the dry summer periods while, in autumn, dead vegetation adds colour, taste, organic breakdown products and bacteria.

Aquatic organisms and algae, in particular, are seasonal and, in surface waters, it is not unusual for algal blooms to occur. These can lead to colourations, taints and, sometimes, toxins. In extreme circumstances, algal blooms can give rise to the presence of complex polysaccharides in water, which produce floc in the end products. Most people would associate algal blooms with late summer and autumn but, depending on the species of algae, blooms can occur at any time of the year.

The yearly agricultural cycles and natural die-back of vegetation means an increase in nitrates entering the water sources in the autumn. This is not just due to the use of nitrate fertilisers, but also to the natural decay processes of plants and vegetation. On the much shorter timescale of a day, there are noticeable differences in dissolved oxygen levels at night and during the day, as the algal production of oxygen changes. Also, any inflows of raw sewage or industrial waste will vary during the day, or a weekend, or industrial shut-down periods.

4.4.3 Processing options
Starting mainly with potable water – or at least water which does not have heavy microbial contamination – only chemical or physical processes need to be used to achieve the desired quality standards. Water treatment processes can be grouped in five main areas, which are: filtration; chemical treatments; ion exchange; physical techniques of electrodialysis and distillation; and disinfection processes.

4.4.3.1 Filtration

Filtration covers a range of techniques, which include physical media, as in sand filtration and diatomaceous earths, adsorption on activated carbon and permeable membranes, as used in reverse osmosis (RO) forward osmosis (FO) and nanofiltration (NF).

Sand filtration

This process is used either post-coagulation or as a standalone unit. The types of material used are varied, and can have some very specific effects. Sand and gravel will filter out suspended matter, and the rate of water flow will depend on the particle size. Other materials, such as manganese dioxide, will remove iron and manganese. Diatomaceous earth or kieselguhr is another silicaceous material, like sand, but with slighter different adsorption properties. Calcium carbonate-based materials can be used to provide pH adjustments or to remineralise softened or desalinated water.

Water is pumped into the top of the sand filter. An inlet baffle deflects the flow so that the water does not directly strike and disturb the surface of the filter bed. The water is, instead, uniformly distributed over the entire filter surface. The filter media gathers particles, and the void space between the particles gets progressively smaller. This debris is removed form the filter by backwashing with water. An under-drain system collects the filtered water and distributes backwash water evenly.

Carbon filtration

Carbon filtration is used mainly to remove chlorine, either before membrane techniques such as RO, or after heavy chlorine disinfection. It will also remove many organic impurities, particularly halogenated materials, such as the trihalomethanes. The process is achieved by passing the water through activated carbon, housed in a pressure vessel, similar in construction to the sand filter but supported on a bed of graded gravel. Activated carbon consists of highly porous particles (grain size usually 1–3 mm), and has an extensive surface area in relation to its volume. The amount of activated carbon used is calculated to give a contact time of about five minutes with the water. This is long enough to remove all the free chlorine, together with some other organic molecules that may remain in the water.

Membrane filtration

This covers a range of separation techniques, which can be graded by the physical or molecular size of the material removed from the water.

Micro-filtration operates in the range of 0.1–5 μm pore size, and is usually specified in the production of sterile water, where either a 0.45 μm or a 0.2 μm filter is used to remove bacteria and other microorganisms. Not all viruses are removed in this way. Though membranes are used for micro-filtration, it is more common to use depth filters in cartridge form.

Ultra-filtration is considered to cover the range of 0.01–0.1 μm pore size and will remove some organics with molecular weights above about 10 000 Daltons. It will not remove inorganic impurities, but it will take out all viruses. Ultra-filtration plant can be used to remove colloidal substances from water which may remain after normal treatment processes such as ion exchange or coagulation. The process will also remove microorganisms and pyrogens, which are the remnants of dead bacteria. Usually, a filter required to remove pyrogens will have a lower molecular weight cut-off, probably in the range of 1000 to 5000 Daltons. The ultrafiltration membranes can be made from both organic and inorganic materials, such as cellulose acetate, polyacrylonitrile, or oxides of aluminium or zirconium. Operating pressures are typically in the range of 2–6 bar. The process may be substituted for the polishing filter in the treated water line of a conventional process.

Reverse osmosis (RO) and nanofiltration (NF)

Nanofiltration tends to operate in the region of removing organics with weights above about 250 Daltons, as well as some inorganic ions. Reverse osmosis will remove most materials with a molecular or atomic weight above about 50–100. Reverse osmosis is a process that will treat water that has a high proportion of dissolved solids, and produces a treated water that can be used directly, or be further treated to meet the desired specifications. 90–95% removal of the total dissolved solids ensures reduction of the cations contributing to hardness, as well as chloride and sulphate removal. All soluble organics of molecular weight above 100, plus bacteria, viruses and pyrogens are also removed. Nanofiltration uses membranes similar to those for RO, but it operates under less pressure, which usually results in a longer membrane life and slightly lower running costs. Depending exactly on the membrane specification, NF will remove medium-to-large organics and some of the larger cations and anions, such as calcium, magnesium and sulphate, but not chloride or nitrate. Thus, NF will reduce hardness and, to an extent, alkalinity in the water.

The process involves exerting a pressure on the feed water through membranes of cellulose acetate or polyamide – either hollow fibre, spiral wound or tubular. Fouling of the membranes is a common problem, and some form of pre-treatment is always necessary, ranging from simple chemical dosing, to complex treatment involving clarification and filtration.

4.4.3.2 Chemical treatments
Coagulation

This is a widely used treatment in large-scale plants that process surface water and effluents, but nowadays it is not particularly useful for soft drinks manufacture. The incoming raw water passes through a reaction vessel, and coagulating chemicals such as ferrous sulphate or aluminium sulphate are added to it. The reaction products and impurities are precipitated, and form a gelatinous sludge blanket, which traps insoluble impurities present in the water. The treated water flows to a clear zone at the top of the vessel. Sludge is drawn off periodically to maintain a constant blanket volume.

Polyelectrolytes
This is a process using chemical agents such as starch or alginates to produce a flocculent precipitate of colloidal impurities, which can then be filtered off. Modern materials may be ionic or neutral in character, and are often based on acrylamide or acrylic acid.

pH adjustment
This is often necessary as a precursor to other treatments, such as RO, or to specifically reduce excess alkalinity or acidity if present. It involves dosing the incoming raw water with a suitable acid, such as sulphuric acid. Hydrochloric acid would be used if there is a risk of insoluble sulphates being formed. If a more alkaline medium is required, sodium hydroxide can be used.

Oxidation
The incoming water is treated either with air or with ozone, which oxidises materials such as iron or manganese to give insoluble oxides. These then precipitate and are filtered out.

4.4.3.3 Ion exchange
This process covers softening of water, dealkalisation, removal of organics and nitrate reduction. Unwanted ions in the water are exchanged for other more acceptable ions, using ion exchange resins based on materials such as zeolites, polystyrene or polyacrylic acid. The resins are made in either a cationic form, with hydrogen or sodium ions, or in an anionic form, with hydroxyl or chloride ions.

Softening
This is a normal base exchange process, using a strongly acid cation exchange resin, and is regularly used to provide softened water for use in many areas such as bottle washers, boiler or pasteuriser feed water, or for the washing of ingredients. The resin (R) is used in the sodium form and is regenerated with brine. The calcium and magnesium ions responsible for the hardness in the water are exchanged for sodium ions:

$$Ca^{2+} + Mg^{2+} + Na_4R = 4Na^+ + Ca.Mg.R$$

This process does not reduce alkalinity.

Dealkalisation
Dealkalisers use weak acid cation exchange resins, based on methacrylic acid, with active carboxylic groups operating in the hydrogen form. Calcium and

magnesium ions associated with alkaline hardness are taken up by the resin, and the corresponding alkalinity is converted to water and carbon dioxide:

$$Ca(HCO_3)_2 + 2RH = R_2Ca + 2CO_2 + 2H_2O$$

In simple terms, this may be explained as:

Calcium alkalinity + Regenerated resin = Exhausted resin + CO_2 and water.

The treated water produced is partially demineralised, with zero alkalinity and a reduction in the total dissolved solids. The total capacity of weak acid cation resins is very high, and they are easy to regenerate. Either hydrochloric or sulphuric acid can be used, and plants are usually controlled and monitored by pH. When the pH of the water rises to 5.5, this is equivalent to an alkalinity of approximately 50 mg/l. The regeneration reactions that take place can be represented by:

$$R_2Ca + 2HCl = 2RH + CaCl_2$$

or

$$R_2Ca + H_2SO_4 = 2RH + CaSO_4$$

Use of sulphuric acid can cause problems due to the precipitation of calcium sulphate, which is very insoluble. The dealkalised water will require sterilisation and filtration before use.

Organic removal

Removal of organic matter by ion exchange was developed in the 1960s, using resins based on a polystyrene matrix. During the 1980s, a range of resins was developed, based on an acrylic structure. These organic scavenger resins are strongly basic anionic exchange resins in chloride form, and will reduce harmful organic substances and silicates in the water. The choice of resin type and organic removal capacity is dependent upon the nature of the organic species to be removed. Typically, around 70% of organic content will be removed.

Nitrate removal

Nitrate concentrations in many waters have increased over the last thirty 30 years, because of the increasing use of nitrogen-, phosphorus- and potassium-enriched fertilisers for agricultural use. Water authorities now treat water to guarantee nitrate level of below 50 mg/l, calculated as NO_3.

Some soft drink manufacturers insist in a lower level than this, particularly where canning operations are involved. High levels of nitrate can cause corrosion problems if there are flaws in the can lining.

A resin was developed in the 1970s that selectively exchanges the nitrate ion. The resin is a modified anion exchange resin, operating in the chloride form. Regeneration is with sodium chloride solution.

4.4.3.4 Disinfection techniques

A wide range of disinfection techniques are available – some chemical and some physical. The chemical processes include the use of chlorine, sodium hypochlorite, chlorine dioxide, chloramines, bromine, bromo-amines, ozone and ionically produced copper and silver ions. The physical techniques include micro-filtration and the action of ultraviolet (UV) light.

Chlorine and sodium hypochlorite

Free chlorine is used as the sterilising agent, and a level of about 6–8 mg/l chlorine is required to be effective. Other impurities in water, such as organic matter, sulphites, nitrites and, particularly, ammonia absorb or react with chlorine, and these reactions must be completed before any chlorine is available for bactericidal action.

Chlorination chemicals are most effective if injected immediately after the treatment process, because direct chlorination of raw water containing humic organic matter produces tri-halomethanes (THMs). These are undesirable substances in treated water and, if present in quantities greater than 100 µg/litre, may impart an unpleasant taint to the treated water.

At the levels used, chlorine will completely oxidise all of the impurities, removing any taste and odour, including those due to phenol and similar substances.

When chlorine is dissolved in water, it reacts to give hydrochloric and hypochlorous acids:

$$Cl_2 + H_2O = HCl + HOCl$$

Hypochlorous acid is a weak acid and dissociates, giving the hypochlorite ion:

$$HOCl = H^+ + OCl^-$$

The non-ionised form is a much more efficient disinfectant than the hypochlorite ion. At pH 5.5, practically all the available chlorine is present as active HOCl. The proportion gradually reduces until, at pH 10, it is all in the OCl⁻ form. Contact time and free residual chlorine required for adequate disinfection are related by the equation:

$$C \times T = 150 \text{ (as a minimum requirement)}$$

Where: T is the contact time in minutes;

C is the concentration of free chlorine in mg/l at the end of the contact period.

Chlorine dioxide

Chlorine dioxide is an important alternative to chlorine, in that it will not react to produce unwanted trihalomethane derivatives. However, chlorine dioxide is a very reactive gas, and has to be made *in situ*. There are two common processes, based either on the reaction of chlorine with sodium chlorite, or the reaction of sodium chlorite with hydrochloric acid and sodium hypochlorite.

$$2NaClO_2 + Cl_2 = 2ClO_2 + 2NaCl$$
$$NaOCl + HCl + 2NaClO_2 = 2ClO_2 + 3NaCl + H_2O$$

Chlorine dioxide is more expensive than chlorine and has a very high toxicity towards micro-organisms, with the advantage of efficacy over a wide pH range. About a one-hour contact time at 1 ppm of chlorine dioxide is required, compared to 5–8 ppm of chlorine for about two hours in the processing of feed water.

Bromine

Bromine is more effective than chlorine against *Psuedomonas* spp. and, in solution, produces hypobromous acid in exact analogy to chlorine. Bromine is rarely supplied in its elemental form – more usually in a powder or tablet form which, when dissolved, produces hypobromous acid either directly by hydrolysis, or in the form of a mixture of sodium bromide and sodium hypochlorite. There is some evidence from the USA that the mixture can produce bromates, which have been recognised as potential carcinogens. These can also be produced by the action of ozone on bromides, and are subject to a WHO-recommended maximum level. Bromine can also be produced, at least in its hypobromous form, from a tableted bromo, chloro hydantoin derivative.

Ozonisation

Ozonisation is a more costly process than chlorination, due to the difficulty of generating the ozone on site and dispersing the gas in the water. It can be generated by a high-voltage silent discharge across a current of clean dry air or oxygen. The electric current dissociates some of the oxygen molecules, which then can recombine as a three-oxygen atom molecule.

$$3O_2 \rightarrow 2O_3$$

Ozonised air is injected into the water, either by a venturi arrangement, through porous diffusers, or by water-spraying. Ozone is sparingly soluble, hence the mixing process must be thorough. Ozone is a very strong oxidant, and can be smelt at about 0.1 ppm. It is also toxic and irritant to the eyes nose and throat, so that extreme care is needed to ensure minimum operator exposure to the material.

However, it has the advantages that it is unaffected by pH, is very rapid in action, and oxidises organic matter, removing some tastes and odours. A residual level of ozone of at least 0.4 mg/l is necessary for at least four minutes to ensure adequate disinfection. Unlike chlorine, trihalomethane by-products are not a problem, but bromates can be formed. Also, low molecular weight carboxylic acids, such as formic, acetic and oxalic acids, can be formed, which could lead to tainting problems. Additionally, other small, readily assimilable organic compounds can be generated, which provide a good nutrient source for microorganisms.

Ionically produced copper and silver ions

The disinfectant properties of silver and copper ions have been known for many years. Research work has shown that very low levels of silver and copper ions (usually in the ppb range) will control bacteria and biofilms (effect of silver) and algae and fungi (effect of copper). Systems of this type can have application in many areas, including drinking water and cooling towers. Because such low levels of ions are required, the equipment is compact, needs little maintenance, and units last several years before being replaced.

Micropore filtration

Filtration through an extremely fine medium may be used to remove microorganisms. Water is filtered through a series of filter elements, and finally through an absolute filter of 0.2 microns porosity. This has the effect of removing bacteria present, and the process is claimed to provide sterile filtration to liquids. It is unlikely that viruses, if present, will be removed, since they are much smaller in size than 0.2 microns and would pass through the filter.

Ultraviolet light

Ultraviolet light treatment is dependent on the passage of UV light through a small depth of water. To be effective, the clarity of the water must be good, otherwise a shadowing effect is created.

The principle depends on the absorption of high-intensity UV light in the range 240–280 nm, which is destructive to the nucleic acids present in microorganisms and inactivates their cells. Various microorganisms have different sensitivities to ultraviolet irradiation (e.g. *Escherichia coli* are one of the most sensitive, while algae are most resistant). The wavelength of the ultraviolet rays can be selected so that not only are bacteria eliminated, but naturally occurring organic matter can also be destroyed.

The flow rate is adjusted so that a contact time of not less than 15 seconds is achieved. Most commercial systems are designed to supply a lamp output much in excess of the international agreed standard of at least 16 milliwatt seconds per square centimetre (mW/cm^2). In the UK, doses as high as 20–30 mWs/cm^2 are normal. Automatic controls of the system are such that, if the lamp intensity drops below an acceptable level, an alarm condition will be indicated on the control panel. There are several different systems, including a low pressure tube similar to a fluorescent tube, operating at about 250 nm which, with average quality water, would be able to treat flows up to about 2 cubic metres per hour. Medium and high pressure tubes operate over a larger spectral band, and give a better bactericidal performance, as well as coping with much higher flow rates.

Additionally, the high pressure tubes are less affected by temperature while, with low pressure tubes, efficiency is temperature-dependent.

Ultraviolet units are generally situated close to the point of use, which may mean several units within a factory. Their compactness is a distinct advantage.

4.5 Analytical and microbiological testing of water

4.5.1 Chemical tests

The amount of chemical testing needed to be done will depend on the source of the water, the treatment it undergoes and the standard with which it has to comply.

4.5.1.1 Source water

In most cases, the water source will be potable water, supplied by the local water supply company. In such cases, a full chemical and microbiological analysis of the water, showing compliance to EU standards, can be supplied upon request from the supplier. The supplier's sampling point will be that nearest to the user's plant but, unless the user's chemical and microbiological testing facilities are comprehensive, it will not be possible to carry out all these analyses at the point of use. If more detailed testing is required on samples at various points through treatment and use, it would, in most cases, be more economical to outsource testing to the water supplier.

4.5.1.2 Water processing

The tests needed to monitor water processing are relatively simple, and do not require an expensive test facility. Some, or all, of the tests listed below should be carried out at the manufacturing plant. If the product being made has special requirements, or the water is from a private source, then other routine tests may needed.

- **pH** – pH meter – gives an indication of acidity and alkalinity.
- **Conductivity** – conductivity meter – gives an indication of solids content.
- **Alkalinity/acidity** – test kits, manual titration or auto-titrator, depending on the number of tests required and their accuracy.
- **Nitrate** – test kit.
- **Chloride and sulphate** – test kit.
- **Residual chlorine** – test kit.
- **Colour and turbidity** – test kit or meter.

4.5.1.3 Effluent and grey water

The company or organisation into whose sewers effluent is discharged will provide discharge consents, based on agreed parameters such as volume, total dissolved solids (TDS), total organic carbon (TOC), pH, temperature, and so on.

If internal testing is required for the above tests, the most difficult and costly to set up is for TOC. For beverage manufacturers, this may not be required, as conductivity value will give an acceptable indication of sugar levels in the waste water.

4.5.2 Microbiological tests

Good standards of hygiene exist in most soft drink manufacturing plants, and many facilities already carry out some microbiological testing, although usually only on end products. Testing required for incoming water often only evaluates total viable counts (TVC) at 22° and 37°C, with additional tests for potential pathogens, such as *E. coli*. For water-bottling operations, more wide-ranging tests are likely to be required. Whether testing is carried out in house, or at an external laboratory, appropriate sampling techniques are required to ensure that there is no adventitious contamination before or during testing. The use of a contract test laboratory is often considered highly desirable, to provide further assurance for microbiological testing.

4.6 Effluents

Soft drink manufacture uses a lot of water, not just in the product, where it could be as much as 99.5% of the product, but in the processes used to manufacture the drinks and to maintain the plant. Because of this, most manufacturers have to treat effluents, in one way or another, to meet local legislative needs. Also, a manufacturer would want to reduce overall water consumption (it costs to obtain and treat the water), and it costs even more to dispose of waste water. Manufacturers are becoming more effective in their use of water. It used to be that returnable glass soft drink manufacture operated on a ratio of about 10 : 1 for water used to product manufactured, while nowadays a PET line could operate on a ratio of less than 2 : 1.

4.6.1 Potential contaminants of water waste

In most cases, the treatment of effluents from a drinks processor will depend not only on the manufacturing processes involved, but also on the other types of effluent generated locally on the site and which, subsequently, mix together prior to the discharge to sewer. The key parameters are pH, solids and BOD. Temperature, volume and flow rates are also considered, as it is assumed that there are relatively few toxic chemicals likely to be added to the effluent stream of a drink manufacturer.

The type and amount of effluent produced, together with the constraints imposed on the effluent parameters, will determine whether a manufacturer will consider treating waste effluent before discharge. For example, within the soft drinks industry, waste streams may have a relatively low level of suspended

Table 4.3 Likely characteristics of effluents from drink manufacturing operations.

Character	Content
Suspended solids	High
Dissolved solids	Variable
Protein	Low
Fats	Very low
Starches	High
pH	High
Colour	Variable
Odour	Moderate
BOD	500–2000

matter, a high pH from bottle washers, if they deal with returnable bottles, and a moderate BOD from residual carbohydrates in the drinks. In this case, the only treatment might be to bulk and mix effluent streams, and remove gross particulates by filtering or the use of a settling tank before direct discharge of the effluent to the sewer system. In order to obtain representative samples, it will be preferable to use and monitor a 24-hour flow proportional sample. Other suitable monitors might be pH or conductivity, to give a measure of caustic and/or sugar being dumped. Table 4.3 gives an indication of the likely characteristics of effluents from drink manufacturing operations.

4.6.2 Use of 'grey' water

Where there may be a need to reduce overall water consumption, waste water may be treated with the intention of reusing it. This may require some, or all, of the treatments outlined below. Treated waste water may be used directly in a secondary capacity, such as cooling. Alternatively, treated waste water may also be directed back into the infeed of the potable water treatment plant, where RO might be used to produce an acceptable quality for product use.

While the use of 'grey water' is acceptable for general purposes where normal potable water is not required, such as flushing toilets and urinals, or vehicle washing, there will be understandable resistance by the general public to its use in products.

4.6.3 Clean-up and reuse of effluents

4.6.3.1 Physical screening

Initially, there is a need to remove particulates by screening or filtering, Settling tanks can be used to remove fine particles, which will either float to the surface or settle on the bottom. Colloidal materials can be removed using alum precipitation. The above treatments will remove suspended solids and reduce the BOD. Subsequent treatment all involve aerobic or anaerobic biological action on the waste.

4.6.3.2 Anaerobic treatment

Anaerobic treatment involves treating the waste with anaerobic bacteria in the absence of air. Carbon dioxide and methane are produced. The methane can be collected and used as a fuel. The residual waste water will have a reduced BOD, and could then be further treated or, if within agreed discharge limits, pumped into the local effluent systems.

Additional treatment following anaerobic digestion is usually in the form of trickling filters, activated sludge tanks, ponds or complete 'wetland' systems. All are systems which expose the effluent to the activity of aerobic bacteria.

4.6.3.3 Aerobic treatments

Trickling filters

Trickling filters allow waste water to interact with aerobic bacteria in very highly aerated conditions. They can consist of a bed of material such as crushed rocks, which have a very high surface area maintained in contact with the atmosphere, or through which air can be blown. The waste water is trickled over the media, which will develop on its surface a very active microbial growth that feeds off nutrients in the waste water. Organic materials in the waste water are oxidised, and the BOD content can be reduced by up to 90%.

Activated sludge tanks

Activated sludge tanks are large volumes of water through which air is bubbled. The tanks will develop a flocculent mass of microorganisms, which feed on the nutrients in the waste water. Like all microbial growths, it may be necessary to add other nutrients, if these are not present in the wastes, to promote activity. The residence time in the tanks is usually longer than for trickle filters, and tanks can be operated in either batch or continuous mode.

Processed water from the above can either be discharged to local authority systems, or processed further using ponds. Here, water is stored in large shallow tanks under aerobic conditions, where the microbial action will reduce BOD further and clarify the waste water. Water which has been treated in this way could, after suitable disinfection (e.g. chlorination) be discharged directly to some suitable water course.

Ponds, reed beds and ecological systems

The above concepts can all be combined and enhanced by the use of complete ecosystems, which can take the waste water from drink production and treat it to render it fit for re-use – if not for primary, then certainly for secondary purposes.

There are several general types of systems, such as reed beds, which can clarify and polish partially treated waste water. Alternatively, ponds and reed beds which consist of a series of pools, with diverse plants and animals, preceding the reed beds, may be used. These systems require minimal operating costs and can give at least 30 days' retention time to produce very clean waste water.

A more complex system of linked tanks, using a very diverse range of self-contained eco-systems, employs bacteria, algae, snails, crustacea, higher plants and fish, which can treat incoming effluent to an extremely high standard, including the removal of pathogens and metals, while having low sludge generation.

Finally, if a zero effluent output is required, the above systems can be used to produce water for re-use in the factory, and with the excess processed water being fed to stands of trees, such as willows, which will take it up and release it to the atmosphere.

Further reading

Franks, F. (2000). *Water, a matrix of life*, 2nd Edition. RSC Paperbacks.
Droste, R.L. (1996). *Theory and Practice of Water and Wastewater Treatment*. J. Wiley and sons.
Binnie, C and Kimber, M. (2009). *Basic Water Treatment*, 4th Edition. Thomas Telford Limited.
World Health Organisation (2004). *Guidelines for Drinking Water Quality*, 3rd Edition.

References

EU Council Directive 98/83/EC of 3 November 1998 on the quality of water intended for human consumption.
SI 2000 No. 3184, Water, England and Wales. The Water Supply (Water Quality) Regulations 2000, Amended 2007.

CHAPTER 5
Other beverage ingredients

Barry Taylor
Firmenich (UK) Ltd., Wellingborough, UK

5.1 Introduction

A visit to any of today's retail food store outlets, either home and abroad, will serve to demonstrate that the wide variety of products to be encountered in the market reflects a background of intense creativity where the characterisation of drink types are concerned. Presentation of the product is, of course, a major factor. Packaging, bottle type and labelling declarations are designed to move the prospective consumer into purchase mode but, ultimately, it is be a favourable response to the nature and condition of the beverage itself that will decide on success or failure of a new brand.

In other chapters, we have considered what might be termed the two major base contributors to be found in beverage formulation – namely, fruit juices and high-intensity sweeteners.

This chapter is engaged with those 'other' ingredients, whose individual contributions serve both to stabilise the drink during its prescribed shelf-life and also ensure that it may exhibit a characteristic identity in its organoleptic performance throughout.

It is useful first to consider the background development of the soft drinks industry and how the term 'additive' found its way into the language.

5.2 Factors influencing development of the industry

Soft drinks, originally designated as 'mineral waters' or 'table waters', appeared as a commercial prospect in the mid-eighteenth century and, in consequence, their development towards full-scale production followed hot on the heels of the Industrial Revolution taking place at that time throughout Europe and the western world.

Chemistry and Technology of Soft Drinks and Fruit Juices, Third Edition. Edited by Philip R. Ashurst.
© 2016 John Wiley & Sons, Ltd. Published 2016 by John Wiley & Sons, Ltd.

Prior to this, while freshly pressed or squeezed juices would have been available for direct consumption to comply with the description 'soft drink', these were inherently unstable if stored, quickly succumbing to yeast attack and resulting fermentation.

It was therefore not surprising that, in the Middle Ages (and before), brewing and wine-making operations were the only reliable methods of producing a stable beverage, as the alcoholic content produced by the fermentation of natural sugars would inhibit further yeast or microbial activity.

Brewing became an art, and most country estates of any substance held their own supplies of alcoholic beverages. In Northern Europe, these were fruit wines, mainly produced from hedgerow stock and, in those countries abutting the Mediterranean, from the ubiquitous grape. Beers, ciders, perrys and so on were also to be found widespread, according to the suitability of locality.

It was not until the discovery of carbon dioxide that a means of stabilising a non-alcoholic drink became attainable. During the eighteenth century, there had been a great move forward in the discovery of gases and the composition of the air we breathe.

Effervescing spa waters and natural mineral waters had been known and taken for some time, carrying with them a decided health benefit connotation, and great scientific interest was shown in the constitution of these waters. In 1741, the identity of carbonic acid gas (or fixed air, *gas sylvestre* – carbon dioxide) in naturally occurring 'waters' was demonstrated by Dr Brownrigg, who was also credited as being the first to accomplish the artificial aeration of mineral water.

Later in the century, Dr Joseph Priestley, F.R.S, was able to claim the discovery – and its practical implications – of the principle and practice of charging, or saturating, water with carbon dioxide, by methods similar to those with which we are familiar today. In the year 1772, Priestly invented 'an apparatus for making aerated water', which he exhibited to the Royal College of Physicians, and upon which the college reported favourably.

In 1794, Priestly and his wife moved to Northumberland, Pennsylvania. While he was acquainted with John Adams and Thomas Jefferson, it seems he did not take up American nationality, and he certainly kept well away from the political affairs of the Nation. He may well, however, have continued his interest in aerated waters for, at about that time, the Seltzer Fountains caught on in the USA, and the emergence of a new industry was about to take place. Mainly fuelled by the Temperance movement, the healthy image of Seltzer waters soon incorporated flavoured varieties, by including botanical extracts, fruit juices and so on.

Soda (Seltzer) fountains were in vogue in America well into the nineteenth century, each fed by a stream of carbonated water to dilute flavoured syrups. The more sophisticated designs would supply a choice of flavours with their drinks, dispensed for direct consumption.

Among the earliest soda water manufacturers in Europe were Jacob Schweppes and Nicholas Paul, who were business partners in Geneva before making their separate ways and moving to England. Schweppes set up in London in 1792, and was making his artificial carbonated waters on what was effectively a factory scale. Paul was also operating commercially in London by 1800, and it is from this period that we can effectively date the origins of soda water as a term for a product, which seems to have been accepted by the turn of that century.

Meanwhile, attention was being given to the problem of keeping the gas in the bottle, for it was well known that corks, unless wet and in contact with the liquid contents, fell well short of requirements. Different designs of bottle and bottle-closure resulted, in order to retain the essential preserving capability of carbon dioxide.

The Victorian era was one of invention and innovation. The new industry in aerated waters was a demanding protégé. Corks, being porous, had to be kept moist in order to prevent gas loss. What better way than to construct a bottle that could not stand vertically? The egg-shaped 'side lying' Hamilton bottle (about 1809) met the requirement of keeping the corks permanently in contact with the liquid, and was successfully used for a time, but it had disadvantages in filling and quantity storage. Glass and earthenware stoppers, and then screw stoppers made of earthenware were tried ($\approx$1843), with varying degrees of success.

Hiram Codd, a Londoner, followed in 1870 with his patent globe-stoppered bottle, constructed so that the pressure of carbon dioxide gas from the 'charged' liquid forced a glass ball against a rubber ring inserted in a groove in the bottle neck. This formed a secure closure in operation, but it needed only a slight downward pressure on the ball to release the pressure and to allow the emptying of bottle contents. A ridge of glass at the base of the bottle neck prevented the ball falling back into the liquid contents. (Figure 5.1)

About this time, the screw (Vulcanite) stoppered bottle, and the 'swing (ceramic) stopper', operated by a wire spring, were also in widespread use. Then towards the end of the nineteenth century, there appeared an entirely new closure concept, the 'Crown cork', first devised in Baltimore in 1889. A patent was later awarded in the UK during April, 1892. This consisted of a thin metal disc, crimped at the edges and originally lined with cork, clamped around the bottle opening using external mechanical force. This type of closure is still in use today, although the cork liners have now been replaced with a soft plastic coating. When introduced, the Crown cork had the advantages of both reducing costs and, because of its 'one-off' usage, the fact it was more hygienic than other types.

This preoccupation with aerated waters was, of course, due to the stabilising effect of dissolved carbon dioxide gas, the first additive of major importance, in inhibiting the growth and effect of spoilage organisms. During the developments taking place around the filling and 'containerising' of the new drinks, there were other changes in formulation that would relate further to the term 'additive'.

Figure 5.1 Effective in retaining carbon dioxide, the Codd's bottle was widely acclaimed, but the drink components it contained proved, at times, cause for concern.

5.3 The move towards standardisation

In its broadest sense, a food additive is any substance added to food, but the term generally refers to any substance, the intended use of which results, or may reasonably be expected to result, in its becoming a component of any food, or otherwise affecting its characteristics,.

According to the Codex Alimentarius Commission (a joint FAO/WHO organisation, active in preparing food standards) and the EEC Commission, a 'food additive' is 'any substance not normally consumed as a food by itself and not normally used as a typical ingredient of food, whether or not it has nutritive value, the intentional addition of which to food for a technological (including organoleptic) purpose in the manufacture, processing, preparation, treatment, packing, packaging, transport or holding of such food results, or may be reasonably expected to result, directly or indirectly, in it or its by-products becoming a compound of or otherwise affecting the characteristics of such foods'. The term does not include contaminants, or substances added to food for maintaining nutritional quality.

Under European Union law, authorisation of food additives, requires, first and foremost, that:
- They present no hazard to the health of the consumer at the level of use proposed, so far as can be judged on the scientific evidence available.
- There can be demonstrated a reasonable technological need (for example, in the processing or preservation of food), and the purpose cannot be achieved by other means which are economically and technologically practicable.
- They do not mislead the consumer.

In practice, it can take ten years or more to obtain approval for a new food additive in the EU. This includes five years to carry out safety testing, followed by two years for evaluation by the European Food Safety Authority, and at least another three years before the additive receives an EU-wide approval in use for every country in the European Union – and, of course inclusion in the official additives listing under a specific E-number .

The road leading to the present rather stringent controls on the use of food additives is littered with countless examples of lack of understanding of the potential risks associated with indiscriminate application of unqualified materials to food. Flavourings, colourings, acidulents and new preservatives have been tried, at times with disastrous results.

There is ample evidence that, with a lack of statutory controls in the early stages, the contents of some of bottled products would have been lethal to the consumer. An early edition of *Skuse's Complete Confectioner*, published about 1890, contained information on cordials and other beverages and, under a section on flavours and colours, seriously admonished the use of chrome yellow (lead chromate), as it had been known for certain confectioners to use a little chrome yellow for stripes in sweets. Heaven forbid its use should also be applied in beverages!

Much of the hazard was created by the presence of impurities in some of the materials used in the manufacture of certain additives. Frequently, the sulphuric acid or vitriol used in the generation of carbon dioxide from whiting, or more effectively from bicarbonate of soda, was contaminated with metallic impurities, including arsenic and nitrous compounds, and care had to be exercised in selection of the best grades.

A recurring problem, again centred on impurities, was the risk of contamination due to the presence of lead and copper in contact with materials used in the preparation of soft drinks. *The Mineral Waters Trade Review and Guardian*, January issue, 1875, dwells on the subject at some length, and a particular case is cited where lemon oils, imported from Messina, Italy, were found to be heavily contaminated with lead, following a period of storage in copper cans. The contact surfaces had been 'tinned', and it was established that the solder used for the purpose contained relatively high levels of lead in its blend. As recently as the 1970s, lemon juice from Sicily has been found with excess levels of lead contamination as a result of old equipment being used in its preparation.

Today, the limits of impurities are well defined in legislation. Under European regulations, non-alcoholic beverages for consumption, without dilution, are given maximum limits for lead (0.2 mg/kg), arsenic (0.1 mg/kg), copper (2 mg/kg) and zinc (5 mg/kg). As regards food and drink additives, custom and use eventually indicated what dosage levels were acceptable but there was no rigid system of assessment for these. In England, the Mineral Water and Food and Drugs Acts were placed upon the statute books in 1875.

These instruments stated very clearly, but nevertheless with some generality, what the industry should not do, but omitted to indicate what might be done without fear of consequences. For example, Section 3 of the 1875 Act stated that 'No person shall mix colour, stain, or powder with any article of food, with any ingredient or material *so as to render the article injurious to health* with the intent that the same may be sold in that state.' There was no restriction on colouring material of any kind, other than that nothing '*injurious to health*' was used.

The absence of standards of quality or composition, apart from those relating to pharmaceutical products (the reference here was the *British Pharmacopoeia*), was a cause for confusion in the early years.

Toxicology was in its infancy as a science, and many of the ingredients used in the manufacture of the beverages, to stabilise and standardise the drink with apparent safety, were subsequently found to be injurious to health at the levels used. It has been said that, in the area of food and drink, 'everything is a poison – it just depends on the dosage or intake'.

The process of assessment and control has continued to the present day such that, by and large, all food ingredients are controlled by legislation. When and where appropriate, these are removed from the permitted list, or limited to an acceptable daily intake (ADI).

Across the globe, most countries employ their own legislative controls for food ingredients, but there are two main regions that exert great influence upon world opinion on this issue: the European system, controlled by the EU parliamentary council, with designated E-numbers for permitted food additives; and the system used in the USA where, at federal level, the Federal Food, Drug and Cosmetic Act (FFDCA) lays down the framework for food safety.

5.4 The constituents of a soft drink

The term 'soft drink' applies to beverages containing flavourings and/or fruit juices, together with other constituents of technological of nutritional value designed to enhance the appearance and stability of the product and to ensure that its organoleptic properties remain intact during a reasonable shelf-life period. The term is generally taken to exclude tea, coffee and dairy-based drinks. These factors are taken into consideration in all development work and, in order to meet the stringent quality and legislative controls of the present day, a new beverage is subjected to extensive trials to assess suitability and performance of all components in its make-up. It becomes essential to arrive at the correct ingredient formulation to achieve a reproducible product.

Table 5.1 lists the principal functional constituents of soft drinks and their typical usage levels. Each category of ingredient, other than fruit juices or carbohydrate and intense sweeteners, is discussed in more detail in the following sections.

5.5 Water

Water, as the main component of a soft drink, usually accounts for between 85% and 99% of the product, and acts as a carrier for the other ingredients. Its quality must conform to rigid requirements and not interfere with the taste, appearance, carbonation or other properties of the drink.

Subject to the location of the bottling plant, the source of water and product specifications, it may be necessary to carry out treatment to improve the quality of water used in the manufacture of soft drinks.

In most urbanised areas of the world, public water supplies can meet consumer requirements of potability but, for the soft drinks manufacturer, this is not always a suitable qualification for use of the water as a raw material. Most soft drinks factories will carry out their own treatments to counteract the likelihood of a possible change in quality. This is most important in areas where changes are brought about as a result of the use of a national grid system for water supply.

In developing and third world counties, water treatment becomes an essential prerequisite where microbial loading could provide cause for concern. It is necessary for a full water treatment to be effective and ensure the wholesomeness of water supplies for beverage production.

5.5.1 Requirements
The quality of water should comply with the following requirements. Water should be free from:
- high levels of elements and mineral salts;
- objectionable tastes and odours;
- organic material.

Table 5.1 Soft drink components, general usage and contribution.

Component	Typical use level
Water (quality must meet rigid requirements)	Up to 99% v/v when high-intensity sweeteners in use
Bland carrier for other ingredients. Provides essential hydration effects to enable body metabolism.	
Sugars	7–12% m/v when sole source of sweetener
Contribute sweetness, body to drink. Act as synergist and give balance to flavour	
Fruit juice	Widely variable usage. Usually up to 10% as natural strength, although some specialised lines in this
Provides fruit source identity, flavour, mouthfeel effects. Also contributes to sweetness and acidity.	
High-intensity sweeteners	Use based upon sucrose equivalence (e.g. aspartame might be employed at 0.40-6% m/v as sole sweetener)
Provide sweetness, calorific reduction. Synergist action. Often used in combination e.g. aspartame with acesulfame K	
Carbon dioxide	0.30–6% m/v
Provides mouthfeel and sparkle to drink (carbonates only)	
Acids (e.g. citric)	0.05–0.03% m/v
Contributes sharpness, sourness, background to flavour. Increases thirst-quenching effects	
Flavours	Nature-identical and artificial: 0.10–28% m/m Natural: up to 0.5% m/m
Provide flavour, character and identity to the drink	
Emulsion (flavour, colour, cloud etc.)	0.1% m/v
Carrier for oil-based flavours or colours. Gives cloudy effect in drink to replace or enhance cloud from natural juices	
Colours (natural or synthetic)	0–70 ppm
Standardise and identify colour tone of drink	
Preservatives	Statutory limits apply (e.g. sorbic acid up to 250 ppm in EU)
Restrict microbial attack and prevent destabilisation of the drink	
Antioxidants (e.g. BHA, ascorbic acid)	Less than 100 ppm, subject to user-country legislation
Prevent oxidation, limit flavour and colour deterioration	
Quillaia extract (saponins)	Up to 200 mg/l (EU), up to 95 mg/l (USA)
Acts to provide heading foam, mainly of use in carbonates	
Hydrocolloids (mucilaginous gums)	0.1–0.2% per GMP, minimum amount required to create desired effect
Carrageenans, alginates, polysaccharides, carboxy methyl cellulose etc. Provide mouthfeel, shelf-life stability, viscosity	
Vitamins/Minerals	ADI applies
Used in 'healthy-living' drinks to provide nutritional requirements	

It should also be:
- clear and colourless;
- free from dissolved oxygen;
- sterile (i.e., free from microorganisms).

Ideally, a non-variable supply of water should be available during all seasons of the year, to allow a standard manufacturing process to be established.

5.5.2 Quality of fresh water

The quality of fresh water supplies will depend upon the geological status of the catchment area. All fresh water is derived from rainfall which, following precipitation, will have filtered through the upper layers of soil, extracting minerals and organic material en route. For example, rainfall on chalk areas will result in a supply of water with a high dissolved solids content, high alkalinity and total hardness, whereas the opposite is the case when precipitation takes place upon hard rocks such as granite.

In marshlands and peaty areas, water may well be a pale yellow in colour, and will contain appreciable amounts of dissolved organic matter. Such waters are sometimes termed 'humic' (because the yellow colour derives from the humic and fulvic acids present), and are likely to possess an unpleasant odour and bitter taste.

5.5.3 Water hardness

The general term 'hardness', for water, refers to the presence of calcium and magnesium salts. 'Temporary hardness' is due to the presence of bicarbonates of calcium and magnesium, and 'permanent hardness' to calcium and magnesium chlorides, sulphates and nitrates. 'Total hardness', as might be expected, is the sum of temporary and permanent hardness. Measurement is expressed as the equivalent concentration of calcium carbonate in mg/l or ppm m/v, and is also termed 'degrees of hardness'.

Approximate classifications are:

Soft	<50 mg/l as $CaCO_3$
Medium-soft	50 – 100 mg/l as $CaCO_3$
Hard	100 – 200 mg/l as $CaCO_3$
Very hard	200 – 300 mg/l as $CaCO_3$

Water for use in soft drinks should ideally be 'soft' or 'medium-soft'.

5.5.4 Water treatment

The standard and well-tried form of water purification involves treatment in a continuous manner with a coagulant (e.g. $Al_2(SO_4)_3$, $Fe_2(SO_4)_3$) and chlorine, together with lime to reduce the alkalinity as necessary (Figure 5.2). A gelatinous precipitate, or 'floc', forms hydroxides of aluminium or iron, which absorb

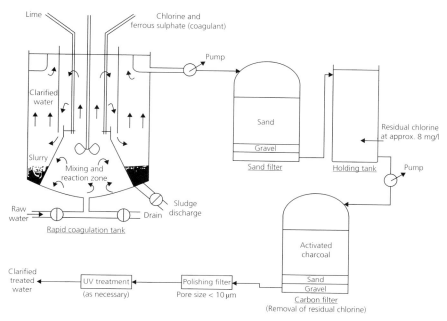

Figure 5.2 Diagram of a water treatment process using rapid coagulation.

foreign organic matter. The chlorine sterilises the water by virtue of its microbicidal and oxidising properties.

After treatment, the water is passed through a sand filter, followed by an activated carbon filter, to remove traces of chlorine, and then through a polishing filter (this is usually a cartridge filter of pore size less than 10 μm).

5.5.5 Water impurities and their effect
5.5.5.1 Suspended particles
These may consist of complex inorganic hydroxides and silicates or, sometimes, organic debris. Particles too small to be easily distinguished can cause difficulties when the drink is carbonated, acting as minute centres of instability and resulting in loss of carbonation, foaming (gushing) at the filler-head and variable fill volumes.

In non-carbonated drinks, there may be visible deposits and, sometimes, a 'neck-ring' in the finished product, caused by agglomeration of smaller particles. Filtration of the incoming water stream, preferably down to at least 0.5 μm, is therefore considered to be essential.

5.5.5.2 Organic matter
This is most likely to be present where the water is from soft regions, and from surface water reservoir fed supplies. The organic material may include humic acid, algal polysaccharides and polypeptides, protozoa and microbial contaminants. The result is unsightly porous crystalline precipitation during storage, as the organic species, notably algal polysaccharides, respond to the lower pH conditions of the soft drink or react with other components in it.

5.5.5.3 High alkalinity

This is due to the presence of bicarbonates, carbonates and hydroxides of the alkaline earth and alkali metals – principally, calcium, magnesium, sodium and potassium. The effect of high alkalinity is to buffer acidity in a soft drink, with the creation of a bland taste. It is essential, therefore, to maintain a consistent alkalinity level, and the majority of manufacturers aim for below 50 mg/l as $CaCO_3$. Alkalinity may be reduced by coagulation treatment or by ion exchange.

5.5.5.4 Nitrates

With modern methods of intensive farming, in which nitrate-based fertilisers are employed, there has been a noticeable increase in nitrate levels from aquifers lying beneath agricultural land. The recommended limit for nitrate has been given as 50 mg/l by the World Health Organisation (WHO). The health risk of nitrates involves a condition seen in infants known as methaemoglobinaemia.

5.6 Acidulents

The use of acidulents is an essential part of beverage formulation. Acidulents perform a variety of functions in addition to their primary thirst-quenching properties, which are the result of stimulating the flow of saliva in the mouth. Because it reduces pH, the acidulent can act as a mild preservative and, in some respects, as a flavour enhancer, depending on the other components present. Most importantly, however, reduction of the product pH to below 4.5, and mostly 4.4, eliminates risk of the presence of pathogens. In addition, by functioning as a synergist to antioxidants such as BHA (butylated hydroxy anisole), BHT (butylated hydroxy toluene) and ascorbic acid, acidulents can indirectly prevent discolouration and rancidity.

In carbonated beverages, there is the additional effect of dissolved carbon dioxide gas. Although it is not officially recognised as an acidulent, the inclusion of carbon dioxide, under pressure, will certainly provide extra sparkle, mouth-feel, flavour and sharpness in a drink. Its inclusion may require the rebalancing of the amount of acidulent added.

Table 5.2 lists the most commonly encountered acidulents.

5.6.1 Citric acid

Citric acid is the most widely used acid in fruit-flavoured beverages. It has a light, fruity character that blends well with most fruit flavours, which is to be expected, as it occurs naturally in many fruit types. Unripe lemons contain 5–8% citric acid. It is also the principal acidic constituent of currants, cranberries and others, and is associated with malic acid in apples, apricots, blueberries, cherries, gooseberries, loganberries, peaches, plums, pears, strawberries and raspberries. It is associated with isocitric acid in blackberries and tartaric acid in grapes.

Table 5.2 Acidulants used in beverages formulations.

Acidulant	Molecular weight	Melting point (°C)
Citric acid, 2-hydroxy-1,2,3-propane tricarboxylic acid, HOOCCH$_2$C(OH)(COOH)CH$_2$COOH	192.1	152–154
Tartaric acid (D-tartaric) 2,3-dihydroxy butanedioic acid HOOCCH(OH)CH(OH)COOH	150.1	171–174
Phosphoric acid orthophosphoric acid H$_3$PO$_4$	98.0	42.35
Lactic acid (DL-lactic) 2-hydroxy propanoic acid CH$_3$CH(OH)COOH	90.1	18
Malic acid (D-malic) 2-hydroxy butandioic acid HOOCCH(OH)CH$_2$COOH	134.1	98–102
Fumaric acid *trans*-butenedioic acid HOOCCH=CHCOOH	116.1	299–300
Acetic acid ethanoic acid CH$_3$COOH	60.0	16–18

Citric acid is a white crystalline solid, and it can be purchased as a granular powder in its anhydrous state or as the monohydrate. Present-day soft drink formulations usually employ the anhydrous form, which may have cost advantages over what was the more 'traditionally' used monohydrate. Citric acid was originally produced commercially from lemons, limes or bergamots by pressing the fruit, concentrating the pressed juice and precipitating citric acid as its calcium salt, from which it was subsequently purified. It is now produced by the action of enzymes on glucose and other sugars.

5.6.2 Tartaric acid

Tartaric acid occurs naturally in grapes, where it is present as the acid potassium salt. During the fermentation of grapes, tartaric acid precipitates from solution as crystals, as its solubility decreases with the increasing alcoholic concentration of the wine. Tartaric acid is also a natural component of numerous other fruits, such as the currants, blackberries and cranberries.

Tartaric acid can be obtained in four forms: dextro, laevo, meso and the mixed-isomer equilibrium, or racemic, form. Commercially, it is usually available as dextro-tartaric acid. This acid has a sharper flavour than citric and it may, therefore, be used at a slightly lower level to give equivalent palate acidity (see Table 5.3). Tartaric acid can be isolated from the crude deposit of tartrates obtained during wine fermentations, in a similar manner to that used for citric

Table 5.3 Palate acidity equivalents.

Acid	Concentration (g/l)
Acetic	1.00
Ascorbic	3.00
Citric	1.22
Fumaric	1.08
Lactic	1.36
Malic	1.12
Phosphoric	0.85
Tartaric	1.00

Note: These concentrations, in water, were considered to be equivalent (tartness, sourness) from taste trials carried out in the laboratories of Borthwicks Flavours (now Danisco (UK) Ltd.), Wellingborough in 1990. Although subjective, they give a proximate comparison of the pure acid effect in solution.

acid – that is, by leaching the deposit with boiling hydrochloric acid solution, filtering it and re-precipitating the tartrates as the calcium salt. The free acid is obtained by treatment of calcium tartrate with sulphuric acid and further purification by crystallisation.

Tartaric acid (dextro-form) is a white crystalline solid with melting point (m.p.) 171–174°C. The acid has a strong, tart taste and it complements natural and synthetic fruit flavours, especially grape and cranberry. If used in beverages, tartaric acid must be perfectly pure and guaranteed for food use. One problem that may need to be addressed is that tartaric acid salts, particularly calcium and magnesium tartrates, have a lower solubility than that of citric acid. There is consequently a tendency for unsightly precipitates of insoluble tartrates to form in products using hard water, and in such conditions it is preferable to use citric acid.

5.6.3 Phosphoric acid

Phosphoric acid is the only inorganic acid to be widely used in food preparations as an acidulent. It does, however, occur naturally in the form of phosphates in some fruits, including limes and grapes. In the soft drinks industry, its use is confined almost entirely to cola-flavoured carbonated beverages, where its special type of 'astringent' acidity complements the dry, sometimes balsamic, character of cola drinks.

Phosphoric acid has a drier, and perhaps sharper, flavour than either citric or tartaric acids, tasting, rather, of flat 'sourness', in contrast with the sharp fruitiness of citric acid. It therefore appears to blend better with most non-fruit drinks.

Pure phosphoric acid is a colourless crystalline solid (m.p. 42.35°C), and it is usually employed in solution as a strong, syrupy liquid, miscible in all proportions with water. It is commercially available in solution concentrations of 75%, 80%

and 90%. The syrupy nature of its solution occurs at concentrations greater than 50%, and is the result of hydrogen bonding between the phosphoric acid molecules.

Phosphoric acid is corrosive to most construction materials, so rubber-lined steel or food-grade stainless steel holding vessels are generally recommended.

5.6.4 Lactic acid

Lactic acid is one of the most widely distributed acids in nature, and it is used to a great extent by the food industry. Its use in beverages, however, is limited. It has a mild taste relative to the other acids, and is used in soft drinks as a flavour modifier or enhancer, rather than as an acidulent.

Lactic acid is supplied commercially as an odourless and colourless viscous liquid. It is produced via the fermentation of carbohydrates such as corn, potato or rice starch, cane or beet sugar, or beet molasses, using lactic acid bacteria.

5.6.5 Acetic acid

Acetic acid has a very limited use in beverages, only finding use where its vinegary character can contribute to a suitable flavour balance in the intended product. It is seldom used in anything except non-fruit beverages.

Pure glacial acetic acid is a colourless crystalline solid (m.p. 16°C), with a suffocating, pungent aroma. It is one of the strongest of the organic acids, in terms of its dissociation constant, and it can displace carbonic acid from carbonates.

5.6.6 Malic acid

Occurring widely in nature, malic acid is closely associated with apples. It is the second major acid, after citric acid, found in citrus fruits, and it is present in most berry fruits. Malic acid is slightly stronger than citric acid in perceived acidity, imparting a fuller, smoother fruity flavour.

Malic acid is a crystalline white solid (m.p. 100°C) that is highly soluble in water. Being less hygroscopic than citric acid, it provides good storage and shelf-life properties. Unlike tartaric acid, its calcium and magnesium salts are highly soluble, so therefore it presents no problem in hard water areas.

The acid finds use in a variety of products, mostly in fruit-flavoured carbonates. It is the preferred acidulent in low-calorie drinks, and in cider and apple drinks, enhancing flavour and stabilising colour in carbonated and non-carbonated fruit-flavoured drinks. Malic acid may also be used to mask the off-taste of some sugar substitutes. Blends of malic and citric acids are said to exhibit better taste characteristics than either acidulent individually.

5.6.7 Fumaric acid

This is not permitted under UK or European legislation for direct use in soft drinks, although it is permitted, under Annex IV of Directive 95/2/EC (modified by directive 98/72/EC), with strict limits, in instant powders for fruit, tea or herbal-based drinks. Fumaric acid finds wide use in other countries as an acidulent, notably in the US market, where it has GRAS status. Fumaric acid is currently manufactured in the US via the acid-catalysed isomerisation of maleic acid. In terms of equivalent palate acidity, it can be used at lower levels than citric acid, and typical replacement is suggested at two parts fumaric per three parts citric in water, sugar water and carbonated sugar water.

The main drawback in the use of fumaric acid is its slow solubility rate compared with citric acid, and special methods need to be employed in its dissolution. It is claimed that fumaric acid and its salts have a tendency to stabilise the suspended matter in both flash-pasteurised and frozen fruit concentrates (McColloch and Gentile, 1958).

5.6.8 Ascorbic acid

This acid, known more familiarly as Vitamin C, is sometimes used as a contributory acidulent, but also as a stabiliser within the soft drinks system, and its antioxidant properties serve to improve the shelf-life stability of flavour components. Many of the ingredients used in flavourings are susceptible to oxidation – particularly aldehydes, ketones and keto-esters. Ascorbic acid shields these from attack by becoming preferentially oxidised and lost, leaving the flavour component unaffected.

It should be noted that, although ascorbic acid acts well as a browning inhibitor in unprocessed fruit juices, its effect can be destroyed, should the juice be subsequently pasteurised or heat-treated. In such cases, ascorbic acid can initiate its own chemical browning reaction.

Another disadvantage of ascorbic acid is its effect on some colours in the presence of light. In the case of azo-colours, such as carmoisine, a light-catalysed reaction occurs, resulting in cleavage of the –N=N– linkage and consequent destruction of the chromophore. This accounts for the disappearance of colour, and bleaching of the characteristic hue associated with some soft drinks.

5.7 Flavourings

It is the flavour of a drink that provides not only a generic identity, but also a unique character. This part of the sensory profile is responsible for pleasing and attracting the consumer. For example, having decided on a cola drink, the consumer will be able to differentiate between colas by virtue of the background flavouring components, which collectively provide a reference point to which the consumer can return, consciously or not, on future occasions, whenever a particular brand of drink is tried.

A flavouring consists of a mixture of aromatic substances that are carefully balanced to convey the right message to the sensory receptors of the consumer. The preparation of such a mixture is a serious matter; the flavourist, like the perfumer, must be well versed in the technique, be creative and be able to translate ideas into a practical solution.

While it is often difficult for the consumer to communicate descriptions of what is being tasted, the flavourist has no such limitations. The flavourist approaches the subject in a professional manner, and is seldom at a loss when describing organoleptic attributes, as a personal library of stored knowledge relating to flavouring substances and types can be called upon. For example, some descriptors that might be applied to a peach-flavoured drink are:

sweet	juicy	fruity	lactonic	astringent
acidic	skinny	floral	estery	aldehydic
ripe	fresh	stewed	jammy	perfumed

Depending on the desired profile, the flavourist may add to, or subtract from, a central theme until an acceptable blend is reached.

Although the art of the flavourist depends largely on individual sensory abilities, it is frequently necessary in present-day flavour work to enlist analytical support at an early stage of the project. Modern instrumental analytical techniques are capable of detection at extremely low levels, but it is still usually necessary to prepare an extract, or concentrated version, of the target flavour before carrying out the analysis. This may be achieved by solvent extraction, distillation, adsorption chromatography, dialysis, headspace concentration and cryogenic or adsorbent trapping, among other methods.

A good gas chromatographic/mass spectrophotometric (GC/MS) system can be used to identify profiles of compounds and individual flavouring substances up to, say, 98% of the target flavour, thus by-passing much of the time-consuming preliminary work associated with organoleptic flavour matching. GC/MS can provide an extremely rapid and reliable assessment, leading to a tolerable flavour match that requires only slight 'tuning' adjustments for completion of the work.

In the creation of a flavouring, there is inevitably a level of comparison against what is already accepted as the generic base. Thus, a strawberry flavour is at once typical, to a marked degree, of the fruit itself; however, on a commercial level, the characterisation of this base flavour into something new will set it apart from the competition and will lead to success in the market. Descriptors such as 'fresh', 'cooked', 'jammy', 'green', 'wild', 'ripe', 'full-bodied', 'creamy', 'estery', 'sweet', 'artificial', 'natural' and many others may be applied in the assessment, as the taster searches for an adjective that best describes what is being conveyed via neurological pathways from taste sensors to brain. At best, the subjective nature of such an assessment will move into a common acceptable pathway across a wide number of tasters. It is at this point that the flavour may be identified as a winner although, even now, success will depend upon the type of application, marketing strategies, and so on.

From the onset of the gas chromatography era, there has been a remarkable move forward in the identification of volatile flavouring constituents occurring naturally, in fruits, botanicals and so on, and in food and drink products. From a list of around 500 in 1955, the current total stands at well over 7000. The flavourist has many ingredients available to chose from although, in practical terms, there is little need to select outside a group of 2000 when creating that extra modification that will distinguish the new flavouring. The majority of flavourings on the market are derived from fewer than 800 or so of these ingredients. A commercial flavouring will typically contain between 15 and 60 components although, among these, there may be natural extracts carrying a composite blend of constituents in their own right.

5.7.1 Flavourings and legislation

Food ingredients in general have been well investigated in terms of use and effect, categorised, and registered under permitted lists as appropriate to local legislation around the world

Flavourings, because of their complexity, have always existed as a separate group when considered as food ingredients, and they are subject to a whole section of legislation in their own right. This legislation, as one might expect, is subject to variation from country to country, and global harmonisation is yet to be completely realised.

Codex Alimentarius is a collection of internationally recognised guidelines relating to food safety, codes of practice and food standards, and it forms a basis for food legislation in many countries. It is recognised by the World Trade Organisation as a reference point for disputes relating to food safety and consumer protection. The Codex Alimentarius Commission was established in 1961 by the Food and Agricultural Organisation of the United Nations (FAO). A first session was held in Rome in 1963, in conjunction with the World Health Organisation (WHO), the objective being to provide clear guidelines ensuring fair practices in the international food trade and protection of the health of consumers. At the 36th session, in 2013, the commission celebrated its 50th anniversary, marking notable achievements made in promoting food safety.

At present (2015) the Commission has 185 member countries, comprising 184 member countries and 1 member organisation (EU). The current Codex guidelines (CAC/GL 66-2008) include revised definitions and terminology on the use of flavourings. The term 'nature identical' (NI) flavouring substance is not retained as such. NI and artificial flavouring substances are now defined as synthetic flavouring substances. Natural flavouring complexes, thermal process flavourings and smoke flavourings have been introduced as new definitions.

Accordingly these changes are now reflected in the International Organisation of the Flavour Industry (IOFI) Code of Practice, published in June, 2010.

One of the most frequently referred registers of flavouring materials appears in the FEMA GRAS listings. These were compiled by the 'Flavor and Extract Manufacturers Association' of the USA, and comprising those substances 'generally recommended as safe' when used in the minimum quantities required to produce the intended physical (i.e. sensory) effect, and in accordance with the principle of good manufacturing practice, each substance being allocated a FEMA number to enable cross-referencing with other listings, such as the Council of Europe (COE), US Food and Drug Administration (FDA) and Chemical Abstracts Service (CAS).

In Europe, flavourings have generally been considered as compound ingredients, and concern has been shown as to the safety of their 'undeclared' components. Following extended interaction with representatives of the European flavour industries, trade associations and so on, a new list was compiled of chemically defined flavouring substances of declared use in the member countries. This list was first published in the Official Journal (OJ) of the European Community on 27th March, 1999. Further work has since been carried out in assessing the level of health risk associated with these ingredients with a view to limitation of use, if and as necessary.

In December, 2008, Regulation EC 1334/2008 was published in the OJ, on flavourings and certain food ingredients with flavouring properties. This updates and replaces Council Directive 88/388/EEC of 22nd June 1988. The new regulation entered into force 20th January, 2009, and has been applicable from 20th January, 2011. Again, as with the Codex, a noticeable revision in the terminology applied to flavourings.

The definitions of flavourings, given under Article 3 of EC 1334/2008, are broadly in keeping with those of the Codex, but have been condensed into two main categories:

- **Flavouring substances** are defined chemical substances, which include flavouring substances obtained by chemical syntheses or isolated during chemical process, and natural flavouring substances.
- **Flavouring preparations** are flavourings other than defined chemical substances, obtained from materials of vegetable, animal of microbiological origin, by appropriate physical, enzymatic or microbiological processes, either in the raw state of the material, or after processing for human consumption.

Definitions are also given for thermal process flavourings and smoke flavourings but, of course, these are not intended for compatibility with soft drink production.

Flavouring substances and flavouring preparations may only be labelled as 'natural' if they comply with certain criteria to ensure that consumers are not misled. If the term 'natural' is used to describe a flavour, then the flavouring components used should be entirely of natural origin and, if the source of the flavour is to be labelled, then no less than 95% of the flavouring component should be obtained from the material referred to.

5.7.2 Flavourings in beverage application

Flavourings for soft drinks are of two main types: water-miscible and water-dispersible. Water-miscible flavourings are formulated to dissolve easily in water, forming a clear, bright solution, at dosages usually in the region of 0.1%. They typically contain mainly oxygenated, highly polar compounds. Water-dispersible flavourings are, strictly speaking, 'insoluble', having in their make-up a relatively non-polar oil phase – usually citrus – which conveys the characteristic zest-like contribution from the peel. This type of flavour is introduced in the form of an emulsion, enabling oil-based flavouring substances to be incorporated in a soluble form.

5.7.3 Water-miscible flavourings

5.7.3.1 Flavouring mixtures

Relatively simple mixtures of flavouring substances are dispersed/dissolved in a suitable carrier solvent system, such as ethyl alcohol or propylene glycol. Such mixtures will tend to exhibit variation in sensory profile following the immediate blending operation, and a short period of maturation is necessary to allow the flavouring to 'settle', as the components may interact and stabilise. These inner workings reflect the reactive nature of the 'oxygenates' responsible for perception of flavour. Aldehydes, alcohols, esters, ketones, lactones, phenols, terpene derivatives, and so on, may interact to form a sort of equilibrium mixture containing acetals, ketals, additional ester types from interactions between some alcohols and acids present in the original mix, and from trans-esterifications, etc. Once 'settled' the new flavouring will remain in a relatively stable condition throughout its allocated shelf-life under recommended storage conditions.

5.7.3.2 Flavouring essence

This is a traditional flavouring product prepared by 'washing' a selected oil blend (predominately citrus oils) with an aqueous alcoholic solvent mixture (e.g. 60% ethanol and 40% water). It is an extraction process in which the aqueous extract phase becomes the flavouring. The process is carried out under cool temperatures (e.g. 5–10°C), either batch-wise or by counter-current extraction. The soluble 'oxygenated' flavouring constituents present in the essential oil blend (e.g. citral in lemon oil) are effectively partitioned between the two phases of the mixture. The low temperatures employed ensure that the transfer of any 'oil' into the hydro-alcoholic phase is minimised, as a poorly processed essence will tend to 'cloud' when used in the drink formulation.

Initial mixing can be vigorous and thorough, to establish a homogenous state that, on settling and separating out, will again form two layers, the lower of which comprises the essence. The separation stage is critical in the traditional 'mix and settle' batch method, and may take several days to achieve.

Table 5.4 Example of a water-miscible flavouring.

Peach flavouring	
Vanillin	50.000
SVR (ethyl alcohol 95%)	95.000
Cinnamic aldehyde	0.225
Terpinyl acetate	0.720
Methyl anthranilate	0.720
Linalyl acetate	0.940
Benzaldehyde	1.200
Oil of neroli bigarade	1.440
Geraniol	2.820
Oil of petitgrain, Terpeneless	3.925
Oil of petitgrain	4.000
Amyl butyrate	4.800
Amyl acetate	4.800
Amyl valerate	9.000
Amyl formate	10.800
Ethyl hexanoate	10.800
Ethyl valerate	26.400
Aldehyde C14	75.000
Propylene glycol (solvent)	600.000
Water (solvent)	100.000

Prepare mixture (in order shown), mixing well at each stage. Leave to stand for 24 hours. Separate from an upper oil phase. Filter to produce a clear, bright product.

Counter-current processes involving, for example, the use of spinning disc or pulse column techniques, do not involve the same degree of pre-mixing, and achieve separation in a shorter period of time. However, as a continuous process, it becomes essential to correctly size the plant required in order to maintain an adequate stock. Unlike the batch process, counter-current techniques are not always easily able to accommodate a sudden escalation in product order requirements.

This important type of flavouring is more commonly referred to as an emulsion, and is designed to introduce oil-soluble flavouring substances to the beverage system.

Beverage emulsions can have the dual role of providing flavour and cloud effects, and are produced by the mechanical dispersion of an oil phase carrying the flavouring components into an aqueous phase containing selected hydrocolloid materials.

5.7.4 Water-dispersible flavourings

A typical sequence used in emulsion manufacture is shown in Figure 5.3. In order to achieve optimum performance of the product, great care is required at the mechanical stage of emulsion manufacture, and the uniformity and size of droplets

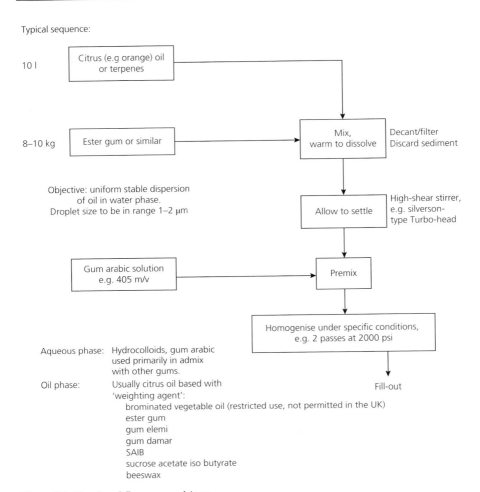

Figure 5.3 Cloud and flavour emulsions.

in the dispersed oil phase is critical. If the emulsion is to be used in promoting a stable cloud in the beverage, and to maximise optical density then, ideally, a droplet of about 1–2 μm diameter should be aimed at. Above these limits, and if the particle distribution curve is not strictly Gaussian, there will be instability, due to coalescence of the oil droplets, and effects such as 'clearing-up', 'clearing-down', 'ringing', 'creaming' and others will be observed, sometimes within a few days, and certainly on storage (see Figure 5.4). Providing the particle or droplet size remains uniform, stable emulsions can be produced, with a particle size average well below 1 μm (say, between 0.3–0.5μm), but there will be a reduced cloud effect. This is not a problem where the dispersion is intended primarily for flavouring purposes.

While mechanical parameters can be established as being important, this is not the only area of control. The selection of the correct mix of stabiliser components

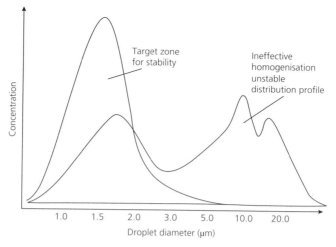

Figure 5.4 Particle distribution in emulsions.

is also critical. Beverage emulsions are essentially different from food emulsions. Their application in a mobile, liquid phase at concentrations in the region of 0.1% results in the formation of a uniform dispersion of the component droplets and, in order to remain stable and to avoid the effects already mentioned, these must remain discrete from each other and also must not interact with other components of the beverage formulation. For example, should another emulsion be present (say, a β-carotene colour emulsion) then, because of the likely differences in particle size, there will be a tendency for attraction between the two types, with resulting instability.

5.7.4.1 Oil phase

Oil phase components comprise the base oil (usually citrus as orange/lemon oils, or terpenes from the same), in which is dissolved a suitable cloudifying/stabiliser agent, such as ester gum or the glyceryl ester of wood rosin (E445), gum damar, gum elemi, sucrose acetate isobutyrate (SAIB) (E444) or beeswax (propolis) (E901). Legislation governing the use of cloudifying agents varies throughout the world, with the most regulated areas being the USA and the EU. Currently, ester gum and SAIB are permitted for use in non-alcoholic cloudy drinks under European law, whereas beeswax is limited to use as a glazing agent for confectionery or in certain colouring materials. Ester gum is permitted in the US, whereas SAIB is not.

The gum exudates, damar and elemi, are effective cloudifiers, but their use in various countries is subject to local interpretation and practice. Of these, gum damar is of most interest, being the only truly naturally sourced cloudifying/stabilising agent of any consequence. Propolis is also used successfully in certain applications although, like damar, it is not officially recognised in Europe or

America. Ester gum is produced by the action of glycerol upon pinewood rosin to produce its esterified form, a mixture of di- and tri-glycerides, then purified by a process of steam-stripping in order to de-aromatise, providing an odourless, tasteless, gum type of melting point 80–90°C.

SAIB is produced by controlled esterification of sucrose, using acetic and isobutyric acid anhydrides, and its composition is consequently dependent upon reaction conditions.

5.7.4.2 Water phase

The aqueous phase is, of course, complementary to the oil phase, in that its hydrocolloidal components tend to provide a protective buffer zone around each oil droplet.

One theory is that the polymeric molecules of the hydrocolloid are drawn to the oil-water interface by charge effects promoted, no doubt, by the oil-soluble cloudifier components referred to above. These micelles will exhibit a composite charge, repelling neighbouring micelles of similar charge, preventing coalescence and maintaining a stable system. It follows that, providing the oil phase droplets are of similar diameters, the associated surface charges will possess similar magnitudes, and the electrostatic repulsive forces between droplets/micelles will prevent any tendency for coalescence between them.

Of the many water-soluble polymers that can act is this manner as emulsifying agents, the most regularly used are acacia gum (also referred to as gum arabic) and the modified starches.

Acacia gum has been employed for many years as an emulsifier in beverage emulsions. Unlike other vegetable gums (e.g. tragacanth, guar, etc.), it has the advantage that it can remain relatively mobile in aqueous solution at concentrations of 30% m/m and above. The gum is obtained as the exudation from selected species of acacia. Upwards of 500 species of acacia are to be found over Africa, Asia, Australia and central America, but only the varieties grown in Africa are of commercial interest and, of these, *Acacia senegal* and *acacia seyal* are the most important. It is *A. senegal* that is the accepted standard for beverages.

The two species are grown widely in a broad band of countries across west Africa, through to the Sudan and Ethiopia. The gum tears are hand-picked and sorted by local people, and transferred to central agencies for grading and distribution on the world markets. Apart from appearance in its raw form and physical quality the main feature of difference is the optical rotation. *A. senegal* is laevo-rotatory, with a specific optical rotation of −25° to −35°, while *A. seyal* is dextro-rotatory, with a specific optical rotation of +30° to +45°.

Acacia gum is a complex polysaccharide with protein groupings, and it is the level of protein present that is thought to be a major factor in its favourable performance as a beverage emulsifier. It has not been unusual for blends of the two types to be offered into the beverage industry as gum arabic, arranged so that the optical rotation would be always laevo-rotatory, but falling short of the optimum required purity. The purity specification issued by JECFA (Joint FAO/WHO Expert Committee on Food Additives) (1995) refers to gum Arabic as the 'dried exudation from stems and branches of *Acacia senegal* (L.) Willdenow or closely related species of *Acacia* (fam. Leguminosae)' An earlier reference (JECFA 1990) to the use of specific rotation and nitrogen content as purity criteria has been dropped, and this now facilitates the use of mixtures of different acacia gums under the description 'gum arabic'

As with all natural products, the quality and harvest yields of gum arabic are subject to climatic conditions. In addition, in the region of Sudan and Ethiopia, availability has also been affected by political factors and, from about 1984, prices for the gum have shown major fluctuation and, at times, have been far higher than other polysaccharides.

Because of this, the starch industry has made great effort to develop replacements. Perhaps the most successful to date, are the starch sodium octenylsuccinates, where native starches have been modified by substitution with 1-octasuccinic anhydride (OSA). The introduction of a hydrophobic grouping onto the starch polymer results in highly effective emulsion-stabilising properties. The conditions of the modified starches during manufacture are controlled, giving some variety in performance to the range available but, in general, a key attribute is the low viscosity of their aqueous solutions, which can enhance formation of uniform droplets at relatively low pressures during the homogenisation stage of emulsion manufacture. Typical concentrations of the OSA starches when used for beverage emulsions are around 10–12%.

Starch octenyl succinates are number registered as E1450 under European legislation, and are generally permitted additives under Annex 1 of Directive 95/2/EC

5.7.4.3 Brominated vegetable oil emulsions

This section has been included more from academic interest than from a practical point of view, as the use of brominated oils has been severely restricted in its absence from European legislation over the past twenty years – although, until comparatively recently, it has been in use in India and parts of Asia. The term 'weighted' is frequently used when describing the oil phase. This alludes to the early use of brominated vegetable oil (BVO) to adjust the density of the oil phase, in order to stabilise the dispersion when in ready-to-drink (RTD) form. BVO emulsions are no longer permitted in Europe, but they are still in use in some parts of the world, particularly in tropical areas, where the improved stability can greatly enhance the storage capabilities of bottled drinks at high ambient temperatures.

The reaction of bromine with vegetable oils, such as peanut or maize oils, involves the addition of bromine atoms at the unsaturated double-bond sites of the oil, to produce a progressive increase in the molecular weight and, hence, the density, which may be thereby easily adjusted to a predetermined value.

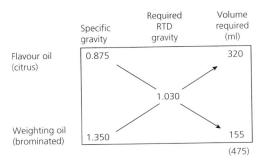

Figure 5.5 Example of the use of weighting oil.

Densities in the region of 1.24–1.4, in accordance with the amount of bromine used, are usually achieved, and these are controlled to meet requirements. For example, the density of an oil phase comprising 'folded' orange oil can be adjusted from 0.875 up to, say, 1.03 by the addition of brominated oil, placing it in the density region of the target beverage and, hence, ensuring a stable system. The amount of weighting oil required can be calculated by the following method, devised by Pearson:

1 Write the target gravity of the finished beverage in the centre of a square (see Figure 5.5).
2 Enter the specific gravities of the essential oil and weighting oil at the upper and lower left-hand corners, respectively.
3 Diagonally subtract the larger from the smaller numbers in each case, and place the resulting difference (omitting the decimal place) in the appropriate right-hand corner, giving the volume ratio required (see Figure 5.5).

Thus 475 parts by volume of the oil mix, to be used as the emulsion 'oil phase', will comprise 320 parts citrus oil plus 155 parts of the weighting oil, giving a gravity of 1.030, identifying with that of the finished beverage.

5.8 Colours

Those of us fortunate enough to possess optical powers capable of distinguishing a variety of colours will appreciate the influence that this particular sensory dimension exercises on our judgement of matters important to our well-being, such as food and drink. The perception of colour gives influence to the taster's reception of the drink and, to this end, there is inevitably some controversy.

One point of view states that colours, which possess no measurable nutritional value, can have no place in food or drink, other than that of deceiving the consumer. To a certain extent, this is true but, to appreciate the full value of colour as a food additive (or, more specifically as a soft drink additive), it is necessary to appreciate the synergy between the sensory responses of sight and taste.

Colour provides a means of correctly presenting a beverage to the consumer, so that the perceived organoleptic attributes are correctly placed in an ordered sequence of appreciation. Both quality and quantity of colour are of importance, and certain colours will provoke, or perhaps complement, a particular taste. Reds will favour the fruitiness of soft drinks (e.g. blackcurrant, raspberry, strawberry, etc.). Orange and yellow tend towards the citrus flavours. Greens and blues reflect the character of peppermints, spearmint and cool flavours, sometimes herb-like and balsamic, and the browns align with the heavier flavours (e.g. colas, shandies, dandelion and burdock). Therefore, the deceit, if ever intended, is aimed at ensuring that the consumer is able to maximise the enjoyment of the beverage concerned.

Where the soft drink is based in part on fruit juices, it may be necessary to restore the appearance of the juice concerned if its natural colours have been destroyed by heat processing, or to intensify such colour when the contribution from the juice is weaker than that normally associated with the effect that the compounded drink is intended to convey. Colour adjustment may also be necessary to ensure uniformity of product, and to offset natural variations in colour tone and intensity associated with the juice type employed in the beverage formulation.

Above all, colour is a major parameter in the assessment of quality, serving at the time of production to standardise the product. It can also give useful information as to quality changes during storage, due to colour deterioration caused by temperature fluctuations or microbial spoilage effects, for example.

The use of food colours is carefully controlled under various legislations (see Table 5.5). There is, at present, no universal listing of colours for soft drinks, and it is necessary to investigate the permitted list to ensure compliance for goods to be manufactured in, or exported to, a particular country.

Both the European Union and the FDA have published lists that are subject to regular review. The greatest concern has been expressed over the use of azo-dye colours, as certain individuals can demonstrate an allergic reaction to some of these. Allergic reactions have been reported most frequently with sunset yellow (E110, FD&C yellow no. 6) and tartrazine (E102, FD&C yellow no. 5).

In recent years, new scientific data on health risks to children exposed to azo-dyes has emerged, which has influenced the European Parliament Committee to adopt a more restrictive approach to their use and move towards better labelling of additives containing azo-dyes. From 2011, foods containing some of those colours (colourings E110, E104, E122, E129, E102 and E124) must be labelled, not only with the relevant E-number but also with the words 'may have an adverse on activity and attention in children'. Additionally, in some countries there are commercial constraints imposed, where some institutional bodies and large retailers may refuse, irrespective of legislation, to accept products containing artificial colours.

Although there are a number of food colours suitable for use in soft drinks, it should be appreciated that the contribution of any one of these cannot be

Table 5.5 Permitted food colourings derived from natural sources (EU Directive 94/36/EC).

Colour	Sources	Shade	E-no.	Stability Light	Heat
Anthocyanins	Grape skins, elderberry, red cabbage, hibiscus	Red-purple-blue, pH dependent	E163	Good	Good
Water-soluble colours; natural indicators; red in acidic solutions (most stable) and bluer as pH increases					
Beetroot Red	Red beetroots (*Beta vulgari*)	Pink to red	E162	poor	poor
Water-soluble, limited stability when exposed to heat, light and oxidation; most stable between pH 3.5 and 5					
Carmine	Cochineal insect (*Dachtilopius coccus*)	Strawberry red, orange/red hues	E120 E160(a)	Excellent Fair	Excellent Good
Soluble in alkaline waters, solubility decreasing with lowering of pH, will precipitate below pH 3					
Annatto	Seeds of annatto shrub (*Bixa orellana*)	Orange	E160(b)	Fair	Good
Bixin is oil-soluble and frequently used for beverages in the form of its emulsion; norbixin is water-soluble in alkaline conditions and will precipitate in acid pHs; versions stabilised to about ph 3 are available					
Beta-carotene	Carrots, algae, palm, synthesised	Yellow to orange	E160(a)	Fair	Good
Oil-soluble; colour sensitive to oxidation; shade varies with concentration; water-dispersible versions available					
Paprika	Red pepper (*Capsicum annum*)	Orange to red	E160(c)	Fair	Good
Oil-soluble colour; sensitive to oxidation; water-dispersible versions available					
Lutein	Aztec marigold (*Tagetes erecta*)	Yellow	E161(b)	Good	Good
Oil-soluble egg-yellow colour; good stability; the xanthophyll lutein is a carotenoid colour which occurs in green leaves, vegetables, eggs and some flowers					
Curcumin	Turmeric (rhizomes of *Curcuma longa*)	Yellow	E100	Poor	Good
The purified colour, curcumin, is not water-soluble, but water-dispersible systems are available					
Chlorophylls	Green-leafed plants	Green	E140, E141	Poor, good	Poor, good
Natural chlorophyll (E140) is oil-soluble but water-dispersible forms are available; copper chlorophylls (E141), effectively chemically modified natural extracts, are water-soluble					

entirely predictable. In any soft drink formulation, the colour component, as with all other ingredients, has to be carefully selected for its performance in the presence of certain acids, flavourings, antioxidants and even preservatives. It is essential, therefore, at all stages of development, that meaningful storage trials are completed to ascertain the real contribution from colour in the newly finished beverage.

Table 5.6 Artificial (synthetic) colours permitted in soft drinks to a maximum level of 100 mg/l[a].

Colour	E-no.	Colour stability			Colour contribution
		Light	Heat	Acids	
Quinoline yellow	E104	Good	Good	V. good	Greenish yellow
Tartrazine (FD&C yellow no. 5)	E102	Good	Good	V. good	Lemon yellow
Sunset Yellow (FD&C yellow no. 6)	E110	Good	Good	V. good	Orange shade (similar to orange peel)
Carmoisine (azorubine)	E122	Good	Good	Good	Bluish red
Ponceau 4R	E124	Good	Good	Good	Bright red
Patent blue FCF	E131	Good	Good	Poor	Bright blue
Indigotine (FD&C blue no. 2)	E132	Fair	Poor	Fair	Dark bluish red
Brilliant blue FCF (FD&C blue no. 1)	E133	Good	Good	Good	Greenish blue
Green S	E142	Fair	Good	Good	Greenish blue

[a] With the proviso that individual levels of E110, E122 and E124 may not exceed 50 mg/l (EU Colour Directive).

Food colours are broadly divided into two classes: natural and artificial. In the USA, these are listed as either 'exempt from certification' or 'certified'.

The natural colours are botanical extracts, with the exception of carmine (a red colour), which should perhaps be termed an entomological extract, as it is obtained from the insect *Dactilopius coccus*, sometimes termed the cochineal beetle, which breeds and feeds on particular cacti indigenous to Central and South America. Table 5.6 lists artificial colours permitted in soft drinks under EU legislation, subject to the controls in use and declaration as mentioned above.

5.9 Preservatives

A preservative may be defined as any substance that is capable of inhibiting, retarding or arresting the growth of microorganisms, or any deterioration of food due to microorganisms, or masking the evidence of any such deterioration, and so on.

In Europe, defined maximum levels of permitted preservatives are given, according to the food substrate concerned. For soft drinks consumable without dilution, the European Directive No. 95/2/EC is as shown in Table 5.7.

The *p*-hydroxy benzoates previously cited in the legislation are no longer permitted for use in soft drinks, although they are still included under certain food uses.

As mentioned previously, carbon dioxide, while not added specifically as a preservative, contributes towards the inhibition of micro-organic growth and,

Table 5.7 Preservative limits under European Directive 95/2/EC.

Preservative	Concentration (mg/l)	E-number
Sulphur dioxide (carry-over from fruit concentrates only)	20	E220
Benzoic acid	150	E210
Sorbic acid	300	E200
Benzoic/sorbic acids in combination	150/250	E210/E200

coupled with other factors (e.g. pH), contributes to the stability of the drink. Carbon dioxide is deemed to be effective at volumes over 2.5 or 3.0 and, for this reason, the incidence of spoilage in carbonated beverages is less than with the non-carbonated versions ('volumes' of CO_2 in general terms refers to the number of times the total volume of the gas, adjusted to 760 mm of Hg and 0°C, can be divided by the volume of liquid in which it is dissolved).

Although preservatives can be used to good effect in beverage formulations, they should never be considered infallible, and there is no substitute for stringent quality and hygiene controls at every stage of manufacture. Within their own product specification, raw materials should be assigned workable limits for microbial activity, so that there is little chance of excessive contamination in the finished beverage product. Equally, all processing plant, machinery or containers likely to come into contact with the product during manufacture should undergo a thorough cleaning (sanitisation) before use.

Certain strains of yeast, moulds and bacteria can survive at relatively low pH conditions and some of these can exist and grow in the presence of certain preservatives, so it is important that everything is done to prevent their multiplying. Under favourable conditions, a typical rapidly growing yeast strain can double its numbers every 30 minutes and, at this rate, in 12 hours one yeast could become 16.7×10^6, providing no inhibitory factor is present.

5.9.1 Microorganisms and beverages

Although there is little evidence of the formation of toxic fermentation products in beverages, the problem of spoilage frequently arises. Because of their utilisation of sugars, yeasts are of most immediate concern.

Yeasts are classified with the fungi, and are unicellular for most of their life cycle. Together with moulds and bacteria, they can bring about a deterioration in flavour, producing taints, off-notes and differences in mouthfeel, and so on. Most yeasts can grow with or without oxygen, whereas most bacteria cannot survive in it. The majority of yeasts thrive in temperatures between 25–27°C, some can survive over 70°C and others can exist, apparently quite comfortably, at 0–10°C.

Bacteria exhibit some similar diversity in their characteristics, with an optimum growth temperature at around 37°C.

Soft drinks provide an ideal growth substrate for many microorganisms, with adequate supplies of the required nutrients. Apart from water, the environmental necessity, typical requirements are sources of carbon (carbohydrates), nitrogen (amino acids), phosphorus (phosphates), potassium, calcium (mineral salts) and traces of other minerals (e.g. sulphur, iron, cobalt and even vitamins). Because of the obvious link with protein formation during cell growth, the presence of combined nitrogen is of particular importance. Also, when it is introduced to beverages via fruit pulp or caramel (colouring), there will be a greater susceptibility to spoilage by certain microorganisms.

Perhaps the most difficult aspect of dealing with microbial contamination in soft drinks relates to the delay factor; an apparently good quality product leaves the bottling line for storage and distribution, only to be returned at a later date (perhaps after several weeks), when severe deterioration in performance has taken place. Fortunately, such occurrences are seldom encountered in today's soft drinks industry but, to any manufacturer, it is a nightmare scenario that must be avoided at all costs.

A bottled drink constitutes a unique system, which can inhibit or enhance the growth of microorganisms. Micro-flora, if present, will enter a dormant stage, during which their chances of survival are assessed in relation to the immediate surroundings. Following this 'lag' stage, while specific micro-flora may adapt to their new environment and start to grow, there is a burst of species-dependent activity, during which the population doubles repeatedly at a steady rate. Since a bottled drink is a 'closed' system, waste products and diminishing nutrients will serve to slow down the growth and will eventually bring it to a standstill when the death rate increases and all activity stops. The product, however, while perhaps not a health hazard, has been spoiled, and can no longer satisfy its intended function.

5.9.2 Sulphur dioxide

Because of the ease with which it can be produced, gaseous SO_2 was one of the first chemical compounds manufactured and used by humans. By Roman times, it was used as a preservative, by burning sulphur before sealing wine into barrels or storage jars. It is one of the most versatile agents used in food preservation, and is well known for its microbiocidal effect on bacteria, moulds and yeasts. Nowadays, it is generally employed in the form of a sulphur dioxide-generating salt. For example, sodium metabisulphite is converted thus in acid medium:

$$Na_2S_2O_5 + H_2O \rightarrow 2NaHSO_3$$
(MW 190)

$$2NaHSO_3 + 2H^+ \rightarrow 2Na^+ + 2H_2O + 2SO_2$$
(MW 2 × 64)

Table 5.8 Preservatives and their salts.

Preservative	E-no.	Alternative form used at equivalent level	E-no.
Benzoic acid (m.p. 122°C) C_6H_5COOH *Benzene carboxylic acid*	E210	Sodium benzoate	E211
		Potassium benzoate	E212
		Calcium benzoate	E213
Sorbic acid (m.p. 133°C) $CH_3CH=CH_2-CH_2=CHCOOH$ *2,4-Hexadienoic acid*	E200	Sodium sorbate	E201
		Potassium sorbate	E202
		Calcium sorbate	E203
Sulphur dioxide (gas) SO_2 *Sulphurous anhydride*	E220	Sodium sulphite	E221
		Sodium hydrogen sulphite, sodium bisulphite	E222
		Sodium metabisulphite	E223
		Potassium metabisulphite	E224
		Calcium sulphite	E226
		Calcium hydrogen sulphite, calcium bisulphite	E227
		Potassium bisulphite	E228

That is, 190 parts of the metabisulphite produces 128 parts SO_2. Table 5.8 lists the various salts of the main preservatives.

The microbiocidal effect increases as the pH falls below 4.0 and, because of this, SO_2 is ideally suited for most soft drink formulations. However, its preserving action is impaired by a tendency to react with many fruit components of soft drinks, to form organic sulphites, in which state the SO_2 is said to be 'bound'. Although the preservative properties are due mainly to free SO_2, it is necessary to analyse for total SO_2 (i.e. free plus bound), as legislation for safe working requirements refers only to maximum total concentrations.

Although SO_2 is used to good effect in the preservation of concentrated citrus juices, with typical concentrations of 1000–2000 ppm m/v, it is now limited under European legislation to no more than 20 ppm in non-alcoholic flavoured drinks containing fruit juice, as carry-over from concentrates only (ref.95/2/EC) (see Table 5.7). There are a number of specific drink products which are permitted to contain higher levels of SO_2 under the same legislation – for example, concentrates/dilutables based on fruit juice and containing not less than 2.5% barley (barley water). However, since this limit has been much reduced from the previous level of 70 ppm, the onus has been squarely placed on manufacturers to attain improved manufacturing practices in terms of plant hygiene.

JECFA has recommended an acceptable daily intake (ADI) of not more than 0.7 mg/kg body weight for SO_2.

Disadvantages associated with sulphur dioxide are that some tasters can detect it as an unpleasant backnote or taint, and it has a tendency to provoke allergic reactions in some individuals. Asthma sufferers tend to be affected by gaseous sulphur dioxide, small traces of which can promote an asthmatic attack. Foods containing sulphites can, therefore, introduce the risk of gas liberation upon swallowing.

5.9.3 Benzoic acid and benzoates

Benzoic acid occurs naturally in some fruits and vegetables, notably in cranberries, where it occurs in amounts of the order of 0.08% m/m (Fellers and Esselen, 1955). It is also found in some resins, chiefly in gum benzoin (from *Styrax benzoia*), and in coal tar. Commercially available benzoic acid is produced by chemical synthesis.

Pure benzoic acid is a white powdery crystalline solid (m.p. 122°C), which is only sparingly soluble in water at normal temperatures. Because of this, it is added to the drink in the soluble form of its sodium or potassium salts. It is normal practice to completely disperse the benzoate during batch make-up, before addition of the acid component, with its resulting pH reduction, to avoid localised precipitation of the 'free' benzoic acid, due its solubility having been exceeded (solubility of benzoic acid = 0.35% m/v at 20°C). It is the free or undissociated form of benzoic acid that exhibits preservative action and, hence, its use is only effective when low pH values are encountered – ideally below pH 3 – at which point, the degree of dissociation has reduced to below 10%.

Benzoic acid is generally considered to exhibit an inhibitory effect on microbial growth, although it is of little use for bacterial control where the greatest problem will occur at above pH 4, and outside the effective limit mentioned above. Improved results are obtained when it is used in conjunction with other preservatives (e.g. SO_2 or sorbic acid), due to synergistic effects. It is interesting to note that the current European Directive, while expressing single limits of 300 mg/l for sorbic and 150 mg/l for benzoic acids in non-alcoholic flavoured drinks, nevertheless permits a 'joint' preservative use of up to 250 mg/l sorbic acid with 150 mg/l benzoic acid. A potential problem that has been reported in some products containing both benzoic acid and ascorbic acid is the potential for the formation of very small amounts of benzene.

Allergic responses to benzoic acid have been reported, particularly among children known to be hyperactive towards other agents, such as tartrazine. Like artificial colourings, benzoates are sometimes effectively banned by retailer.

The maximum ADI for benzoic acid, recommended by JECFA, is 5 mg/kg body weight.

5.9.4 Sorbic acid and sorbates

Sorbic acid is found naturally in a number of fruits and vegetables, notably in the juice of unripe mountain ash berries (from *Sorbus aucuparia*), where it occurs together with malic acid. Sorbic acid and its salts are some of the most widely used preservatives in the world. In soft drinks, the most commonly used form is potassium sorbate because, like benzoic acid, there are problems in preparing its solution (solubility of sorbic acid = 0.16% m/v at 20°C).

In common with benzoic acid, as a microbial inhibitor, sorbic acid and its sorbates show reduced effectiveness with increase in pH. Although activity is greatest at low pH values, sorbates have the advantage of being effective at pH values as high as 6.0–6.5, in contrast to benzoic acid, for which the comparative range is pH 4.0–4.5. The undissociated form, as with benzoic acid, is primarily responsible for its preservative action. Sorbates can be metabolised by some moulds, and the generation of 1,3-pentadiene as a breakdown product has been known. The presence of this substance causes unpleasant taints in a product.

In the United States, sorbates are classified as GRAS, and have maximum permissible levels in various foods of 0.05–0.3%. The EU permits levels within the range 0.015–0.2% although, in soft drinks, the limit is 0.03% (300 ppm m/v).

In addition to being less toxic than benzoates, sorbates seem to be less obtrusive in terms of taste detection by certain individuals and allergenic reactions. Overall, sorbates are considered to be one of the safest food preservatives in use, and WHO has set the ADI for sorbate at 25 mg/kg body weight.

5.10 Other functional ingredients

The functionality of the ingredients already discussed tends to be self-evident as contributing to the main identity of the drink. Except for the preservatives, these ingredients are used primarily for taste and colour. There are others whose contribution is to improve performance and further characterise the drink.

5.10.1 Stabilisers

We have already referred to the use of stabilisers in the production of water-dispersible flavourings and/or emulsion-based clouding systems. Such additives, as well as contributing to stability, are also used, where appropriate, in soft drink formulations, to impart stability to natural clouds (e.g. dispersion of fruit solids), and to improve mouthfeel characteristics by increasing the viscosity of the drink.

In European legislation, there are over 50 E-coded materials with stabilising properties for food use, although perhaps there are no more than ten are used on a regular basis in soft drink formulations. These include the alginates, carrageens, vegetable gums, pectin, acacia, guar, tragacanth, xanthan and carboxy methyl cellulose. Also included under number E999 is extract of quillaia, which is permitted specifically in soft drinks and, apart from its use as an emulsifier, is valued for its foam-stabilising properties.

5.10.2 Saponins

Saponins occur in the roots of many plants, notably the genus *Saponaria*, whose name derives from the Latin *sapo*, meaning soap, because of lather-like reaction that occurs when parts of these plants are soaked in warm water. This ability to generate foams finds use in beverages such as ginger beer, shandy, cream soda, cola formulations and so on, to improve and standardise heading foam characteristics.

Saponins for beverage use are sourced from quillaia bark (*Quillaia saponaria* Molina) and the yuccas. Of the latter species, two main varieties are used in the United States for production of the water extract – the Mohave yucca (*Yucca mohavensis*) and the Joshua tree (*Yucca brevifolia*).

At the levels used, these additives are colourless and tasteless; the dried extract, however, possesses an acrid, astringent taste. Permitted limits are quoted in terms of the dry weight of the extract. In the European Union, subject to Directive 95/2/EC, quillaia is permitted only in non-alcoholic drinks to a maximum level of 200 mg/litre.

5.10.3 Antioxidants

Perhaps the most common problem encountered during storage of a beverage relates to the oxidation effects involving certain ingredients. Both flavour and colour components can be subject to deterioration in the presence of dissolved oxygen, to the detriment of the product. Antioxidants are, therefore, included in those formulations containing ingredients most vulnerable to oxidation. Oxidation can frequently be attributed to the oxygen permeability of the plastic materials used in container manufacture, but it is essential that the oxidation process should not start at the production stage of the drink or any of its ingredients.

Citrus-flavoured drinks, notably lemon drinks, are particularly susceptible to oxidation, so antioxidants may feature in their formulation.

Oil-based, water-dispersible flavours (emulsions) are protected by the addition of oil-soluble antioxidants, such as butylated hydroxy anisole (BHA) and butylated hydroxy toluene (BHT), to the oil phase before the emulsification process. 1000 mg/l is the typical usage level in essential oils.

Since the flavour emulsion will be used at the rate of about 0.1%, the level of antioxidant in the finished beverage will be of the order of 1 mg/l, which will safely comply with an ADI of 5 mg/kg body weight for either additive.

Increasing use is being made of natural and nature-identical antioxidants as, in many countries, the usage of BHA and BHT continues to be restricted on health grounds. Ascorbyl palmitate (6-O-palmitoyl-l-ascorbic acid) and its sodium and calcium salts, natural extracts rich in tocopherols (vacuum-distilled from soya-bean oil, wheat germ, rice germ, cottonseed oil, for example) and synthetic α-, γ-, and δ-tocopherols, are used to good effect in preventing oxidative deterioration in oil-based systems. In combination, ascorbyl palmitate and α-tocopherol (Vitamin E) synergise to exhibit enhanced antioxidant properties.

5.10.4 Calcium disodium EDTA

This is a mixed salt of ethylene diamine tetra-acetic acid (EDTA), and it is prepared by reacting the acid with a mixture of calcium and sodium hydroxides. It acts as a sequestrant, its binding action removing traces of metal ions present in raw materials or process water. These metals (e.g. iron), can destabilise the beverage by a tendency to catalyse the degradation of flavouring components, causing oxidation and off-notes. Their removal serves to maintain stability of the products during storage and to increase shelf-life.

Under European Directive 95/2/EC, calcium disodium EDTA is permitted only in a limited number of foods, including some canned and bottled products, with maximum levels specified in each case.

In the United States (Code of Federal Regulations), it is permitted to a level of 33 ppm in canned carbonated soft drinks, to promote flavour retention.

5.11 Food safety

Although the primary function of food for humans is survival, it now has additional associations with health, enjoyment and acceptability. Today's consumer looks to suppliers and manufacturers for a product with which there is no associated risk in consumption, and one that is marketed in accordance with strict observance of the laws governing food safety.

Although microbial contamination must always be a point of concern for food, it is much less of a potential hazard for beverages, where the lower pH conditions (usually below 4.0) make the survival of pathogenic species virtually impossible, and the likelihood of food poisoning equally remote. In addition, any attack from, say, yeasts and moulds would manifest itself as spoilage and would be detectable either visually or organoleptically well before any lasting danger can occur. However, to be assured of complete safety, it is necessary to look further, below the superficial and into the actual ingredient make-up of the drink itself.

The safety of food additives and other ingredients is monitored according to guidelines issued by the Foods and Agriculture Organisation (FAO) and the World Health Organisation (WHO) Joint Expert Committee on Food Additives (JECFA). Knowledge of health safety is gained primarily as a result of animal feeding trials, coupled with relevant short- or long-term toxicological investigations. In later stages of testing, humans may also be included in the studies, to ascertain that their physiological reactions are similar to those found in animals.

Because of potentially longer-term ingestion, the standards applied to food additives in feed trials are significantly higher than those applied to pharmaceuticals – for example, where, in order to treat successfully some conditions of illness, certain side-effects caused by the treatment can be tolerated. The Acceptable Daily Intake (ADI) is an estimate of the amount of a food additive, expressed in mg/kg body weight per day, that can be consumed safely over an entire lifetime. ADI levels are set by JECFA after considering the result of various feeding trials. As a general rule, the ADI value is set at one hundredth of the intake, which produces virtually no toxicological effects in long-term animal feeding trials.

The registering of a 'new' ingredient, or perhaps the retention of an existing one, under the European Miscellaneous Additives Directive, is a complicated, very expensive and long drawn-out procedure, not just because of the strict protocols employed in the assessment itself, but also because of the accompanying exchanges taking place between all interested parties representing the member states in setting up the application. These include agricultural and trade associations, consultancies, research establishments, among others, whose opinions may be called upon in the preparation of a suitable dossier laying out the justification for inclusion of an additive. The associated costs are formidable. Currently, starting from scratch, the costs involved in preparing a monograph on the constituent and physical properties of a proposed additive, together with a full toxicological study, will be of the order of at least £250 000 ($287 500). Thus, it is little wonder that new additions to the list are few in number.

An ongoing issue, periodically provoking media coverage, is that of intolerance by members of the population to certain food ingredients. This is of most concern where serious allergenic reactions occur. The apparent growing prevalence of severe allergic reaction to peanuts during the 1990s has been a case in point, and has initiated, through the manufacturing industries, a series of controls in raw material supply line, to ensure that products are, where appropriate, free from nut components and labelled accordingly. In 1996, the European Scientific Committee for Food (SCF) issued a report on 'Adverse Reactions to Food and Food Ingredients'. The level of intolerance to food additives was indicated in the report of 0.026% – equivalent to about three persons per 10 000 of the population being implicated

5.12 Future trends

As we move further into the 21st Century, the food industry in general will have access to an ever-increasing market. Global forecasts place the world's population, now already over six billion, in the region of nine billion by the year 2050. There should be no shortage of outlets for soft drinks.

The challenge facing the soft drink industry will remain largely unchanged, although it will perhaps become increasingly more complex. To provide products that are wholesome, durable and acceptable in character, and of interest to an ever-growing market, has been the objective for many years. However, with increasing realisation that almost anything we consume can have detrimental as well as beneficial effects, the focus, as today, will be very much on additives, processing aids and functional ingredients.

In the western world, there is increasing concern over processed food products and what goes into them – a trend partly fuelled by increasing obesity among members of the population (in many cases due to a change to a more sedentary lifestyle), and those who are genuinely concerned about the nutritional value of many foods, including beverages, on the market today. The term 'processed' gives rise to suspicion among consumers where food products are concerned and, despite not yet having entered common parlance where drinks are concerned, the discerning customer will always show preference for the use of natural ingredients in the make-up of the beverage. Consequently, it should be expected that the market will, in time-honoured fashion, be forced to adapt itself accordingly.

The term 'functional drinks' is now part of the vocabulary, and this is an area where there will be ample scope for future development. The general aim of a functional drink is not just to alleviate thirst, but also to contribute to a sense of well-being in the consumer by assisting with diet, metabolism, improved lifestyle, and so on.

Healthy lifestyle drinks, drinks with added vitamins, drinks with pro-biotic properties and isotonic drinks have already been seen in the market, and will provide a basis for more innovation. Recent examples of such include drinks claiming to have a 'thermogenic effect', whereby the use of combinations of ingredients such as caffeine, green tea, amino acids and ginger stimulate the body to generate more heat to digest food and temporarily elevate the metabolism, thereby 'burning off' calories as an aid to slimming. Products of this type have been used by athletes for many years, but we may now expect to see increased interest in thermogenic drinks aimed at the general consumer.

Such claims for drinks as being rich in fibre, fat free, sugar reduced, beneficial in calcium adsorption by the body, improving intestinal flora, probiotic, with ascorbic acid, energy reduced, and so on, will no doubt continue to be seen, and will have to be supported with sound evidence in order to comply with labelling regulations.

Another worldwide concern to be addressed is the consumption of energy associated with the manufacture and marketing of soft drinks. Although there will be little room for manoeuvre in terms of processing, because of raw material cost restrictions on beverage ingredients, it is likely that methods of bulk distribution and transportation will change to accommodate energy savings. The main area energy saving will surround the manufacturing costs of soft drink containers. Collection and recycling of plastic bottle and drink can materials is already in practice, and the trend is bound to continue and to become more efficient in the years ahead.

During the past decade, legislative controls have become more stringent, with the gradual phasing out or reduction in use of many of the additives that have hitherto been considered an essential part of the production formula, and this process is bound to continue. In conjunction with this, improved packaging, aseptic processing techniques and the selection of top-quality raw material ingredients will be in vogue, creating a highly specialised system of presentation designed at all times to satisfy the consumer – for this is what the successful marketing of soft drinks ultimately depends upon.

Further reading and references

Barnett and Foster (1916). *Bottling Plant Catalogue, Publication No.179.*
Bernard, B.K. (consulting ed, 1985). *Flavor and Fragrance Materials.* Allured Publishing Corp, Wheaton, IL, USA.
Code of Federal Regulations (1989). 21CFR170 to 199, Office of the Federal Register, USA.
European and Council Directive 95/2/EC (as amended) concerning food additives other than colours or sweeteners.
European Parliament and Council Directive 88/388/EEC (as amended) on the use of flavouring substances.
European Parliament and Council Directive 89/107/EEC (as amended) on the approximation of laws of the member states covering food additives for use in foodstuffs intended for human consumption.
Fellers, C.R. and Esselen, W.B. (1955). *Cranberries and cranberry products.* Agricultural Experiment Station, Amherst, Bulletin no. 481, 62 pp.
Hanssen, M. (1987). *E for Additives,* 2nd edition. Thorsons Publishing, Wellingborough, UK.
McColloch, R.J and Gentile, B. (1958). *Stabilization of Citrus Concentrates.* US Patent 2,845,355.
Regulation (EC) No. 1333/2008 of the European Parliament and of the Council of 16 December 2008 on Food Additives.
Saltmarsh, M. (ed,(2000). *Essential Guide to Food Additives.* Leatherhead Publishing, Surrey, UK.
Stevenson & Howell Ltd. (1926). *The Manufacture of Aerated Beverages, Cordials, etc.,* 7th edition. Own trade publicity, London.
The Commission of the EC (1999). *Official Journal of the European Communities. L84, Vol. 42.,* Register of Flavouring Substances, per Regulation (EC) No. 2232/96.

CHAPTER 6
Non-carbonated beverages

Philip R. Ashurst

Dr. P R Ashurst and Associates, Ludlow, UK

6.1 Introduction

Non-carbonated beverages represent an important segment of the market for soft drinks, but they present some special technological issues for product developer and manufacturer alike. The principal groups of non-carbonated beverages are as follows:
- dilutable drinks
- ready-to-drink pre-packaged beverages
- fruit juices and nectars.

As indicated elsewhere in this volume, the technical issues relating to tea, coffee and milk-based drinks are not considered, except insofar as they relate to the marketplace, or where they are used in soft drinks as ingredients.

Soft drinks are low-pH beverages that are based mostly around fruit-derived ingredients, or incorporate fruit flavours. They are an important source of hydration, but are usually selected on the basis of pleasant taste and convenience of use. There are some soft drinks, of which cola-flavoured beverages are the most prominent, which do not rely primarily on fruit flavours. Some of these non-fruit-flavoured products are, almost invariably, produced only in a carbonated form, whereas others (such as elderflower) appear as both dilutable and carbonated products.

The particular issues that must be addressed when non-carbonated beverages are to be produced relate mainly to the prevention of microbial spoilage, the deterioration of product taste and appearance as a result of oxidation, and the enhancement of flavour. Carbonated beverages use the presence of carbonation to boost flavour characteristics and to provide palate stimulation. Thus, there are different considerations to be applied to formulation and packaging when non-carbonated beverages are produced.

6.2 Dilutable beverages

6.2.1 Overview

Dilutable beverages have been widely used for many years as a low-cost, convenient means of producing soft drinks on the consumers' own premises. The product formulation is often broadly similar to that for a syrup produced by a carbonated beverage manufacturer, who subsequently dilutes this intermediate with carbonated water before packing it into the containers sold to consumers.

Dilutable products, on the other hand, offer the consumer a number of advantages, including the ability to use differing syrup : water ratios, the ability to produce variable volumes of end-product, and the possibility of using different diluents such as water, alcoholic drinks or milk.

Although syrups in various forms have been around for many years, a particular milestone was the production and use of concentrated orange juice as a dilutable 'syrup' in the United Kingdom during the Second World War.

That product was a 60°Brix concentrated orange juice, packed in eight fluid ounce (200 ml) 'medical flats' – glass bottles with flat sides and rolled-on metal caps, with cardboard inserts. The product was produced for and distributed by the UK government as a means of enhancing the nutritional intake of babies and young children, particularly in respect of their vitamin C needs.

Although this product was available following the war years, it spawned the development of a variety of other products, such as whole-fruit drinks, 'squashes' and cordials, which became, and remain, the mainstay of the dilutables market in the United Kingdom and Commonwealth countries.

6.2.2 Nomenclature

Dilutable products were given the particular product designations used above and, to ensure consistency between manufacturers, the United Kingdom introduced legislation in 1964 which defined, for the first time, specific compositional requirements for these products. These regulations, known as the 1964 Soft Drinks Regulations, were revoked some 31 years later, when it was recognised that many factors rendered these compositional constraints unnecessary. In particular, this move coincided with a change of emphasis in legislation towards improved labelling and consumer choice. For example, the imposition of regulations in the United Kingdom, requiring percentages of fruit ingredients to be declared (Anon, 1998), was one of a number of factors that facilitated such a change of emphasis.

An outline of the principal fruit component compositional requirements of the UK 1964 Soft Drinks Regulations compared with those of today is shown in Table 6.1.

In addition to reserved descriptions for these products, the 1964 regulations also defined minimum levels for carbohydrates and imposed various compositional constraints. It is perhaps noteworthy that, at the time of writing, the only compositional constraint that is legally required in the United Kingdom is that

Table 6.1 1964 UK reserved descriptions for dilutable fruit drinks.

Product description	Juice/fruit minimum content 1964	Juice/fruit minimum content post-1995
Squash	25% v/v cloudy juice	No minimum
Cordial	25% v/v clear juice	No minimum
Whole fruit drink	10% w/v comminuted citrus fruit	No minimum
X flavour drink or cordial	No minimum	No minimum
Citrus barley water	15% citrus juice	No minimum

Table 6.2 Principal ingredients of dilutable soft drinks.

Nutritional	Non-nutritional
Fruit components	Preservatives
Carbohydrate syrups	Colours
Acidulants	Emulsifiers and stabilisers
Other nutritional components, e.g. vitamins and minerals	Antioxidants
Acidity regulators	Intense sweeteners
	Flavourings
	Clouding agents

for a quinine content (57 milligrams/litre) to enable a product to be described as 'Indian Tonic Water'. Tonic water is well known in many other countries, where different minimum, and sometimes maximum, levels of quinine apply.

Despite the removal of compositional legislation, the descriptions of dilutable soft drinks in the United Kingdom are still widely used today. Consumers and enforcement authorities alike still have an expectation that a product described as a 'squash' will be a cloudy product, containing a significant proportion of fruit juice. Similar expectations still apply to the other products mentioned above.

6.2.3 Ingredients

Other chapters of this volume deal in more detail with the ingredients of all soft drinks, and readers requiring more information should refer to them. However, it is appropriate here to make reference to special issues concerning ingredients, insofar as they relate to dilutable soft drinks. The main ingredients of dilutable soft drinks are set out in Table 6.2.

6.2.3.1 Fruit components
Concentrated juices

It will be evident, from the section on nomenclature, that the principal fruit components that are used in dilutable soft drinks are fruit juices (both clear and cloudy) and whole-fruit preparations – the so-called comminutes.

Fruit juices and comminutes that are added to dilutables (and other non-carbonated drinks) may be either freshly pressed or in the form of a concentrated juice. If a significant proportion of juice (25%, for example) is required in a dilutable drink, the level required may be difficult to achieve, unless a concentrated juice is used. In practice, therefore, most non-carbonated beverages use concentrated juices and comminutes to obtain the required level of fruit components.

The concentration of most fruit juices is conveniently measured in degrees Brix, although the strict interpretation of this measure refers to pure solutions of sucrose in water (e.g. 10°Brix is 10% w/w sucrose in water). For juices with a high proportion of sugars to acids, such as orange, pineapple and apple, this is a useful and convenient means of measuring concentration. In some instances, a correction factor may be introduced to take account of the acidity (see Chapter 10). Brix measurement is simply related to refractive index, and there is a slightly different relationship between the refractive index and the concentration of citric or malic acids and that of simple sugars.

The observed Brix (and acidity) of a given freshly pressed juice will vary over a limited range, depending on a number of plant variables, such as seasonality, variety and location. However, concentrated juices are produced to an industry standard, so there will be slight variations in the degree of concentration required to achieve the standard of concentrated juice.

For example, frozen concentrated orange juice (FCOJ) – the industry standard material for orange – is traded as 65–66°Brix concentration. Oranges that are used may, on pressing, yield a juice of variable Brix – say, from around 10° to as much as 14° or 15°. Thus, the degree of concentration required to produce 65–66°Brix concentrate will be different for a 10°Brix juice, compared with a 13°Brix juice.

Most countries, therefore, adopt a 'standard' Brix for juices, in order to facilitate the production of a comparable product when juices are reconstituted. In most European countries, this standard is 11.2°Brix for orange juice. A similar approach will be adopted when calculating the amount of a concentrated juice required to deliver, say, 25% juice in a dilutable drink.

The typical concentrated juices and comminutes used by the industry for manufacturing dilutable drinks are shown in Table 6.3.

Table 6.3 Typical concentrated juices and comminutes for manufacturing dilutable drinks.

Fruit	Typical concentrate	Comminute
Orange	Frozen concentrated juices 65–66°Brix	30–60°Brix (3 : 1 to 6 : 1)
Lemon	Frozen juice or sulphite- preserved 4 : 1 or 5 : 1 clear or cloudy product	25–40°Brix (4 : 1 to 6 : 1)
Lime	450 g per litre citric acid = 6 : 1 to 3 : 1	Not normally produced
Apple	69–70°Brix concentrate	Not available

Comminutes

The process of comminution refers only to citrus products, where the oils that reside in the flavedo (coloured peel) have intense flavour characteristics. At its simplest, comminution involves taking a complete orange (or other citrus fruit) and making a pulp from it. This pulp will have a much more intense flavour than juice alone but, because of the presence of much peel and albedo (pith), it would be unacceptable in taste to most consumers. Thus, the process of comminute production, developed in the immediate post-war years, is typically as set out in Figure 6.1.

In the manufacturing process for comminutes, components of the citrus fruit are separated into the principal products: citrus oils, juice and residual peel and pith. After concentration of the juice, the oils and some peel and pith will be recombined, usually in different proportions, and the whole mixture will then typically be finely milled and homogenised, before being pasteurised.

Finished comminutes are usually available at strengths between about 35° and 60°Brix. When used for making whole fruit drinks, apart from a more intense fresh flavour, they deliver cloud and colour.

Other fruit components

Other fruit components that may be used in the manufacture of non-carbonated beverages, particularly dilutables, include pectins and aroma substances obtained during the concentration of the fruit juices. These components do not normally count towards the fruit content of products, and they are usually classified as types of permitted additives.

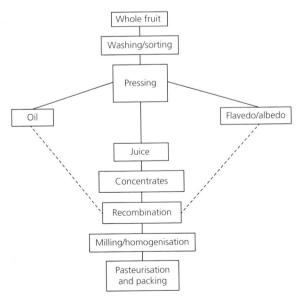

Figure 6.1 Simplified outline process for citrus comminute production.

6.2.3.2 Carbohydrates

Carbohydrates still feature as important components of many non-carbonated beverages, and they are particularly important in the manufacture of dilutable drinks. Historically, the UK Soft Drinks Regulations of 1964 required dilutable drinks to have a minimum level of 22.5% w/v carbohydrates, unless they were declared to be 'low calorie'. The regulations assumed a five times dilution factor (one part dilutable plus four parts water), and thus a minimum carbohydrates level of 4.5% w/v in finished drinks.

Today, manufacturers in most countries can choose how much carbohydrate (if any) to use, the information being passed to the consumer by the product label.

Sweeteners generally are dealt with in more detail in Chapter 5.

Sucrose

The preferred carbohydrate for most manufacturers is still sucrose, although its 2004 price within Europe is so artificially high that other alternatives are often sought and are increasingly used. Sucrose is readily available as a bulk dry solid, or as a 67°Brix syrup, and it is in this latter form that most manufacturers will use it.

Invert sugar

Invert sugar, sometimes referred to as partially inverted refiner's syrup, is produced by acid or enzymic hydrolysis of the disaccharide sucrose into its component parts of fructose and dextrose (glucose). Invert syrups usually contain a mixture of sucrose, fructose and dextrose. The main advantage of such a syrup is the reduced likelihood of crystallisation and an increase in osmolality, which may be useful in reducing spoilage risk.

Because of the development of fructose-containing glucose syrups, invert sugars are little used now. Some product formulators maintain that the sweetness of invert syrup is marginally greater than that of sucrose at the same strength.

Glucose syrups

Glucose syrups are a group of industrial syrups manufactured from starch – usually corn starch (maize). The starch may be hydrolysed by either acid or enzymic hydrolysis or, more usually, a combination of the two. Glucose syrups are normally referred to as having a dextrose equivalent (DE), which broadly relates to the percentage of dextrose in the mixture of carbohydrate produced on hydrolysis. In general, the DE also gives an indication of the sweetness of the syrup. Typical glucose syrups that are commercially available include 42DE and 63DE syrups. Products available from the hydrolysis of starch include pure dextrose, glucose syrups with a range of carbohydrate components, and maltodextrins.

Table 6.4 Comparison of degrees Brix, degrees Baumé and physical characteristics of carbohydrate syrups.

Degrees Brix	Degrees Baumé	Refractive index at 20°C	Specific gravity 20°/20°C
0	0	1.3330	1.0000
5	2.79	1.3403	1.0197
10	5.57	1.3478	1.0400
15	8.34	1.3557	1.0610
20	11.10	1.3638	1.0829
25	13.84	1.3723	1.1055
30	16.57	1.3811	1.1290
35	19.28	1.3902	1.1533
40	21.97	1.3997	1.1785
45	24.63	1.4096	1.2047
50	27.28	1.4200	1.2317
55	28.54	1.4307	1.2451
60	32.49	1.4418	1.2887
65	35.04	1.4532	1.3187
70	37.56	1.4651	1.3496
75	40.03	1.4774	1.3814
80	42.47	1.4901	1.4142

Glucose syrups are often used in energy drinks, where a high level of carbohydrate is required (e.g. 20% at drinking strength), but without the sickly sweetness that this strength of sucrose would bring. There can also be commercial advantages in using glucose syrups, as the solids levels are usually around 80% w/w, compared with the maximum of 67% w/w for sucrose syrup. One particular technical disadvantage is that glucose syrups are often extremely viscous and, if allowed to cool to below 30°C, can become very difficult to handle. The solids levels in glucose syrups are often measured in degrees Baumé, rather than degrees Brix, and some examples of the relationship between degrees Baumé, degrees Brix and physical characteristics are shown in Table 6.4.

Modified glucose syrups
An important development in the production of alternative carbohydrate sources for beverage and other food uses has been the production of fructose-containing glucose syrups. One such product is known as high-fructose glucose syrup (HFGS). It is widely used in the United States, and to a lesser extent in Europe, where there are fewer commercial advantages of use. In these products, starches (usually corn starch) are hydrolysed to dextrose syrup. A further enzymic modification then takes place, whereby a proportion of the dextrose present is converted to fructose. Depending on the proportion converted, the resulting level of fructose can reach up to 100% or more of the dextrose level, to give a product that is chemically similar to invert sugar syrup and has similar technical

and organoleptic properties. Syrups with a lower proportion of dextrose converted to fructose have also been found to be a useful carbohydrate source for beverage manufacture. The approximate comparative sweetness values of various carbohydrate sweeteners are shown in Table 6.5.

Fructose syrup

In addition to the glucose/fructose syrups mentioned above, a fructose syrup has been produced using inulin as a source. Inulin is the fructose analogue of starch, and the chicory root is the standard source for commercial hydrolysis. Fructose syrups are usually too expensive for routine use in beverage production, but they have been employed where a particular claim is to be made for fructose. They have also been used for the adulteration of fruit juices, as they are chemically difficult to detect. Detection is possible at the sub-molecular level by techniques such as stable isotope ratio measurement. Fructose is also manufactured using sucrose as a starting material.

6.2.3.3 Intense sweeteners

No chapter dealing with beverage manufacture would be complete without mentioning these important ingredients. A comparative picture is shown in Table 6.6.

Table 6.5 Comparison of carbohydrate sweeteners.

Carbohydrate	Approx sweetness compared with sucrose	Typical form
Sucrose	1.00	67°Brix syrup or solid
Invert sugar	1.00–1.1	Syrup
Glucose syrup	0.4–0.8	Syrup
High-fructose glucose syrup	1.0–1.1	Syrup
Glucose/fructose syrups	0.8–0.9	Syrup
Fructose syrup	1.05–1.1	Syrup

Table 6.6 Comparison of intense sweeteners.

Sweetener	E-number	Typical sweetness factor compared with sucrose	Maximum UK permitted use (mg/l)
Sucralose	(955) provisional	450–500	300
Saccharin	954	450–550	80 (as imide)
Aspartame	951	160–200	600
Acesulfame K	950	160–200	350
Cyclamic acid	952	30–40	400
NeoHDC	959	200–300	30

Saccharin
This substance was employed as a sugar substitute during the Second World War and for many years was used, together with sucrose, as a mainstay of beverage sweeteners. Saccharin, by experiment, has a sweetness factor (compared with sucrose) of 450 for the soluble form (sodium saccharin dihydrate) and around 550 for the much less water-soluble imide form. Despite commercial advantages, saccharin is little used now because of its bitter aftertaste.

Aspartame
Aspartame is a widely used intense sweetener that has excellent taste characteristics. It is a peptide made from two amino acids – phenylalanine and aspartic acid – but, in an acidic beverage medium, it will slowly hydrolyse to its components. The fact that aspartame is a source of phenylalanine is of concern to consumers with certain complaints, and suitable label declarations are now required by law. Technically, this slow hydrolysis brings about loss of sweetness.

Acesulfame
Acesulfame K has similar taste characteristics to aspartame, but without the disadvantages of hydrolysis causing loss of sweetness. The product has found wide use in beverages.

Cyclamic acid
Cyclamic acids, in the form of cyclamate salts, were in wide use during the 1965–75 period but, because of a sudden scare that they could be a cause of certain cancers, they were removed from the marketplace. Cyclamates were re-permitted on a limited basis in Europe in around 1995, but have found little commercial use since then.

Neohesperidin dihydrochalcone
This sweetener (NeoHDC) is a substance of natural origin that has been chemically modified. It has found little use in the beverage or food industries.

Sucralose
Sucralose is a chemically modified sugar with a very high sweetness factor, comparable with that of saccharin, but without the unpleasant aftertaste. The sweetness profile of sucralose is claimed to be excellent, and it has already found wide use in the beverage and food industries.

Stevia
Extracts of the stevia plant, Stevia rebaudia Bertoni, provide a glycoside that is about 250–300 times sweeter than sugar. The substance is now permitted for use in the EU under regulation EU 1131/2011, and is finding many applications as a zero-calorie natural sweetener. Its use is permitted in flavoured drinks at a level of up to 80 mg per litre (or per kg) of steviol glycosides

6.2.3.4 Other ingredients

Acidulants

The preferred acidulant for dilutable (and other) soft drinks is citric acid. This is readily available both as a crystalline solid (citric acid anhydrous) and as a 50% w/w solution in bulk. Other acidulants that are used in specific products include malic acid, lactic acid and tartaric acid. Phosphoric acid, until recently permitted only in cola drinks, is now available for use in the United Kingdom, but has so far found little, if any, use in dilutable products. Acids other than citric acid are usually employed only where a slightly different taste profile is needed. Ascorbic acid is usually employed as an antioxidant, rather than as a direct acidulant.

Preservatives

Despite the requirement that most dilutable drinks should be pasteurised (see Section 6.4), the use of chemical preservatives in these products is, in most situations, almost essential. The main reason for this is that dilutable products are used over a period of time, during which the container will remain partially full. The storage period will vary from user to user, and may be as short as a few hours from first opening, to several weeks, or even months. During this time, the consumer expects a product to remain free from fermentation, mould growth or other microbial development, and to retain an acceptable taste. Preservatives permitted in the United Kingdom include benzoic acid, sorbic acid and sulphur dioxide (in limited situations). Dimethyl dicarbonate (velcorin) is permitted, but is little used in dilutables.

Sulphur dioxide remains a key preservative in dilutables containing fruit components, where it is permitted (at least in the United Kingdom) at a rate of 250 mg/l. This preservative, which is a gas in solution in the product, will diffuse into the product headspace and help to minimize microbial development.

It is normal to use, in addition, a mixture of both benzoic and sorbic acids, added as their sodium and potassium salts, respectively. Current UK preservative regulations permit a maximum level of 300 mg/l of sorbic acid and 150 mg/l of benzoic acid, both at drinking strength. For this reason, it is normal to suggest, on the product label, a dilution ratio, which can then be used as a factor in calculating the amount of these preservatives to be used. An example is set out in Table 6.7.

In most dilutables, these levels would be more than adequate to deliver enough preservation after pasteurisation. A typical preservative mix for a dilutable containing up to 25% fruit juice might be as follows:
- Sulphur dioxide 150 ppm (parts per million)
- Benzoic acid 500 ppm
- Sorbic acid 800 ppm

The actual additions would be sodium metabisulphite (solution or solid) and aqueous solutions of sodium benzoate or potassium sorbate. Convenient conversions for these materials are shown in Table 6.8.

Table 6.7 Preservative levels.

Preservative	Max. ready-to-drink level mg/l	Dilution recommended for product	Max. level in dilutable
Benzoic acid (sodium benzoate addition)	150	1 + 4 (i.e. 5 × dilution)	750
Sorbic acid	300	1 + 4 (i.e. 5 × dilution)	1500

Table 6.8 Convenient conversions for preservative materials.

Preservative	Molecular formula and weight	Added form	Molecular formula and weight	Conversion factors	
				Salt to preservative	Preservative to salt
Sulphur dioxide	SO_2 MW 64	Sodium metabisulphite	$Na_2S_2O_5$ MW 190 (+2 × SO_2)	0.674 (128/190)	1.484 (190/128)
Benzoic acid	$C_7H_6O_2$ MW 122	Sodium benzoate	$C_7H_5O_2Na$ MW 144	0.847 (122/144)	1.1803 (144/127)
Sorbic acid	$C_6H_8O_2$ MW 112	Potassium sorbate	$C_6H_7O_2K$ MW 150	0.747 (112/190)	1.696 (190/112)

Because of the limited solubility of benzoic and sorbic acids in water, great care must be exercised during the manufacturing process of dilutables to ensure that acidification does not result in precipitation and loss of the preservatives.

There is a small but growing market for high-value dilutables that are declared to be free from preservatives. These products must be adequately processed, using in-bottle pasteurisation or aseptic packaging, and the label clearly marked with the need for time-limited storage in refrigerated conditions.

Flavourings

Flavourings are widely used in dilutable soft drinks, to boost or substitute those occurring naturally. There are other publications that deal with this topic in more detail but, in brief, it is necessary to ensure that appropriate beverage flavours are selected to produce adequate solubility. Most manufacturers of dilutables will use either natural or nature-identical flavours.

Colourings

Most dilutable beverages are formulated with added colourings although, depending on the fruit preparation used, many products will have a significant level of colour delivered by the fruit components. In the United Kingdom, artificial colours are now little used in dilutable beverages, except for lime juice cordial, which is usually marketed as a clear product with a synthetic lime-green colour that is difficult to achieve by means of natural ingredients.

The available natural colourings offer a limited range of yellow, through orange, to red/purple colours for products. The most common natural colours used in dilutables include β-carotene, apocarotenal, curcumin and anthocyanins.

To obtain maximum colour stability, a careful balance must be achieved between sulphur dioxide and ascorbic acid contents, in order to avoid bleaching the colours.

Remaining additives

Various other additives are employed in dilutable soft drinks manufacture, including antioxidants, acidity regulators, emulsifiers and stabilisers. Stabilisers are particularly important for ensuring physicochemical stability of the product to avoid unsightly oil ring formation or undue sedimentation of fruit components. Cloudy agents are often used to boost the turbidity of natural fruit components. These ingredients can also be made to incorporate citrus oils and colourings, by creating an oil-in-water emulsion, using a mixture of permitted emulsifiers and final emulsification of cloud, to a low particle size (<10 μm).

Compound ingredients

Compound ingredients were widely used at one time for the manufacture of dilutables. They are still available today, but are much less widely used. This probably reflects a number of factors, such as the disappearance of many small soft drinks manufacturing companies and the need for precise fruit component content. A compound would typically contain all the components to make a dilutable, except for the water and carbohydrate. Thus, a manufacturer would purchase, for example, a ten-fold orange squash compound. By adding the required amounts of sugar and water, the manufacturer would make ten times the volume of compound (i.e. 25 litres of ten-fold compound would make 250 litres of orange squash) into a product that would contain the legal minimum fruit content.

6.2.4 Manufacturing operations

The manufacture of dilutables is essentially a very simple process, with the required ingredients being mixed in order in a large vessel. After checking the final volume for process variables the mixture is then flash pasteurised and filled into the required containers.

The process is diagrammatically summarised in Figure 6.2.

6.2.4.1 Ingredients

Addition of the ingredients in the correct order is essential in order to avoid production problems. The normal order starts with the presence of around 30–50% of final product volume of process water, to which preservatives (other than sulphur dioxide) are first added. This volume should be as large as possible, to

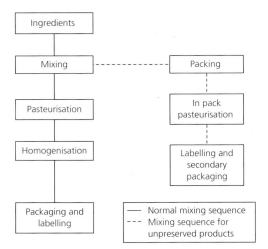

Figure 6.2 Process for manufacture of dilutables.

allow the addition of carbohydrates and fruit components, which follow in that order. At this point, the volume should be approaching 90% of final volume, to allow the dilution of preservatives. Acidulant is then added, followed by colourings, flavourings and all other components.

The last ingredients to be added are cloud emulsions. Sulphur dioxide, if used, should ideally be added after the final make-up water, to avoid loss to atmosphere of the gaseous preservative.

6.2.4.2 Mixing

A mixing vessel, fabricated from high grade stainless steel, with some form of cover that allows access for ingredient addition, is an ideal unit for mixing dilutables. The vessel is normally fitted with a stirrer, the power and design of which take account of whether sugar is to be added as a crystalline solid (thus needing dissolution) or added as a syrup. Either a top-mounted propeller stirrer or a side-entry unit will mix components adequately, especially if the inside surface of the vessel is fitted with fixed baffles. The use of a stirrer that creates a sufficient vortex to draw in air should be avoided.

High-shear stirrers can be a useful way of mixing components, but they often draw in air and can have an adverse effect on added emulsions. Mixing can also be done through an external circulating loop, with an in-line pump or emulsifying mixer. All systems should ideally be connected to a clean-in-place (CIP) system.

6.2.4.3 Pasteurisation
Flash pasteurisation

For a normally preserved dilutable soft drink, the typical conditions for flash pasteurisation are 85–90°C for 30–60 seconds. The actual conditions should be determined by reference to the quality of ingredients used, although pasteurisation must never be used as a means of employing sub-standard components. Products containing particulate, such as fruit cells, should be pasteurised in a plate pasteuriser with 3–4 mm spacing, or in a tubular pasteuriser. Great care must be taken to avoid microbial contamination downstream of the pasteuriser. Pasteurised product should be stored in a very clean (or even aseptic) bulk buffer tank prior to filling.

In-pack pasteurisation
In-pack pasteurisation is normally reserved for dilutable products that are made without preservatives. Unless there are particular circumstances that demand a preservative-free product (e.g. manufacture of a certified organic product or other marketing considerations), preservatives should always be used in dilutables, because of the way the drinks are used and stored. A dilutable without preservatives is very vulnerable to microbial contamination, which can lead to fermentation, and possibly, bottle bursting. Dilutables without preservative must be labelled to encourage refrigerated storage and short shelf life.

In-pack pasteurisation normally demands large and expensive tunnel pasteurisers, which have several stages. Bottles are introduced into a pre-heating stage (typically around 40°C) to reduce thermal shock, and then into the pasteurisation zone, which will normally be at 70°C, for some 20 minutes. Following this are hydro-cooling zones. The first of these reduces product temperature to around 40°C, and the second to ambient temperature. Recovery of heat is essential to an economically viable operation.

Final product temperature should ideally be below 20°C, to avoid the phenomenon of 'stack burn', where packed and palletised product that is not adequately cooled will effectively be 'slow cooked'. This can result in excessive browning, and the development of a cooked taste.

6.2.4.4 Homogenisation
Some manufacturers homogenise all cloudy dilutable products, to obtain maximum physical stability for the product, but others achieve the same result by careful ingredient selection. If homogenisation is to be used, a piston-type unit is preferred, with an operating pressure range of around 50–100 bar. As with all beverage manufacturing plant, effective cleaning is essential.

6.2.5 Filling and packaging
Gravity fillers are normally employed for dilutable products, and filling speeds tend to be fairly slow, as container sizes are relatively large. For most dilutables, the smallest container is usually 0.7 litres, with sizes up to 3 or 5 litres being common.

Most manufacturers now use polyethylene terephthalate (PET) bottles, which provide a good degree of protection from oxygen ingress, but without the weight disadvantages of glass packs. For PET bottles, closures are normally moulded polyethylene (LDPE or HDPE), whereas manufacturers packing in glass will normally use roll-on pilfer proof (ROPP) caps made of aluminium.

6.2.6 Product range

The dilutables of the 1960s were fairly limited in range – orange squash, whole orange drink, lemon squash, lemon barley water, lime juice cordial and blackcurrant cordial being a typical product spectrum. Seasonal products that were added to the range included ginger and peppermint cordials.

Over the past 20 years, the UK market for dilutables has developed and grown to include high-value products that have become a niche market. The range of dilutable products available today is much wider, and includes products such as elderflower cordial, summer berry fruits, lime and lemongrass, to name but a few.

6.3 Ready-to-drink non-carbonated products

6.3.1 Overview

There has always been a market for ready-to-drink (RTD) non-carbonated products, but it was a difficult market to develop, because consumers needed persuading away from making their own RTD products using dilutables. The market for these products is now well established.

The market was once seen as one of low-quality products, with an overriding convenience factor but, over the last 25 years, packaging developments and increasingly affluent consumers have encouraged the development of this market. Early products in the market were packed in pre-formed plastic cups, pouches or early Tetra Paks, but today there is a wide variety of packing options available.

6.3.2 Formulations

RTD non-carbonated drinks are usually made by formulations and processes that are identical to the manufacture of dilutables, except that the dilution takes place at the manufacturer's premises rather than the consumer's. However, because many non-carbonated drinks often sell in low-unit-price markets, many of the early formulations used contained little, if any, fruit components. As mentioned above, a market for higher unit value products, in more expensive packaging formats, is now well developed

6.3.3 Special problems

There are particular problems in the manufacture of non-carbonated RTD beverages that are not aseptically packed, and these relate to microbial contamination. Products that have no carbon dioxide in their head space are particularly vulnerable to contamination by moulds and certain types of bacterial infection.

For many years, it was possible to control such potential contamination by the use of low levels of sulphur dioxide (50 ppm). Changes in European Preservative Regulations now make the use of this preservative in RTD formulations (but not dilutables) illegal, unless it is 'carried over' from a fruit component, when up to 20 ppm SO_2 may be present. Even at this level, the gaseous preservative is rapidly lost and is quickly ineffective.

To avoid such microbial problems, manufacturers must either employ aseptic packing lines, which are very capital-intensive, or use flash pasteurisation and scrupulous downstream hygiene, and close control over formulations.

One significant difference in these RTD products is that levels of preservatives will normally be raised to close to the permitted maximum, to gain maximum benefit. Even preserved products are particularly vulnerable to the development of mould spores and, when sorbic acid is used, a particular problem can arise as a result of its metabolism by a mould. The production of 1,3-pentadiene can occur, giving the affected product an unpleasant off-taste and aroma.

A further potential problem with non-carbonated RTD products is that they invariably contain atmospheric air in their headspace, as there is no carbon dioxide to displace it. This often leads to undesirable oxygen levels in the product, with resulting flavour and colour deterioration in a short time period.

Accordingly, it will often be necessary to adjust the product formulation to incorporate appropriate levels of antioxidants, such as ascorbic acid, and to use flavour and colour preparations that are stable to oxidation.

6.3.4 Manufacturing and packing

The normal manufacturing and packing sequences for both aseptic and non-aseptic products are shown in Figure 6.3.

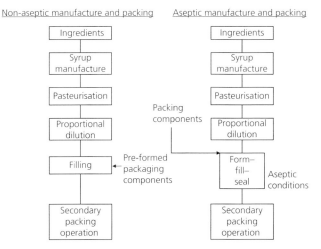

Figure 6.3 Normal manufacturing and packing processes for aseptic and non-aseptic products.

Various alternative configurations can be used and, in particular, some manufacturers employ non-aseptic form-fill-seal operations, which usually produce either in-line cup packs, or cartons such as Tetra Pak or Combibloc packs.

6.3.5 Packaging types

Many non-carbonated RTD products that are not pure fruit juices or nectars are packed in either pre-formed or form-fill-seal plastic packages, although an increasing number are now packed in PET bottles. Flexible pouches have also been used by a number of manufacturers.

Depending on the shelf life required, the use of some form of barrier in the packaging is highly desirable or even essential, depending on the shelf life required. Much of the packaging used today will be based on a rigid or semi-rigid container, employing polystyrene as a major component. The incorporation, often in the form of laminated structure, of a barrier plastic, can significantly enhance the product shelf life by reducing the rate of oxygen transfer. Some mechanical flexibility can be introduced into containers by incorporating polyethylene into the laminate.

The increased use of PET bottles for packing non-carbonated RTD drinks probably reflects the availability and convenience of this form of packing, coupled with the low oxygen transfer rate.

6.4 Fruit juices and nectars

This chapter will not provide the background to the production of fruit juices and nectars, as that is dealt with elsewhere in this volume. However, fruit juices and nectars represent the largest volume of non-carbonated beverages that are sold in almost every marketplace. It is therefore appropriate that some aspects of these products, particularly those relating to processing and packaging, are mentioned here.

6.4.1 Processing

Fruit juices and nectars are highly susceptible to fermentation and other forms of microbial spoilage and, with few exceptions, it is essential that some form of pasteurisation is employed when these products are packaged. The exceptions that are seen usually relate to freshly squeezed orange (or other) juices that are processed directly from fresh fruit and packaged immediately. These products have a very short shelf life – usually a few days – and are maintained by storage at temperatures between 0–5°C.

There is also a market for reconstituted fruit juice made from concentrate and not further processed, but maintained, during its short shelf life, by refrigeration.

For all other fruit juice and nectar products, either frozen storage or in-pack pasteurisation will be used, although some manufacturers employ a hot fill process.

6.4.1.1 Flash pasteurisation

Typical flash pasteurisation operations for fruit juices and nectars will employ a plate pasteuriser, with heat recovery and final product cooling. Typical flash pasteurisation conditions will use temperatures between 85–95°C, with holding times varying between 15–60 seconds. Selection of the appropriate conditions will depend on the product, including the level of microbial load pre-pasteurising. If enzyme deactivation is required as well as microbial removal, then a temperature between 90–95°C will normally be used. At these temperatures, holding times are normally reduced to around 15 seconds.

Juices containing cells, particulate material or products that are particularly viscous, such as some of the tropical juices, may be pasteurised in tubular units, or plate pasteurisers with wide (3–5 mm) spacing.

For aseptic packaging operations, flash pasteurisers are often linked integrally with the aseptic packaging unit, either directly or via an aseptic buffer tank. When flash pasteurisation is used, care should be taken to minimise product recirculation when the pasteuriser is in divert mode. Excessive recirculation can lead to thermal damage to the product, resulting in unpleasant cooked flavours and product browning.

6.4.1.2 In-pack pasteurisation

In-pack pasteurisation is often regarded as a foolproof operation, although product integrity will ultimately rely on the seal provided by the pack closure, and the maintenance of the required temperature over the whole pasteurising area.

For small-scale operations, in-pack pasteurisation can be achieved at very low cost by simply immersing the bottled product, with closures tightly applied, in tanks of heated water. A pre-heat tank at around 40°C should be employed to minimise thermal shock to the containers, and the main pasteurising tank will be held at around 70°C. A single container, into which is inserted a remote temperature probe, should be used to ensure that the whole contents of the bottles reach pasteurising temperature.

The normal means of achieving in-pack pasteurisation is to use a tunnel pasteuriser. These are large, capital-intensive pieces of plant, and they require significant floor space and provision of services. Most units work by using water sprays in a pre-heating zone, pasteurising zone and cooling zone(s). Some form of heat recovery is almost essential if a tunnel pasteuriser is to work economically.

After containers leave the pasteuriser, they should be air-dried and then labelled. Typical pasteurising conditions will be 70–75°C for up to 20 min.

6.4.1.3 Hot filling

Hot filling provides a further means of ensuring the microbial integrity of fruit juices and nectars. However, because of the time delay in applying caps to filled containers, and cooling that rapidly occurs, hot filling should be regarded as a higher risk process than in-pack pasteurisation. The bulk product is heated to the required temperature, then filled into containers and the closure applied. If glass bottles are used, they should be pre-heated (e.g. by a warm water spray) before filling, to minimize thermal shock. Following filling, containers are usually rotated through 360° to ensure contact between the hot liquid and the whole inside of the container and cap.

Depending on the fill temperature, which is usually around 70–80°C, the filled containers will be held for the required time before being placed in a hydro-cooler. Containers should be cooled to below 25°C before being stacked. This will avoid further low-temperature 'cooking' of product inside a stack of containers. Labelling is carried out after air-drying the containers.

6.4.1.4 High-pressure pasteurisation

Claims have been made for the successful high-pressure pasteurisation of fruit juice in containers. The equipment required is expensive, and the process is carried out as a batch operation, which tends to be both slow and ineffective. It may, in future, provide a very interesting means of low-temperature pasteurising of fresh juices, thus retaining all the flavour characteristics of the product.

6.4.2 Packaging

6.4.2.1 Boxes

Most fruit juices for retail sale are now in cartons, a high proportion of which will be aseptic packs. Cartons are formed, filled and sealed in a single operation, which will either be clean or aseptic, depending on the product and shelf life sought. Typical packs include Tetra Paks, Combibloc and Elopak.

The long shelf life packs for aseptic products are often made of a board, foil, plastic (polyethylene) laminate, which gives protection from oxygen ingress and light, as well as providing mechanical strength and an excellent surface for printed material. Cardboard packs for short shelf life products are often a simpler laminate, excluding foil.

6.4.2.2 Bottles

The selection of a container for fruit juices will often be based on a combination of the technical, cost and marketing needs. Many outlets for fruit juices require relatively small unit packs, and these will invariably be glass bottles. In the United Kingdom, most fruit juice in bottles is limited to these small units (e.g. 200 ml), whereas many European markets prefer larger (up to 1 litre) glass bottles, with wide necks. Glass bottles will normally be pasteurised using either the hot fill method or in-pack pasteurisation.

There is some limited use of plastic bottles for juice packing, mostly related to short shelf life products sold from the chill cabinet. PET containers and plant are now available to support aseptic filling, but they have found limited support for the sale of fruit juices.

6.4.2.3 Cans

The sale of fruit juice in cans to either the retail or industrial markets has largely died out with the availability of other forms of packaging and storage, except in specialised markets such as transport catering. Some juices from developing country suppliers (e.g. mango juice) are still supplied in cans containing around 5 kg. This reflects the technology and packaging available in the supply country.

6.4.2.4 Bulk packs

Juice for industrial use has, over many years, been packed in a wide variety of drums. Typical drums are open-head steel containers, with the juice packed inside several plastic bags. This package is the usual container for frozen juices and typically contains around 200 litres.

Plastic drums have also been widely used over many years, without the need for plastic liners, but they are less suitable for freezing, as the plastic has a tendency to become brittle and may rupture. Plastic drums usually contain 200–250 litres, although larger containers have been very successfully used, especially for chemically preserved juices. The Israeli manufactured 'Rotoplas' container was probably the best-known example. This container typically held around 1300 litres.

Aseptic bulk packing has now become a well-established means of packing concentrated or RTD juice. Containers are available from as little as a 5 l bag-in-a-box, to be dispensed from a bar, up to a 1000 l bin in a 1 m^3 pallet box.

Finally, the transportation of fruit juice in temperature-controlled bulk road tankers of up to 25000 litres is well established, as is intercontinental transfer by shipping tankers.

Further reading

Arthey, D. and Ashurst P.R. (eds, 2001). *Fruit Processing*, 2nd edition. Aspen Publishers Inc., Gaithersburg, MD.

Ashurst, P.R. (ed, 1995). *Production and Packaging of Non Carbonated Fruit Juices and Fruit Beverages*, 2nd edition. Blackie Academic & Professional, Chapman & Hall, London.

Ashurst, P.R. and Dennis, M.J. (eds, 1996) *Food Authentication*. Blackie Academic & Professional, Chapman & Hall, London.

Ashurst, P.R. and Dennis, M.J. (eds, 1997). *The Analytical Methods of Food Authentication*. Blackie Academic & Professional, Chapman & Hall, London.

European Union Council Directive 2001/112/EC.

UK Fruit Juices and Nectars Regulations 2003, Statutory Instrument No. 1564, HMSO, London.

UK Soft Drinks Regulations 1964 (as amended), Statutory Instrument N.

CHAPTER 7
Carbonated beverages

David Steen
Casa Davann, Murcia, Spain

7.1 Introduction

Naturally occurring carbonated mineral waters have been known for a long time. These effervescent waters exist as a consequence of excess carbon dioxide in an aquifer dissolving under pressure. Although claims for the medicinal properties of these mineral waters have been grossly exaggerated, the presence of carbon dioxide does make aerated waters and soft drinks both more palatable and visually attractive: the final product sparkles and foams. The first noncarbonated soft drinks appeared during the seventeenth century. In 1767, Joseph Priestley produced the first man-made, palatable carbonated water. Three years later a Swedish chemist, Torbern Bergman, invented a process that produced carbonated water from the reaction between chalk and sulphuric acid, allowing the commercial production of aerated mineral water. In 1783, Jacob Schweppes, a young watchmaker and amateur scientist, perfected an efficient system for manufacturing carbonated mineral water and founded the Schweppes Company in Geneva. He relocated to Drury Lane, London, England in 1790. Since then, the addition of flavourings to aerated waters has seen the development of major soft drinks brands throughout the world. To meet the need for carbonated soft drinks, the soda fountain was developed by Samuel Fahnestock in the United States in 1819. The patenting of the Crown cork by William Painter in 1892 and the automatic production of glass bottles using a glass-blowing machine by Michael J. Owens in 1899 were notable achievements that at last allowed carbonated soft drinks to be successfully bottled without significant loss of carbonation. Since then, developments in closure technology, polyethylene terephthalate (PET) bottle production, can design and manufacture, syrup making methods, carbonation technology and filling machine manufacture have led to the worldwide beverage industry as we know it today.

Chemistry and Technology of Soft Drinks and Fruit Juices, Third Edition. Edited by Philip R. Ashurst.
© 2016 John Wiley & Sons, Ltd. Published 2016 by John Wiley & Sons, Ltd.

7.2 Carbon dioxide

Carbonation is the impregnation of a liquid with carbon dioxide gas. Carbon dioxide is a non-toxic, inert gas that is virtually tasteless and is readily available at a reasonable cost. It is soluble in liquids (the degree of solubility increasing as the liquid temperature decreases) and can exist as a gas, liquid or solid. When dissolved in water it forms carbonic acid. It is carbonic acid in combination with the product that produces the acidic and biting taste found in carbonated waters and soft drinks. Above a certain level of carbonation, carbon dioxide also has a preserving property, which is a bonus from its use.

Carbon dioxide gas is heavier than air; it has a specific gravity of 1.53 under normal conditions of temperature and pressure. It has a molecular weight of 44.01 and does not burn, although it will support the combustion of magnesium. It is a fairly stable compound that decomposes into carbon and oxygen only at very high temperatures. It can cause death by suffocation if inhaled in large amounts. The gas is easily liquefied by compression and cooling. When liquid carbon dioxide is quickly decompressed it expands rapidly and some of it evaporates; this evoporation removes sufficient heat that the remainder cools into a solid. Carbon dioxide is part of the atmosphere, making up about 1% by volume of dry air. In various parts of the world it is formed underground and issues from fissures within the earth. This happens notably in Italy, in Java and in Yellowstone National Park in the United States. Carbon dioxide is a known contributor to the greenhouse effect; the proportion in the atmosphere increases each year, thus disrupting the natural carbon dioxide cycle.

The phase diagram in Figure 7.1 shows the effect of temperature and pressure on the state of carbon dioxide. At the triple point, carbon dioxide can exist in the three states as a solid, a liquid or a gas by just a small perturbation. All phases are in a state of equilibrium at the triple point, which is at 5.11 bar and −56.6°C.

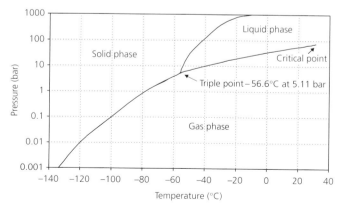

Figure 7.1 Carbon dioxide phase diagram.

Above 31°C, it is impossible to liquefy the gas by increased pressure; this is termed the critical point. At normal temperatures and pressures carbon dioxide is a colourless gas; at high concentrations it has a slightly pungent odour. Carbon dioxide cannot exist as a liquid at atmospheric pressure. Liquefaction can be achieved by compression and cooling between the pressure and temperature limits at the triple point and the critical point. Above the critical point of 31°C it is impossible to liquefy the gas by increasing the pressure above the corresponding critical pressure of 73 bar. When liquid carbon dioxide under pressure is released into the atmosphere it will be as a gas and a solid only, in the form of a dense white cloud due to the solid content and the condensation of atmospheric moisture at the low temperatures obtained. The solid will fall to the ground as snow, which, when compressed, forms a translucent white solid known as dry ice.

7.3 Carbon dioxide production

Several methods of carbon dioxide production are in commercial use. These include the reaction between sulphuric acid and sodium bicarbonate, the combustion of fuel oil, the extraction of carbon dioxide from the flue gas of a boiler or similar heating facility, the distillation of alcohol and the fermentation of beer; carbon dioxide is also a byproduct of fertiliser manufacture. Following manufacture the gas must be cleaned to ensure it is free from impurities and is fit for purpose. Two typical processes are described below.

7.3.1 Fermentation

When a sucrose- or other simple carbohydrate-based solution is mixed with yeast and oxygen in a fermenter, carbon dioxide vapour and alcohol are produced. The carbon dioxide can then be passed through a separator to remove any trace carry-over of foam. Once the foam has been removed the carbon dioxide is compressed. It is then scrubbed with water in a packed tower, removing water-soluble impurities such as alcohol, ketones and other aroma chemicals produced during fermentation.

7.3.2 Direct combustion

A hydrocarbon fuel such as light oil or natural gas can be burned specifically in order to produce carbon dioxide. The flue gas from this process, which contains less than 0.5% oxygen by volume, is cooled and scrubbed to remove any impurities that may be present. The resultant gas is then passed through an absorbent tower, where it comes into contact with a carbon dioxide absorbing solution. The absorbing solution, now rich in carbon dioxide, is pumped to a stripper tower, where the heat from the combustion of the fuel is used to release the carbon dioxide in vapour form. The absorbing solution is then recycled and reused. The resultant carbon dioxide vapour is then cooled and further treated to meet the requirements for use within a beverage.

Large manufacturers of aerated waters and carbonated soft drinks produce their own carbon dioxide on site using packaged systems, thereby reducing their operating costs and minimising logistical problems.

7.3.3 Quality standards

The European Industrial Gases Association working with the Compressed Gases Association of America and the International Association of Beverage Technologists has prepared a specification for liquid carbon dioxide for use in foods and beverages. This is shown as Table 7.1. It is to this minimum standard that all carbon dioxide to be delivered to soft drinks and aerated mineral water bottles is manufactured.

7.3.4 Delivery to the customer

Carbon dioxide is delivered as a liquid, most frequently by road tanker. It is then transferred to pressurised vessels of 5–50 tonne capacity and held at a pressure in the region of 20.5 bar at −17°C, the temperature being maintained by the use of a small refrigeration unit. This is shown schematically in Figure 7.2. To change the carbon dioxide from the liquid to the gas phase all that is required is to vaporise the liquid by heating it using either steam, water or electricity. This

Table 7.1 Commodity specification for carbon dioxide (CGA/EIGA limiting characteristics).

Component	Specification
Assay	99.9% v/v min.
Moisture	50 ppm v/v max. (20 ppm w/w max.)
Acidity	To pass JECFA test
Ammonia	2.5 ppm v/v max.
Oxygen	30 ppm v/v max.
Oxides of nitrogen (NO/NO$_2$)	2/5 ppm v/v max. each
Non-volatile residue (particulates)	10 ppm w/w max.
Non-volatile organic residue (oil and grease)	5 ppm w/w max.
Phosphene[a]	≤0.3 ppm v/v
Total volatile hydrocarbons (calculated as methane)	50 ppm v/v max. of which 20 ppm v/v max. non-methane hydrocarbons
Acetaldehyde	0.2 ppm v/v max.
Benzene	0.02 ppm v/v max.
Carbon monoxide	10 ppm v/v max.
Methanol	10 ppm v/v max.
Hydrogen cyanide[b]	<0.5 ppm v/v
Total sulphur (as sulphur)[c]	0.1 ppm v/v max.
Taste and odour in water	No foreign taste or odour

[a] Analysis necessary only for carbon dioxide from phosphate rock sources.
[b] Analysis necessary only for carbon dioxide from coal gassification sources.
[c] If the total sulphur content exceeds 0.1 ppm v/v as sulphur then the species must be determined separately and the following limits apply: carbonyl sulphide, 0.1 ppm v/v max.; hydrogen sulphide, 0.1 ppm v/v max.; sulphur dioxide, 1.0 ppm v/v max.
Source: With thanks to Messer Gases.

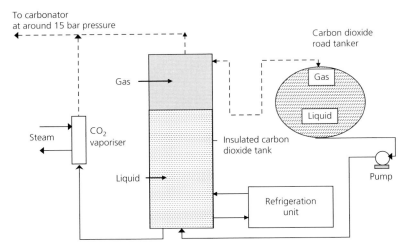

Figure 7.2 Carbon dioxide delivery process.
Note: No valves are shown.

is normally done in tubular heat exchangers. The most common system is to use steam, which has traditionally been available already in soft drinks plants for bottle washing, pasteurisation and factory heating. The use of water is becoming more common as it is cheaper and the process actually reduces the temperature of the water, which saves subsequent product cooling costs as well as the cost of steam generation. A typical 2 tonne/h carbon dioxide vaporisation system requires a constant flow rate of 50 m^3/h of water at ambient temperature to be effective. Electric vaporisers are expensive and rarely used. Yet another method is the use of ambient air blown across evaporator coils. It is claimed that this achieves significant energy savings over conventional steam vaporisation. Considering Figure 7.2, it is important to ensure that both the gas phase and the liquid phase in the tanker and the receiving tank are connected. This allows the gas displaced by the liquid delivered to the storage tank to flow into the tanker and thus balance the pressure between the two vessels. Non-return check valves must be incorporated to ensure no backflow occurs. The system needs to be installed in accordance with the recommendations of the gas supplier, and regular planned maintenance on the system is necessary to ensure any risks are minimised.

7.3.5 Precautions

The user must take great care when using carbon dioxide to ensure that it is fit for purpose. Scares such as the detection of residual benzene in CO_2 and the risk of contamination from nuclear plants have to be considered. All batches supplied must have a certificate of conformance. The supply chain must be regularly audited, including the actual carbon dioxide manufacturing plant, storage and distribution. A full hazard analysis and critical control points (HACCP) survey of the site-installed system is required to ensure any inherent risks are minimised. A list of possible impurities is given as Table 7.2. Depending on the method of

Table 7.2 Possible trace impurities by source type (excluding air, gases and water).

Component	Combustion	Wells/geothermal	Fermentation	Hydrogen or ammonia	Phosphate Rock	Coal gasification	Ethylene oxide	Acid neutralisation
Aldehydes	✓							
Amines	✓			✓				
Benzene	✓	✓	✓	✓				✓
Carbon monoxide	✓	✓	✓	✓		✓		✓
Carbonyl sulphide	✓	✓	✓	✓	✓	✓	✓	✓
Cyclic aliphatic hydrocarbons	✓	✓		✓		✓	✓	
Dimethyl sulphide		✓	✓		✓			✓
Ethanol			✓			✓	✓	
Ethers		✓	✓	✓		✓	✓	
Ethyl acetate		✓	✓	✓		✓	✓	
Ethyl benzene		✓	✓	✓		✓		
Ethylene oxide						✓	✓	
Halocarbons	✓					✓		
Hydrogen cyanide						✓		
Hydrogen sulphide	✓	✓	✓	✓	✓	✓	✓	✓
Ketones	✓	✓	✓	✓	✓	✓	✓	
Mercaptans	✓	✓	✓	✓		✓	✓	
Mercury	✓	✓				✓		
Methanol	✓	✓	✓	✓		✓	✓	
Nitrogen oxides	✓					✓		✓
Phosphine					✓	✓		
Radon					✓	✓		✓
Sulphur dioxide	✓				✓	✓		✓
Toluene	✓	✓	✓	✓		✓	✓	✓
Vinyl chloride						✓	✓	✓
Volatile hydrocarbons		✓	✓	✓		✓	✓	
Xylene		✓	✓			✓	✓	

Note: Examples of specific analytical methods are given in the ISBT document. *Analytical Methods for CO_2 Analysis.*

Source: As prepared by an ad hoc working group of the European Industrial Gases Association working in conjunction with the Compressed Gases Association of America and the International Society of Beverage Technologists (ISBT), EIGA Document 76/01.

carbon dioxide production, the certificate of conformation for each batch must include analytical checks on the relevant listed compounds. These analytical checks need to be carried out to the standards agreed by the International Society of Beverage Technologists (ISBT). The risk of contamination is very real. If too much gas is withdrawn from the top of the tank then contaminants can build up due to distillation of the liquid carbon dioxide within the tank. Other possible risks are oil contamination from the transfer pump between the delivery tanker and the user's tank, degradation of the delivery hose, contamination of the hose connections by particulates, water, oil or mud, and contamination by the backflow of cleaning fluids from the user's process. Dedicated tankers are required that are not used for non-food applications including nuclear plants or where a risk of cross-contamination exists.

7.4 Carbonation

7.4.1 Basic considerations

For a liquid–gas mixture in a sealed container, equilibrium is said to exist when the rates of gas leaving and entering the liquid solution are equal. Take any PET bottle of carbonated soft drink and shake it: the liquid–gas interface will initially fob, but after a short while the equilibrium condition will have been reached. Fobbing is a term used within the carbonated beverage industry to denote product foaming. If the cap is then opened and some of the contents poured out, the cap replaced and the shaking repeated, the bottle will go from being limp before the shaking to being rigid. Gas has come out of solution to attain the equilibrium condition. This state is just stable. Any decrease in pressure or increase in temperature will render the mixture metastable, that is, supersaturated, so that the temperature/-pressure combination is insufficient to keep the carbon dioxide in solution. If this occurs, gas is spontaneously released, giving rise to fobbing. If the mixtures are agitated or some irritant, such as small particulates, is added to the mix, the rate of gas release will be even more pronounced. This is due to nucleation sites being generated by the presence of these particulates or other gases, such as air.

Any carbonated product that is kept in a container that is open to the atmosphere will gradually lose carbonation. This is due to the gas being liberated into the atmosphere as the liquid–gas interface continually strives to achieve the equilibrium condition. In a closed container the gas fills the container headspace, thus increasing the headspace pressure. This happens quickly at first and then slowly as equilibrium is approached. The rate of transfer of gas from the product to the headspace depends on the proximity of the headspace pressure to the equilibrium pressure, the temperature of the liquid, the nature of the beverage, the extent of any agitation and the presence of any irritants. A quiescent, stable product will take many hours to reach equilibrium when not subjected to any external forces such as agitation, movement, temperature or pressure change. However, the same product roughly shaken will take only seconds to achieve the equilibrium condition. The faster the rate of change towards the equilibrium condition, the sooner this condition will be reached. For a given volume, the amount of carbon dioxide that can be retained in solution depends on the temperature and pressure. The lower the temperature, the greater the amount of carbon dioxide that is retained. Conversely, the higher the temperature, the greater the pressure required to maintain the carbon dioxide in solution. Henry's law was postulated by William Henry (1774–1836) and states that 'The amount of gas dissolved in a given volume of solvent is proportional to the pressure of the gas with which the solvent is in equilibrium.' Charles's law (Jacques Charles, 1746–1823) states that 'The volume of an ideal gas at constant pressure is directly proportional to the absolute temperature'. These two laws can be combined to form the universal ideal gas law:

$$pV = mRT$$

where p is the absolute pressure, V is the volume, m is the number of moles of gas, R is the gas constant (for that particular ideal gas) and T is absolute temperature. A mole is that quantity of a substance which has a mass numerically equal to the molecular weight of the substance. For carbon dioxide, the molecular weight is 44.01 and R is 0.18892 J mol^{-1} K^{-1}. From this relationship the carbonation chart shown as Figure 7.3 can be deduced. Here the concept of carbonation volumes is introduced. Volumes 'Bunsen' measures the gas volume at atmospheric pressure (760 mm of mercury) and the freezing point of water (0°C). It is defined as the number of times the total volume of dissolved gas can be divided by the volume of liquid in the container. As an example, a product with 4 volumes carbonation will contain a volume of carbon dioxide four times the volume of the beverage. A 1 l container carbonated to 3.5 volumes would contain 3.5 l of carbon dioxide, and likewise a 3 l container carbonated to 4 volumes carbonation would contain 12 l of carbon dioxide. One volume 'Bunsen' is equivalent to

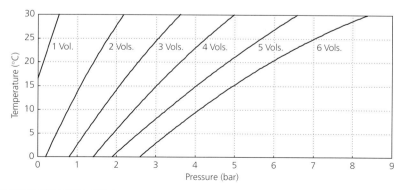

Figure 7.3 Carbonation chart.

1.96 g carbon dioxide per litre. This is often simplified to 2 g/l. For PET bottles, the smaller the container normally the higher the carbonation volumes, as the rate of loss of carbon dioxide by permeation due to a high surface-to-volume ratio is large. A 2 l bottle can easily meet a shelf-life requirement of no more than 15% carbonation loss in 12 weeks. This will reduce to around 9 weeks for a 500 ml bottle and some 7 weeks for a 250 ml bottle. The light weighting of PET bottles gives rise to thinner wall thicknesses and hence greater permeation and a shorter product shelf life. Cans have carbonation levels up to a maximum of 3.5 volumes; above this, the internal pressures that can be generated during expected use would cause can rupture or deformation. Glass bottles can be designed to accommodate higher pressures for products such as tonic water, which is traditionally a high volume carbonation product. The level of carbonation will depend on design and wall thickness.

7.4.2 Carbonation measurement

The measurement of carbonation is carried out using a device similar to that shown as Figure 7.4. It consists of a jig in which the container can be restrained, and a piercer attached to a pressure gauge. The container is placed in the jig, and is first of all pierced, then shaken before the pressure is measured. The release valve is opened until the pressure gauge reads zero and all the gas has been exhausted from the container headspace. The release valve is then closed and the container shaken again. The pressure is re-taken. The container is released from the jig and the temperature of the contents measured. A carbonation chart is then used to determine the volumes of carbonation.

The pressure must be released from the container before the pressure reading is taken because of the problem of air inclusion in the beverage. This gives a

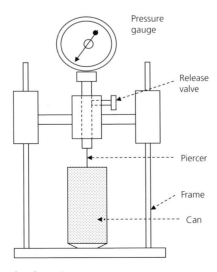

Figure 7.4 Measurement of carbonation.

two-gas system of air and carbon dioxide, and it is necessary to release the air to determine how much carbon dioxide is present. Air has approximately one-fiftieth the solubility of carbon dioxide in a liquid. Hence any air contained within the beverage will exclude some 50 times its own volume of carbon dioxide. Under Dalton's law of partial pressures, the pressure of a mixture of gases is equal to the sum of the pressures of the individual constituents when each occupies a volume equal to that of the mixture at the temperature of the mixture.' The main constituents of air are around 79% nitrogen and around 21% oxygen, ignoring for simplicity the presence of the inert gases. So in any carbonated mixture carbon dioxide, nitrogen and oxygen will be present. Owing to the differing solubility and proportions of oxygen and nitrogen, the dissolved air actually contains 35% oxygen and 65% nitrogen as the solubility of nitrogen is low. (This oxygen enrichment can give rise to spoilage problems if care is not taken to minimise the amount present.) The presence of air will give rise to a higher pressure and hence a false reading of the volumes carbonation from the carbonation chart; thus the amount of air present clearly has to be minimised when taking carbonation measurements. If we consider a bottle with a gas headspace of 5% of the bottle volume, on the first snift the gas loss would be 5% of the bottle volume. On the second snift we would lose a further 5%. If carbon dioxide only were present in the headspace, we would expect to lose 5% pressure on the first shake and some 7% by the second shake. If other gases were present, we would lose more pressure. Thus, the amount of air present in the product can also be estimated during carbonation measurement. If excess air is found (often caused by air entrainment or poor sealing with the filler bowl) then action needs to be taken to minimise its presence. This can be seen from an example where, by volumetric analysis, it has been found that the headspace of a carbonated drink container contains 90% carbon dioxide, 3.5% oxygen and 6.5% nitrogen. Considering the properties of perfect mixtures, which is only an approximation of the actual situation, it follows from Dalton's law that if

$$xi = n_i / n$$

where i corresponds to on individual component, n is the mole fraction defined as $n = m/M$, m is the number of moles and M is the molecular weight, then

$$pi = x_i p$$

For our example, if the mixture pressure is taken as 4 bar, we can develop the following table:

	M_i	V_i/V	$(V_i/V)M_i$	m_i/m	p_i
Oxygen	32	0.035	1.12	0.026	0.14
Nitrogen	28	0.065	1.82	0.043	0.17
Carbon dioxide	44	0.90	39.60	0.931	3.60
Total		1.00	42.54	1.000	4.00

156 Chapter 7

The carbon dioxide pressure is only 3.6 bar. From the carbonation chart, assuming a temperature of 20 °C, the volumes carbonation at 4 bar would be 4.3 but only 4.1 at 3.6 bar. Hence, with the presence of air we have lost 0.2 volumes carbonation. This is the maximum tolerance normally given for volumes carbonation during the manufacture of soft drinks.

7.5 Syrup preparation

Most products are traditionally prepared as a syrup-plus-water mix, in a ratio of some 1 part (volume) syrup to between 3 and 6 parts (volume) water. This allows a concentrated batch of syrup to be made and then proportioned with water to form the final product. For a sugar-based product the syrup would typically consist of 67°Brix sugar, citric acid, flavourings, colourings, preservatives and water. The ingredients are carefully weighed out and added to the mixing vessel. The syrup is pre-prepared and fully tested before being sent to the proportioner for mixing with water and subsequent carbonation. This is carried out in the syrup room as a batch process, allowing the multitude of soft drink flavours to be catered for.

Various methods exist to accurately proportion syrup and water, though the most popular current system uses flow meters. The syrup is usually dosed though a mass flow meter and the water is dosed volumetrically using a magnetic induction flow meter. This allows for density variations within the syrup to be accounted for to give the required Brix of the final product, since a mass flow meter works on the same Coriolis principle as a densitometer, although the degree of accuracy of measuring density using a mass flow meter is an order of magnitude less than if a densitometer is employed. (The Coriolis principle is an effect whereby a mass moving relative to a rotating frame of reference is accelerated in a direction perpendicular both to its direction of motion and to the axis of rotation of the frame. It explains why water flows down a plughole clockwise in the northern hemisphere and anticlockwise in the southern hemisphere, and is named after G.G. de Coriolis (1792–1843).) The density of water, within the range under consideration, does not vary significantly and hence the simpler volumetric flow meter can be used. The latest adaptations of these proportioners allows for the final product either to be collected in large vessels of some 30 000 l capacity or greater or to be fed direct online to the carbonator, with the syrup being individually proportioned as a premix and online metering of sugar, citric acid and other components. The accuracy of mass flow meters ensures the product is produced at the required Brix, thus ensuring conformance to specification, tight cost control and minimum wastage. A typical system is sketched as Figure 7.5. The older Mojonnier system uses a fixed orifice to meter the syrup component whereas the water is fed through a variable orifice, both operating under a constant head pressure as shown schematically in Figure 7.6. Other systems in use employ volumetric dosing pumps.

Carbonated beverages

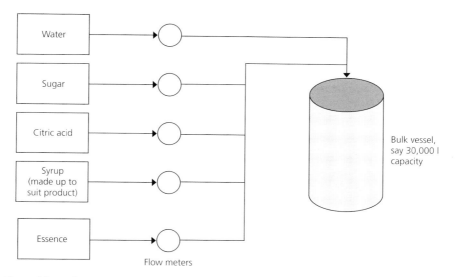

Figure 7.5 Product preparation using flow meters.
Note: It is normal to water-flush between introducing each component to ensure the feed lines are clear of the previous ingredient.

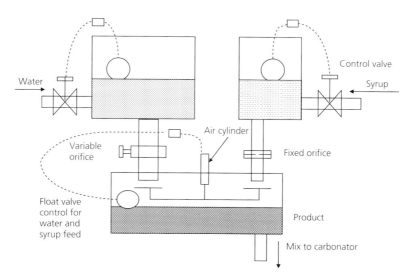

Figure 7.6 Mojonnier proportioning system.

7.6 De-aeration

Why de-aerate? As discussed earlier, the presence of air in a product causes product deterioration, as well as giving a false reading of the level of carbon dioxide present due to the partial pressures involved. Experience has shown that the aim should be to reduce the level of air within a product to below 0.5 ppm wherever possible. In this way the product will be at minimum risk from deterioration due to the presence of oxygen; hence shelf life will be improved and filling problems minimised. The presence of air and carbon dioxide causes nucleation sites within the products, giving rise to the phenomenon known as fobbing. The higher the air content the more difficult it is to hold carbon dioxide in solution. Two main methods of de-aeration exist, vacuum and reflux, both of which are normally applied to water before mixing with syrup rather than to the final product. By de-aerating only the water no product contamination of the equipment occurs and consequently less product risk is involved. Also the frequency of cleaning is decreased. In modern plants, it is normal to de-aerate all the water used in the product, including the water used to make up the syrup content, thus minimising the amount of air that is present in the final product. Simple methods such as introducing all liquids into mixing vessels through the base of the vessel will minimise the entrainment of air, as will careful mixing of the product.

The most effective method of de-aeration is to atomise water into a vessel held under a vacuum. In this way air is stripped out as the atomised water is exposed to the vacuum. Alternatively, if a positive carbon dioxide pressure is applied in a sealed vessel (reflux de-aeration) the air attaches to the carbon dioxide, in a process known as nucleation, and is then driven off through a vent. Often these two processes are combined, such that the effective use of vacuum de-aeration followed by reflux de-aeration will give an air content of less than 0.5 ppm in the water and of order 0.5 ppm in the final product, as the syrup will also have been produced from de-aerated water.

7.7 Carbonators

Carbonators take the form shown schematically in Figure 7.7. The final product is fed to a vessel pressurised with carbon dioxide gas. The rate of flow and the pressure of the carbon dioxide are critical to ensure the correct carbonation level. The greater the liquid surface area exposed to the carbon dioxide, the higher the rate of absorption of the carbon dioxide by the liquid. The carbon dioxide is often sparged into the liquid under pressure; this allows small bubbles of gas to be formed which can be easily absorbed by the liquid. The higher the pressure, the smaller the gas bubbles formed at the sparger and the greater the gas bubbles' surface area made available for the gas to be absorbed by the liquid. Early carbonators used refrigeration to carbonate at around 4°C. A typical system is sketched as Figure 7.8. Water alone is often carbonated to ensure minimum contamination of the system by syrup. The product is spread over chilled plates, such that the product runs down the plates as a thin film. This is carried out in a constant-pressure carbon dioxide atmosphere, the lighter, displaced air being

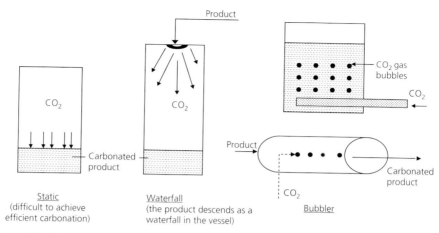

Figure 7.7 Carbonation methods.

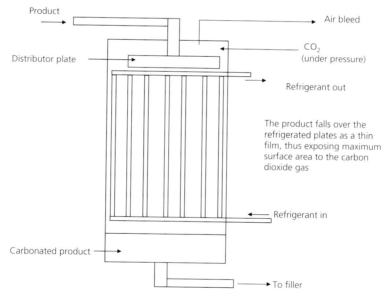

Figure 7.8 Refrigeration carbonation.

bled off. Chilling the product as a film maximises the surface area available to the carbon dioxide, thus promoting effective carbonation. This has the added benefit that at a lower temperature the carbon dioxide stays in solution more easily, thus minimising future filling problems. However, the energy usage is high and packaging problems are created due to condensation within shrink-wrapped packs as the temperature reaches ambient conditions. This is especially a problem for steel cans, where corrosion can easily occur. To overcome this problem the packaged product must be warmed to ambient conditions, thereby further increasing the energy load.

7.8 Filling principles

A carbonated product made to specification has then to be filled into the required container at a commercially viable filling rate. This is achieved under gravity, the rate of flow being dependent on the head difference between the filler bowl and the container. The rate of flow will increase if an overpressure is introduced. With reference to Figure 7.9, the pressure from the top of the filling bowl to the outlet of the filling valve provides the driving force to fill the container. The example shows a bottle, but the principle is the same for a can or carton. The rate of flow to fill the container is a function of the overpressure applied to the top of the filling bowl (p), the viscosity of the liquid to be filled (μ), the diameter of the filling tube (d) and the length of pipe (l). For simple flows this can be expressed by the Poiseuille formula as

$$\text{Volumetric rate of flow}, V = \frac{\Pi p d^4}{128 \, \mu l}$$

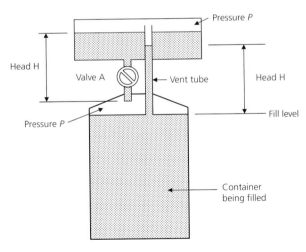

Figure 7.9 Principles of gravity filling.
1 Seal container to the filler
2 Both the container and the header tank are under the same pressure P due to the connecting vent tube.
3 Open the liquid valve A.
4 Filling starts.
5 Filling stops when the vent tube is covered by the liquid and the pressure heads H equate.

This is a much simplified equation for very-low-speed laminar fluid flow. In actual practice, the rate of flow would probably be turbulent and proportional to $\sqrt{p}$ rather than to the pressure alone. It does demonstrate that for viscous liquids it is necessary either to increase the filling tube diameter or to increase the driving pressure to maximise the flow rate through an orifice. Considering the process in more detail reveals some of the problems facing the filler designer, especially with regard to how the process is controlled. It is simple to envisage how a container is filled under gravity alone: it is the same as filling a bottle from the kitchen tap. To control the process under pressure with carbonated product is more complex. However, if the pressure in the container and the pressure of the gas in the filler bowl headspace are the same, gravity filling conditions will apply. This is exactly what is done.

7.8.1 Gravity filler

It is first necessary to seal the container to the filler bowl such that no leakage can take place around the seal. This applies whether the container is glass, plastic or metal. The filler bowl is filled to a given level, which is maintained within close tolerances by means of float valves. This ensures a near-constant pressure head during the filling process. In terms of Figure 7.9, with the container sealed to the filler bowl, valve 'A' is opened and filling commences. By means of a vent tube the gas within the container is expelled, the rate of flow of liquid into the container being proportional to the rate of flow of gas displaced. When the liquid reaches the vent tube it will fill the tube until the pressure within the vent tube equates to the filling tube pressure. When this equilibrium condition is achieved, the liquid flow stops and the filling valve can be closed. As the container is lowered from the filler bowl the liquid within the vent tube will drain back into the container. Small amounts of liquid are left in the vent tube due to surface tension effects that are dependent on the characteristics of the liquid being filled. The cycle is shown in Figure 7.10.

This process requires certain standards for the container as well as the filler. A bottle must have a standardised neck finish to allow it to seal effectively at all times with the filler bowl. It must also have sufficient top load to withstand the forces involved during the filling process. Most modern bottle fillers now lift PET bottles by the neck to overcome possible deformation problems during the process. This also allows light weighting of the bottle for both environmental and commercial reasons. To achieve commercially acceptable filling speeds the majority of fillers are rotary. Bottles are fed into the filler by conveyor to an infeed worm and star-wheel in single file. This mechanism incorporates a clutch-operated bottle stop which, under normal conditions, rotates freely. Should the filler be stopped for some reason, the bottle stop will engage automatically. From the star-wheel, bottles are fed to a bottle-lift stirrup sited below an individual filling valve and lifted by the neck to the seal with the filler bowl. These filling valves are sited at equal intervals around the base of the filling bowl. The filling

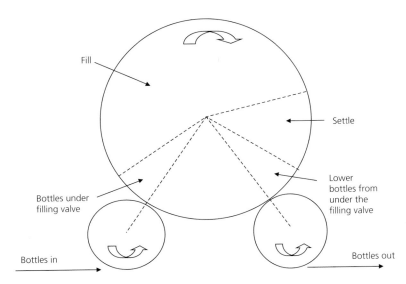

Figure 7.10 Gravity filling cycle.

bowl itself is annular to allow all the feed pipes and central drive to function. The rotation of the filling bowl gives rise to a centrifugal force, and the flow through apertures into individual valves is very complex. The flow paths are short, such that the flow through a valve is not fully developed. The closer the filling valves on a small-pitch filler, the more the flow into individual valves is influenced by the flows into adjacent valves. Consequently, design is based on experimental results and experience with fillers in practice rather than using computational fluid mechanics.

For each filling valve a centring bell acts as the seal for the bottle, this seal being made of food-grade rubber. The hardness of the rubber, any defects on the bottle sealing surface, the state of the sealing rubber and the sealing pressure all affect the performance of the seal, which is critical to good filling. Each filling valve is opened and closed by an actuating lever from cams fitted to the stationary filling frame. The valves are actuated when the filling bowl, and hence the valve-actuating lever, pass the cam. If no bottle is in position, this is sensed because the centring bell has not been lifted into position, and no filling takes place. The actual flow through a valve is complex. For example, there could be some 100 valves sited around the base of the filling bowl. The liquid has to flow through each of these orifices and then into the container via an annulus between the vent tube and the side of the valve. A spreader is usually fixed to the vent tube to deflect the product on to the container side wall to minimise turbulence conditions in the liquid within the bottle. Each bottle has a defined filling level, which should be as close to the sealing surface as possible commensurate with the expansion of the liquid during its potential lifecycle as a result of temperature variations. Once filling is completed, a short settling period is allowed before the bottles are lowered from the filling bowl and passed to the capper to be sealed.

7.8.2 Counter-pressure filler

The process above forms the basis of counter-pressure filling of carbonated liquids into a container. To achieve gravity filling the pressure within the filling bowl and the container must be equated. A typical process cycle is shown in Figure 7.11. Once the container is sealed to the filler bowl, the gas valve is opened and the gas within the filling bowl headspace flows under pressure into the container displacing air at atmospheric pressure. In modern fillers this air is vented to the atmosphere, but on older fillers it is often vented to the filler bowl headspace, causing the problems associated with gas mixtures discussed previously. In practice, however, since air is lighter than carbon dioxide, the majority of the air can be vented off. The example shown in Figure 7.12 illustrates how evacuating the air from the bottle prior to filling can reduce the air content. It can be seen that it is not feasible to completely eliminate all the air within a bottle prior to filling. However, the air within the bottle can be minimised if it is evacuated away from the filler and not into the filler headspace.

Once the pressures are equated, the gas valve is closed and the liquid valve opened. Flow commences, stopping when the pressures are equated. A short settling period is then allowed before the liquid valve is closed and the gas within the container headspace is snifted (vented) off. This settling period, and subsequent

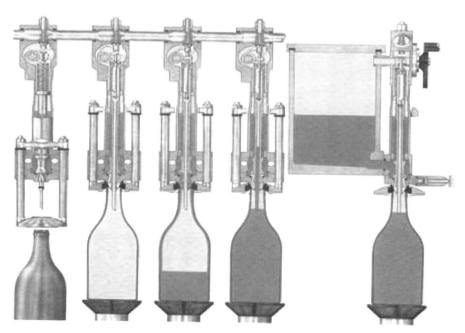

Figure 7.11 Counter-Pressure filling cycle.

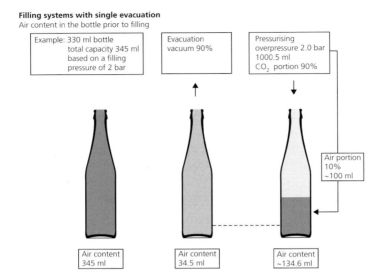

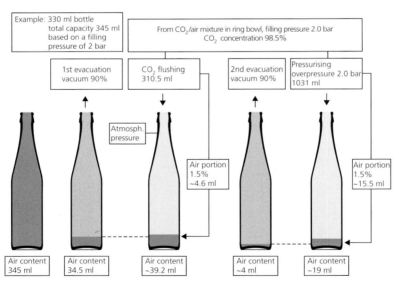

Figure 7.12 Bottle evacuation. *Source*: By courtesy of Krones (UK) Ltd.

snifting are required because otherwise, when the container is lowered from the filler bowl, the pressurised gas within the headspace will be exposed to the atmosphere. This would result in severe fobbing of the product within the container until equilibrium conditions were reached. To overcome this potential problem, the pressure within the headspace is gradually reduced by snifting the gas off externally to the filler. Often this process is repeated several times in short bursts allowing a short rest period between snifts. The rate of flow of the gas is controlled by the diameter of the snift valve orifice. Snifting is the most difficult part of the process to control, especially if the product itself is lively – cream soda, say. Care when snifting off to atmosphere has to be taken as the snifted gas often contains a product mist, which can cause contamination as it condenses on the surface of the filler and surrounding plant. A conventional filling cycle is illustrated in Figure 7.13. A diagrammatic representation of how bottles are fed through the system using an infeed worm and star-wheels is shown in Figure 7.14.

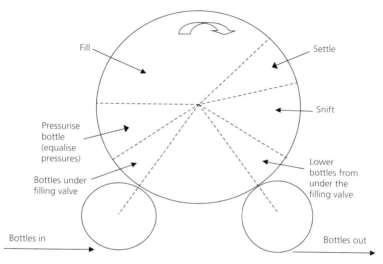

Figure 7.13 Counter-pressure filling cycle.

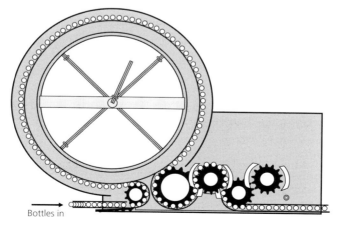

Figure 7.14 Filler layout showing infeed worm and star-wheels.
Note: Each circle represents a container.

Considerations during the process include ensuring that filling is as quiescent as possible, otherwise 'fobbing' will occur. The term 'fobbing' is generally applied to a carbonated product that is still 'lively' and has not reached a state of equilibrium. Fobbing can be induced by excessive agitation during filling and too fast a snifting process. For this reason great care must be taken to ensure that the process is under control at all times. There is an increasing tendency for carbon dioxide to come out of solution with higher temperatures as well as with greater volumes of carbonation. Most modern fillers can operate at up to 22°C, though any liquid temperature above 20°C will tend to increase the risk of fobbing. Most fillers to operate at around 14°C, which has been found to be a good compromise temperature when considering energy consumed to cool the product and the efficiency of the filling process, as the lower the temperature, the more easily carbon dioxide will stay in solution. Filling down the insides of a bottle will reduce the level of fobbing as will a carefully designed container. The former is achieved using swirl-type valves that impart a tangential force to the liquid, thereby forcing it to the sides of the bottle. Any discontinuity along the side wall of a container will cause the liquid to 'jump' off the inner surface side wall and fall directly on to the product in the base of the bottle, thus increasing the level of agitation and hence causing fobbing. Likewise, too high a shoulder on a bottle will not allow the liquid to flow along the inside walls as early as possible, giving rise to increased agitation of the liquid and loss of carbon dioxide from solution. In addition, the shape of the bottle shoulder can exaggerate any inconsistency of fill level control, especially with tapered bottle shoulders. Bottle top load is a critical factor. On older fillers that use bottle-lifters whereby the bottle is lifted by the base, any bottle with a low top loading ability will tend to buckle during the snifting process, thereby causing some loss of product, especially if fobbing is induced, and hence a reduced fill level in the container. When filling carbonated products with fruit cells it is usual to employ siphon-type valves, which allow a smoother passage through the filling valve, though with a loss of performance and a requirement for a lower carbonation level.

The choice of a filler depends on the number of bottle sizes being considered for use on it. For example, if 3 l PET bottles are to be filled, a 126 mm pitch between filling valves is required. If only small bottles up to 500 ml are to be filled, then a 70 mm pitch will suffice. This has a direct effect on the size of the filler and its footprint on the factory floor. Unless the filler bowl level is kept within tightly controlled limits, pressure head variations will affect the rate of flow into containers. Systems such as that shown in Figure 7.15 need to be employed. Wherever possible there should be minimal contact between any instrumentation and the product. Conventional float valves should be avoided and simple capacitance probes, which are easily cleanable but small and very effective, should be used. It is not uncommon in older fillers to only have one float valve. This often gives rise to filler bowl flooding, which may lead to inconsistent fill level and poor counter-pressurisation of the container pre-filling.

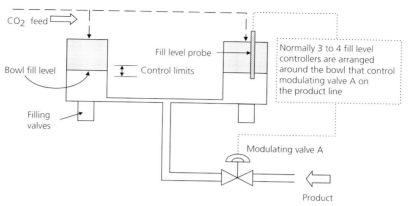

Figure 7.15 Filler bowl level control.

A filler is usually constructed of 316 stainless steel to ensure all surfaces can withstand chemical attack from both the product and cleaning processes, as well as being easily and effectively cleanable. The annular bowl is supported on a central frame, which also includes the drive motor and pipework supports. Each valve is individually piped to the central frame, connected via a rotary valve. Most drives are now invertor controlled through a process logic controller (PLC), giving a wide range of possible operating outputs. Valve design has improved greatly over the last few years, so that swirl valves now have sufficiently large passageways to enable many fruit- or cell-based drinks to be processed, instead of using the slower siphon valves, with their tendency to give rise to fobbing. Even so, owing to the nature of the design of counter-pressure fillers, the vent tube will always contain some product, which will inevitably be carried across to the next fill cycle unless it is blown out prior to the new cycle commencing. The cutoff for the vent tube occurs at the full filling rate, and so is never perfect because of the momentum of the liquid product during filling. The traditional filling valve with springs can thus be easily contaminated since the springs are submerged in the product itself. This can cause cross-contamination from one product to another following a flavour change unless care is taken to ensure an effective cleaning regime.

7.8.3 Other filler types

Many modern types of filler are electronically controlled. They fall into two main categories, volumetric and capacitance probe fillers. The capacitance probe filler uses a capacitance probe to detect the point at which filling should cease. The probe is sited in the centre of the filling aperture of the valve, and is calibrated along its length to detect when the fill level is being approached and then progress until the actual shut-off at the end of filling. This sophisticated type of filler allows the initial fill through the swirl valve to be slow, to minimise the effect of fobbing, then fast until the fill level is approached. As the fill level is approached the probe senses this and allows the fill rate to be reduced, allowing a very accurate end-of-fill cutoff to occur. The use of electronic valves within the system ensures that the whole operation can be controlled electronically. The snift can be more finely controlled than in counter-pressure fillers by applying a small back-pressure for the initial snift, followed by a normal snift operation to atmospheric pressure. Owing to the fact that the capacitance probe filler is more complex than traditional fillers, it does require a fuller understanding of the physics of the operation to achieve the best results. A typical electronic probe filler is shown in Figure 7.16.

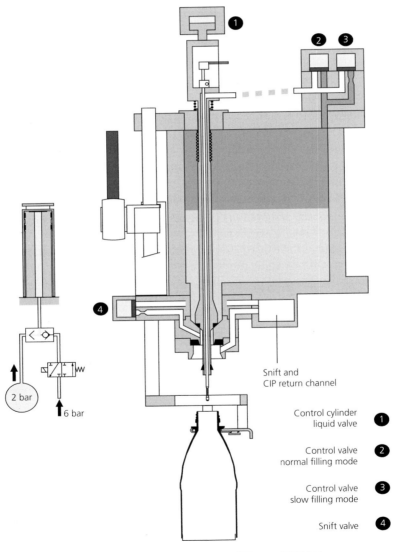

Figure 7.16 Electronic probe filler. *Source*: By courtesy of Krones (UK) Ltd.

The volumetric filler is also electronically controlled, ensuring an accurate dose of product to each container. For PET bottles, which 'creep' under carbonation pressure, there is the drawback that they will have differing fill levels over a filling cycle due to the fact that no two bottles are ever exactly identically blow-moulded. With light weighting, this problem is exacerbated.

Two main types of volumetric filler exist. In the first type, a predetermined volume of liquid can be dosed using either a magnetic inductive volumetric flow meter or a mass flow meter. Alternatively, measuring cylinders can be used that incorporate a level probe. The product is fed to a set level in the metering cylinder, at which point the flow valve is closed off from the supply tank. A typical can filler operating at 1,500 cans of 330 ml capacity per minute will fill as per the supply contract within ±2.5 ml. However, a standard deviation for a volumetric filler as low as ±0.58 ml has been quoted. As with the probe filler, these fillers are fitted with electro-pneumatic valves, enabling accurate control of the process. Care has to be taken to ensure these valves are kept clean and in good condition as they are critical to the operation. Any sluggishness in a valve's operation will affect performance. Two typical volumetric filler operations are shown in Figure 7.17.

7.8.4 Clean-in-place systems

All modern fillers are designed for clean-in-place (CIP) to ensure the sterility of the system. The CIP process operates from a centrally located system that is piped to the filler. A return cycle to the CIP set from the filler is included, the temperature of the return liquor being sensed at the filler outlet. Specifically designed CIP cups are attached to each filling valve to allow the CIP process to take place entirely through the filler bowl and filling valves and all associated pipework at sufficient velocity and contact time to ensure effective cleaning is achieved. A typical operating cycle would be

1 Rinse the filler and filling valves for a set time using rinse water to drain. This removes particulates. The rinse water used is often the final rinse water from the preceding CIP operation so as to conserve water.
2 Circulate hot caustic soda solution ($c.1.5\%$ w/v) at $80°C$ (as recorded at the filler outlet) for some 15 min. The caustic soda is returned to the hot caustic tank through filters.
3 Circulate hot water at around $75°C$ for 15 min to allow the interface with the caustic to flow to the drain. A conductivity probe is included within the circuit to ensure that all the caustic lye has been removed. The hot water is recirculated to conserve energy.
4 Use cold water as a final rinse.

Careful records of the CIP process need to be maintained to ensure the process used has been effective. A typical CIP process is shown in Figure 7.18.

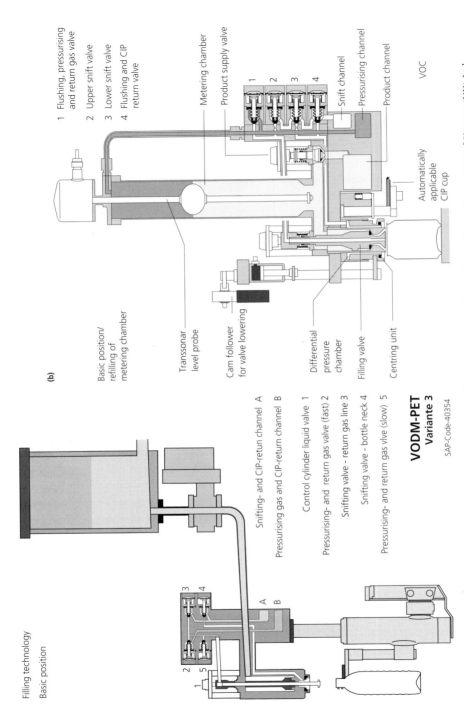

Figure 7.17 (a) Volumetric filler using flow meters. (b) Typical volumetric filler using a measuring cylinder. *Source*: By courtesy of Krones (UK) Ltd.

Filler VK(2)VCF
CIP cleaning system

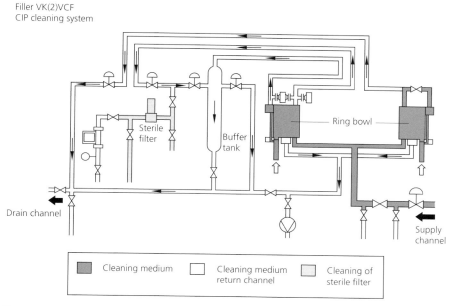

Figure 7.18 Typical CIP system. *Source*: By courtesy of Krones (UK) Ltd.

7.9 Process control

Process control is the essence of modern factory systems. It is imperative at all times to have the process fully under control and for full records to be kept.

Online instrumentation and feedback control is the key to a successful operation. The syrup premix is produced as a batch offline. All the weighed ingredients are logged on a computer-controlled weighing system that can feed back data to the central factory system. The level of de-aeration of the water can be continuously logged using an online probe, and the water quality itself must be regularly analysed. The treated water is monitored for turbidity problems online, with automatic shut-off should a problem arise. The carbon dioxide and all materials delivered should have a certificate of conformance to an agreed, signed specification. The premix, water, acid and sugar, if required, are fed by flow meter, with online Brix control feedback, either to a bulk vessel or direct to the carbonator and are checked against an agreed recipe within the PLC of the dosing system. At all times, records need to be computerised and backup copies kept. Automatic shutdown occurs should a problem be found by the online instrumentation. The effectiveness of the carbonation process can be detected online by taking regular samples and checking pressure and temperature against the specification. The feedback control system will regulate the process within agreed limits. The filler itself is fully PLC controlled, often using a touch-screen control system. This allows production to be carried out with Brix and carbonation levels controlled

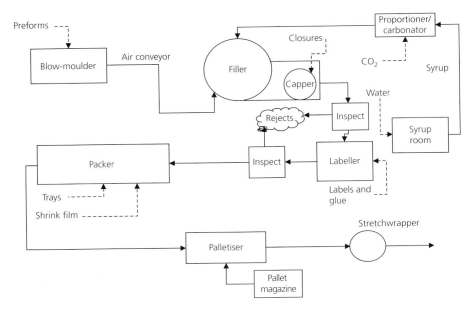

Figure 7.19 Typical PET filling line.
Note: For high-throughput items, display trays can be used.

within ±0.05 Brix and carbonation values of the set-point. It should be remembered that the filler is just part of the filling line process. This is illustrated in Figure 7.19 for a typical PET bottling line. The line is usually controlled by the blow moulder, but for small bottle lines silos are used to buffer the bottles and allow the blow moulder to operate as efficiently as possible. The overall line efficiency is the multiplication of all the individual unit efficiencies comprising the line. Even the conveyors between the machines have a mechanical efficiency. However, if sufficient bottle accumulation between machines is provided by the conveying system this will improve line efficiency by providing a buffer to overcome short duration machine stoppages. Such accumulation systems are of three minutes or more in duration and are designed to overcome typical machine stoppages of the order of 30 seconds.

7.10 Future trends

It is difficult to imagine what the next major step forward in beverage technology will be. We already blow-mould PET bottles direct online to the filler via an air conveyor. These bottles are manufactured and filled in just over 2 min, unless a silo system is used to store the bottles and act as a buffer between the blow-moulder and filler and hence increase operational output. Such silos, which tend to be used for PET bottles up to 500 ml capacity, also offset the usual efficiency imbalance between the blow-moulder and the less efficient filler. One obvious trend is the rationalisation of bottle shapes and sizes, such that common diameters, heights and fill heights are used. This reduces the number of size changes required on the blow-moulder and filling line, thus eliminating setup and startup time losses. Further light weighting of PET containers will continue, especially as the cost of PET resin is directly proportional to oil prices. This means that the filling process needs to handle flimsier containers. This will preclude many of the older fillers from still using full bottle-lifts rather than not neck handling. Rapid size changes will become even more important, as will the dedication of filling lines, wherever possible, to one bottle size and if possible one product. The light weighting of cans will continue, though at a lower rate than the major reductions seen in the past 20 years. This will imply more careful control of the pressure compensation to ensure effective sealing of the can with the filling bowl without any can crush occurring. More processes will become PC rather than PLC controlled, thereby facilitating easier programming. The snifting operation will incorporate more sophisticated instrumentation to minimise the risk of fobbing and will be carried out in as short a time as possible to ensure the filling-cycle operation itself is optimised. More flexible operations will be required as marketing staff try to gain increasing market share for their products. This will see the emergence of 'cottage industry' lines as separate cost centres within large dedicated-line factories, thus allowing the true cost of producing minor flavours and specialised containers with short run lengths to be determined. The real flexibility will come from packaging changes rather than improvements in carbonation practices and filling techniques. The increasing sophistication of consumers will require even more aseptic and semi-aseptic filling operations. For non-carbonated products fully aseptic processes are now operational whereby the product being filled does not come into contact with the filler bowl. This situation cannot be achieved for carbonated products; so the aseptic state can only be approached but not actually achieved with current technology. It is in this area that further developments will take place, driven by consumer demand.

Further reading

Giles, G.A. (ed.) (1999) *Handbook of Beverage Packaging*, Sheffield Academic Press, Sheffield.

Mitchell, A.J. (ed.) (1990) *Formulation and Production of Carbonated Soft Drinks*, Blackie, Glasgow and London.

Rogers, G.F.C. and Mayhew, Y.R. (1964) *Engineering Thermodynamics, Work and Heat Transfer*, Longmans, London.

Steen, D.P. and Ashurst, P.R. (eds) (2006) *Carbonated Drinks Formulation and Manufacture*, Blackwell Publishing, Oxford.

CHAPTER 8
Processing and packaging

Robert A.W. Lea
GlaxoSmithKline, Weybridge, UK

8.1 Introduction

This chapter covers juice extraction, blending, processing, factory layout, filling and packaging. The purpose of processing and packaging is to produce a product that is wholesome and refreshing in a pack of convenient size for the consumer. The topic is extensive and so only the main points are covered in this chapter. However, equipment suppliers provide a good source of free information and their names can be found from trade magazines, trade associations, etc.

The designer of a blending/filling system must have a full understanding of the process: its limitations, the raw materials, the packaging materials and their levels of contamination and the means of controlling the contamination to enable wholesome product with adequate shelf life to be produced.

8.2 Juice extraction

The details of the juice extraction process will depend on the fruit, but this section outlines a typical method. Some processors size the extraction plant to handle all the juice they will require for the year within the harvesting period; others will freeze or otherwise store the fruit and extract juice as it is required on a much smaller plant.

The fruit is normally delivered in large containers (or lorry loads) to a reception area where it is washed and screened to remove foreign bodies. The fruit is then milled, heated, mixed with enzyme to break it down and held for a period (enzyming is a time/temperature process). The juice is separated using either centrifuges or fruit presses and processed for aseptic storage as single-strength juice (see Section 8.4) or concentrated by evaporation of water and stored. Many soft drinks are made from bought-in concentrates produced by a specialist supplier.

8.3 Blending

There are three main blending routes, which can be used individually or in combination to form a product. Traditionally this is carried out in the 'syrup room' as it used to be normal to produce the drink as a syrup and dilute it prior to filling (before that it was usual to fill it as two separate phases: syrup and carbonated water). The three routes are
1 batch
2 flip-flop
3 continuous.

Figure 8.1 shows a typical combined process flowsheet.

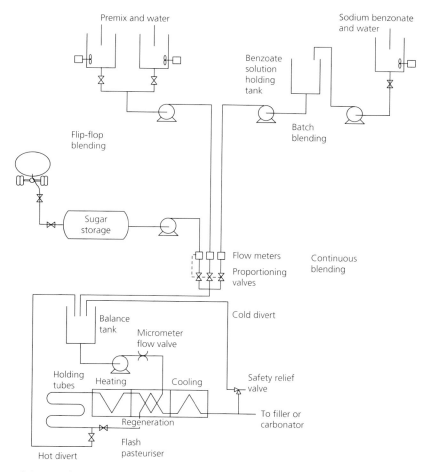

Figure 8.1 Typical process flowsheet.

8.3.1 Batch blending

The ingredients, usually powders or liquids, are mixed together in a large batch tank. Bulk ingredients can be metered or weighed from storage. The storage tanks or silos are filled by bulk tanker. The minor ingredients are often combined in a smaller premix tank prior to addition to the main mix tank. Some of the ingredients cause an endothermic or exothermic reaction on solution and may need heating or cooling to maintain temperature. There is a variety of stirrers and mixers available for the mix and premix tanks, ranging from simple propellers to specially designed high-shear mixing heads.

Powders are dissolved either directly in the main mix tank or premix tank or indirectly using a vortex-type mixer (Figure 8.1) where powder is dropped into the vortex of a horizontally mounted pump head recirculating the fluid from and to the batch tank. Some specialised versions of this mixer can handle very viscous blending applications (50,000 cP or more).

8.3.2 Flip-flop blending

This is really a special case of batch blending. In batch blending systems, once the batch has been made and checked for quality it is pumped into a holding tank before being sent to the line. In flip-flop systems a batch is made in one mix tank and then fed directly to the line while the next batch is made in another mix tank. The second batch is sent directly to the line when the first batch has finished: batches flip-flop between the tanks. Flip-flop batching is often used for premixing in batch blending or when mixing time is tight (i.e. the time taken to pump the batch out of the mix tank to storage is too long); however, batch mixing systems are generally cheaper as only one mix tank is required.

8.3.3 Continuous blending

In continuous blending, streams of raw materials and premixes are blended into one another on a continuous basis. The proportions are closely controlled either using a system of positive displacement proportioning pumps (either the stroke or speed is accurately controlled) or using flowmeters to control the throughputs of each stream via speed-controlled pumps or proportioning valves to throttle the flows. The metered liquids then pass through an in-line mixer to complete the blending; this mixer can be static (a series of blades creating turbulence) or powered to create high shear. Continuous systems are used when the number of streams is limited; for example, water, sweetener, premix 1 and premix 2 (assuming stable premixes can be made). Systems with more than eight streams are unusual because of difficulties in accurately controlling the streams and validating the process.

8.4 Processing

In this section, we consider the means by which the product is rendered microbiologically secure. This is usually accomplished by the addition of heat; although chemical means are also used in some circumstances (i.e. dimethyldicarbonate, which breaks down into water and carbon dioxide and is thus safe for the consumer). There are five main processes for juices and soft drinks:
1. flash pasteurisation
2. hot filling
3. in-pack pasteurisation
4. aseptic filling
5. chilled distribution.

The choice of process depends on the level of microbiological contamination of the raw materials and packaging, and the ability of the product to withstand growth of micro-organisms (e.g. preservative levels and sugar content). Also of great importance is the ability of the product to withstand heat.

The theory of pasteurisation is very simple: the product is heated and kept at or above the pasteurisation temperature for a certain time. It is the responsibility of the processor to specify the temperature and time. For pasteurisation to be microbiologically safe, it is essential that the correct conditions are maintained during the complete period of operation and that reinfection is prevented.

8.4.1 Flash pasteurisation

The raw juice is normally passed through a balance tank (or feed tank) before being fed to the pasteuriser.

The drink is generally heated by hot water in a plate or tubular (spiral) heat exchanger to the desired pasteurisation temperature and held at that temperature for the specified time in a holding tube before being cooled to the filling temperature (usually ambient) using chilled water. Normally, flash pasteurisers have a regenerative section (see Figure 8.1); this is an energy-saving feature whereby the incoming raw product is initially heated by the hot product returning from the holding coil, which in turn is cooled. The energy used for heating is regenerated. To prevent reinfection the equipment must be sterilisable using culinary steam or hot water and be impervious to microbiological spores (i.e. there should be no risk of contamination from unpasteurised product or product trapped in 'dead legs' or equipment that is not bacteria tight, e.g. valves and pumps). In some countries regulations require that pasteurised product is always at a higher pressure than the raw product or heating or cooling medium on the other side of the plates. This is usually achieved using a boost pump between the incoming regenerative section and the heating section.

The equipment should be designed such that the treatment time is correct. This is usually achieved by controlling the flowrate and passing the product through a holding tube of known volume. The flowrate and temperature (at the end of the holding tube) are critical to the process and should therefore be monitored.

In the basic system described above, if the temperature drops below the pasteurisation temperature or the flowrate exceeds that for the correct holding time there is no other choice than to shut down the process and to clean and resterilise the equipment before production can be restarted. The consequences of this may be limited by diverting the flow of insufficiently pasteurised product back to the balance tank: forward flow may be resumed once correct conditions are restored. The divert valve should be placed sufficiently downstream of the temperature monitoring probe that the system response time (probe, controller and valve) is less than the time taken for the unpasteurised product to reach the valve.

Pulsations and vibrations should be kept to a minimum (e.g. pumping and valve switching) because they reduce the life of plates by mechanical damage leading to fatigue failure. To ensure product security, plates should be crack tested on a regular basis (annually).

Pasteurised product is either sent to a proportioner and diluted to the desired strength (if produced as a syrup) and then carbonated and mixed with carbon dioxide or sent direcly to a filling machine; excess product is diverted (cold divert) back to the balance tank of the pasteuriser. The system is typically sized for a flow of 5–10% in excess of that required; the positive forward flow maintains the sterility of the system so that bacteria, etc. cannot pass up the cold divert line against the flow.

8.4.2 Hot filling

In hot filling product is heated (in a heat exchanger), sent to the filler hot and filled into containers. The containers are closed and held at or above the required temperature for a specified time prior to being cooled, usually in a tunnel with water sprays. In this system not only has the product been heat treated but so has the container.

A typical regime would be to heat the product to 87°C (monitored at the heat exchanger) and send it forward to the filler bowl. The temperature is monitored in the filler and should it fall below a preset temperature, say 85°C, filling is inhibited, the filler emptied and its contents replaced with fresh hot product. Emptying the filler is often by means of a pump; care is needed in the selection of the pump as it is operating at high temperature, increasing the risk of cavitation (localised boiling in the pump suction due to reduced pressure). The product is then filled into the containers, where the equilibrium temperature is typically 83–84°C. If glass containers are being filled, hot water at, say, 60°C should be used to rinse the containers thus limiting the effect of thermal shock of hot liquid on cold bottles. The containers should then be held on a conveyor for the predetermined time, say 2 min, at above 80°C. This time and temperature are the microbiological control requirements and are thus important parameters that should be specified (the other values are not the controls for the process; however, the filling temperature and conveyor speed are the engineering parameters that are controlled to achieve the specifications). For some applications, it may be necessary to tilt the product so that the hot liquid wets the cap and neck area of the container. However, this is not often required: the head space typically reaches 75°C with the above regime and this temperature is normally adequate. Even with gable-topped cartons, in which there are four triangular areas surrounded by ambient air, if these areas do not reach an elevated temperature, experience has shown that product is not put at risk (this has been a challenge tested by the author). After the holding section, the containers are cooled to below 30°C (to limit thermal degradation).

8.4.3 In-pack pasteurisation

Generally, in-pack pasteurisation is the severest and also the most microbiologically secure form of heat treatment. The completely filled closed pack is put into a tunnel pasteuriser. The treatment is via water spray at various controlled temperatures. The pasteuriser is divided into zones. First there are a heating zone and a 'superheat' zone; these raise the temperature of the container and product to the desired temperature. Next there is the pasteurising zone, where the product is held at the pasteurising temperature for the specified time. Finally the product is cooled to below 30°C. To verify correct pasteurisation a detector is sent through the machine regularly. This consists of a temperature sensor mounted within a 'dummy product' and a recorder to produce the time–temperature profile.

A special case of in-pack pasteurisation is a retortable process in which the packs can be treated to elevated (above 100°C) temperatures in a retort and cooled.

8.4.4 Aseptic filling

The processing for aseptic filling is a special case of flash pasteurisation (often using a higher temperature profile). The plant is not equipped with a hot divert as this allows the forward flow to stop when the divert is actuated. Should pasteurising conditions be lost (flow or temperature) then the plant is stopped and resterilised. Great care is needed with any equipment downstream of the holding tubes as it must be microbiologically impervious to maintain asepticity (e.g. valves should have steam barriers on shaft seals) and should be kept to an absolute minimum.

The aseptic processor either feeds the filler directly, with excess flow returning to the balance tank (and positive flow maintained under all circumstances), or feeds directly to an aseptic storage tank, which acts as a buffer between the processing and filling operations.

For successful aseptic filling, clean containers, clean product, clean head space and clean closures need to be brought together in an environment that prevents recontamination. This operation normally takes place in an enclosure overpressured with sterile air. The air is sterilised using either heat or filtration of 0.2 μm. (The filter will need sterilising periodically using either steam or chemicals.) It is recommended that the filling equipment is located in a separate room that is in turn overpressured with clean filtered air (typically 2–4 mm water column) and has 10–15 air changes per hour. The air quality is defined in terms of the number of particles per cubic metre, that is, a class 100 room would have 100 particles per cubic metre.

The cleaning regime specified for the containers will depend on their initial level of contamination. Normally heat (steam or hot air) or chemical means (hydrogen peroxide or oxonia) are used; sometimes these are used in combination with ultraviolet radiation.

Laminate board packs are made either from flat reels of material or from preformed blanks. These systems are generally 'tied', that is, the laminate supplier also makes the filling equipment. The fillers are well proven microbiologically and do not need to be placed within a special environment.

Bottle systems are more varied, whether for glass, polyethylene terephthalate (PET) or other plastic. Bottles are rinsed with oxonia solution and then sterile water prior to filling. The filler is generally of a non-contact type (it does not touch the bottles) and product is either weighed in or measured volumetrically. Caps are also chemically sterilised (unless a foil closure is used) and applied on a capper monoblocked with the filler, enclosed in a high efficiency pure air (HEPA) filtered enclosure. The filler and final rinser are in a class 100 room and the operator wears full protective clothing to prevent infection of the product.

There are some more specialised systems for PET bottles, cans, other plastic bottles, form-fill-seal packs and returnable PET or glass bottles for still and carbonated drinks (generally high acid). These are material-dependent solutions and each arises from the limitations and properties of the materials and the way they are formed into containers.

PET bottles are made from resin chips in a two-stage process (even if made on one machine). A preform is injection moulded and is sterile when formed; this is then blown into the final shape. If the machine is enclosed in an HEPA-filtered environment and the bottles blown with sterile air, the machine produces a sterile bottle. The bottle can be transferred to the filler within an enclosed sterile air conveyor and filled without further treatment, thus eliminating the need for rinsing and chemical treatment. Carbonated products can be filled if the carbon dioxide is sterilised.

Cans can be sterilised easily with super-heated steam or hot air as they are stable above sterilising temperatures.

Other plastic bottles can be blown on a sterile bottle blower and be supplied closed (i.e. material seals the bottle above the screw thread or neck). The bottles are rinsed on the outside before entering the aseptic filling zone, the excess material is cut off and the bottle filled. (This process is not suitable for PET bottles because of the need to form the screw thread as part of the injection moulding process.)

Form–fill–seal packs are also suitable for aseptic production. The pack is blown aseptically in a mould and filled through the soft neck. Once filling is complete the fill tube is retracted and the neck sealed with heat and pressure.

Returnable bottles are generally subjected to a three-part process. Bottles are washed in a bottle washer, as they would be for non-aseptic production; however, the final rinse is with hot water, which has an initial bacteriocidal effect. The bottle rinser and filler are enclosed in an HEPA-filtered chamber; the rinser uses hot water, maintaining the decontamination; finally the filler injects steam into the bottle just prior to filling. The total process leads to a high log reduction in bacterial contamination, producing commercially sterile products. This process has been used for carbonated drinks.

8.4.5 Chilled distribution

If products are filled cold into clean bottles on clean (but not aseptic) fillers, stored in refrigerated warehouses and sold to the customer from chill cabinets then quite long shelf lives can be obtained. The product is generally flash pasteurised to ensure its cleanliness prior to filling.

8.4.6 Summary

Many products will go through a combination of these processes depending on the product, preservative system, preferred packaging and shelf life requirements. Some products will develop a cooked taste if exposed to a full in-pack pasteurisation and if for marketing or legal reasons preservatives cannot be used, aseptic filling is the choice. However, the capital cost of aseptic equipment is high and so it is only used for products at the premium end of the market. Some packages cannot be exposed to elevated temperatures (e.g. PET) and some are not suitable for carbonated products (e.g. laminated cartons). The choice of the packaging and process is an important technical decision and requires full understanding of the limitations of materials while still meeting the customer's need.

8.5 Control of process plant

In the last 10–15 years tremendous strides have been made in improving the control of process plant; however, fewer improvements and innovations have been made in the field of hygienic instruments. Ideally these should be noninvasive and should be cleanable in place. All controls should be defined in a functional specification of how the plant is to work. Each part of the process must be described in detail, defining its route, start, stop and alarm conditions and its cleaning procedures.

8.6 Factory layout and operation

In addition to process design it is essential to design the drinks factory such that it can be operated economically and cleanly. It is also important that its design is of minimal capital cost and consistent with production to desired quality standards.

Prior to considering the factory layout the designer must understand the risks to the production environment. These come from factors external to the plant, transmission media into the plant and internal factors resulting from operations within the plant. Table 8.1 details some examples of risk factors. Measures must be designed into the factory to eliminate or control all the risks to acceptable levels.

A sensible layout can be achieved through consideration of the main flows within the facility. Material flow, both ingredients and packaging, is important and should follow in a logical order. Clean operations should be separated from dirty operations, and wet operations from dry. Figure 8.2 shows a typical

Table 8.1 Examples of risk factors in the production environment.

External factors	Transmission media	Internal factors
Birds	People	Unpacking materials
Insects	Raw materials	Processing
Rodents	Packaging	Generation of waste
Other vermin	Pallets	People movement
Site ground	Engineering parts	Material movement
Surface contamination	Canteen materials	Air movement
Surrounding atmosphere	Other consumables	Drainage
		Cleaning
		Maintenance
		Exterior packing

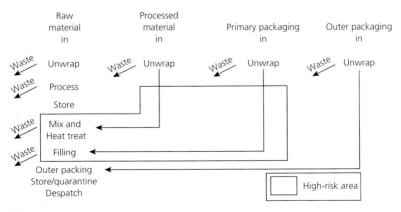

Figure 8.2 Typical process flow.

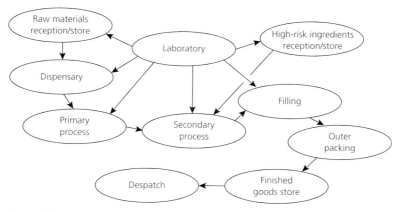

Figure 8.3 Bubble diagram.

process flow for materials; as can be seen there is an area where the risk of contamination is high. Generally, unwrapped packaging or processed material and the operations of mixing, processing and filling are high risk and care should be taken to ensure appropriate levels of cleanliness. Another flow within the factory is people; they can be a source of contamination or cross-contamination from an unclean operation to a clean area. Care must be taken to ensure that people can only access high-risk areas and operations through adequate and appropriate routes via cleaning and changing areas. A useful tool in understanding required people movements is a bubble diagram (Figure 8.3): each bubble represents an area of the factory and they are connected by arrows representing required access for personnel. The third important flow is waste. The process flow (Figure 8.2) shows waste being generated by a number of operations. The flow of waste and its exit from the factory needs careful consideration and its flow should not cross incoming material flows to avoid cross-contamination. The ideal factory layout can be developed from these flows.

Figure 8.4 shows a layout suitable for a complex aseptic production facility; each of the operations is carefully controlled to minimise the risk of cross-contamination. The material, people and waste flows have been superimposed on this layout in Figures 8.5–8.7. In many cases this layout is too expensive and elaborate for the production of preserved high-acid beverages and many of the separations between operations can be eliminated. However, for some products these precautions are essential. It should be noted that high-risk ingredients are those that are added to the product without further processing (e.g. aseptically packed fruit materials that can be added aseptically to the process flow; further treatment, although microbiologically desirable, might affect the flavour).

184   Chapter 8

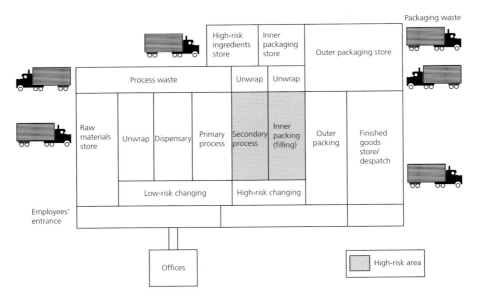

Figure 8.4 Typical factory layout.

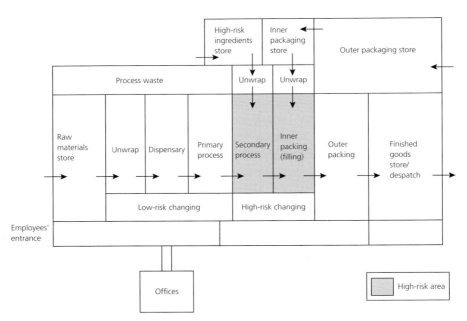

Figure 8.5 Factory layout showing material flow.

Processing and packaging

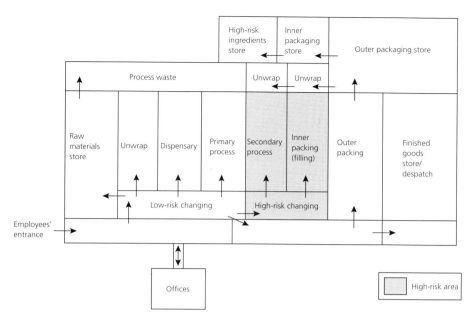

Figure 8.6 Factory layout showing people flow.

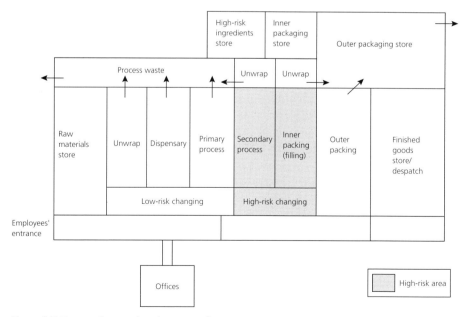

Figure 8.7 Factory layout showing waste flow.

8.7 Hazard Analysis Critical Control Points

Once the process has been specified and the layout designed, the system should be subjected to a Hazard Analysis Critical Control Points (HACCP) study. This is an analytical tool to ensure continual cost-effective product safety. The steps involved in the operation are systematically assessed and those that are critical to the safety of the product are identified. Operational targets and corrective actions are defined with monitoring procedures. Any design modifications can be easily made at the planning stages of the factory; the operation should have had a HACCP study and once the factory has been built a further study should be conducted to ensure compliance. Any planned changes to the use or operation of the facility should also be analysed. The HACCP concept allows movement to a preventative approach to quality issues whereby potential hazards are identified and controlled prior to causing product failure. As with any steps in the validation of a process the HACCP study should be fully documented and linked with verification procedures and supplementary tests.

8.8 Good manufacturing practice

The plant should be run under the principles of good manufacturing practice (GMP). These are defined as all those management activities, including the effective planning, design, maintenance and control of buildings, equipment processes and personnel, which together ensure integrity of the product in compliance with the law.

In the United Kingdom adoption of HACCP and GMP can be used to show that due diligence has been exercised in terms of the Food Hygiene Regulations 1990 or trade standards such as those required to meet the British Retail Consortium (BRC) criteria. Other countries have similar legal requirements. To date due diligence has been used to show mitigating circumstances in the event of a breach of the regulations rather than as a defence to a prosecution in court.

All equipment should be designed and constructed in a manner that prevents the retention of moisture (unless it is designed to deliberately stay full of sterile water), ingress and harbourage of vermin and soil and to facilitate inspection, servicing, maintenance and cleaning.

The following need special attention:
- materials of construction and cleanability
- surface finish
- joints and fasteners
- drainage
- internal angles and corners
- dead spaces
- bearing and shaft seals
- methods of cleaning.

8.9 Cleaning in place

Cleaning in place (CIP) is a method of cleaning plant by circulating detergent solutions and thus obviating the need for stripping down and manual cleaning. There are two main types of plant to be cleaned within a drinks plant: pipelines and vessels. For pipeline cleaning hygienic design is vital and correct CIP velocity is critical. This is typically 1.5 m/s, an empirical value based on achieving turbulent flow to achieve a scouring action (i.e. Reynolds number above 2500). For vessel cleaning, two systems are available depending on the soil to be cleaned: high pressure/low flow, which relies on jet impingement to remove soil, or low pressure/high flow (e.g. spray balls), which relies on there being a continuous film of detergent solution on the vessel walls. Cleaning solution must be continuously removed from the base of the vessel (known as scavenging).

A typical CIP cycle consists of:

1 pre-rinse to remove loose soil and reduce the work of detergent;
2 detergent recirculation to remove remaining soil – there are three main variables to be considered: chemical strength, cleaning temperature and exposure time;
3 intermediate rinse to remove detergent with fresh water until a neutral pH is achieved;
4 sanitisation to destroy remaining microbial contamination using hot water above 90°C recirculating for more than 20 min.

In addition to internal cleaning of the equipment, effective external cleaning of the plant and environment is required to remove pathogens and spoilage organisms introduced during normal operation. This must include regular cleaning under the equipment and cleaning the drains. Gel cleaning is the only real innovation in plant cleaning since the introduction of foam cleaning in the 1970s. Its main advantage is the prolonged contact time. The detergent increases viscosity on dilution; hence the neat product is easy to handle and contact times of 45 min are possible compared with around 10 min for foams. Additionally, reduced application pressure reduces the production of aerosols on impact with the surface to be cleaned; these can spread contamination over a wide area.

Automatic fogging systems are available to disinfect rooms. A dense fog is quickly achieved by atomising disinfectant solution through nozzles.

Some construction materials, for example, polyvinylidene fluoride (PVDF), are bacteriologically resistant and it has been shown that simple rinsing without detergent is capable of cleaning them to a state suitable for further drinks production. Such materials are suitable for the production of pipes, coating vessels, etc. and will probably be used more widely in the production of process plant in the future.

188 Chapter 8

8.10 Packaging

Packaging machinery is as varied as the many packages available on the market: some kinds of machinery form generic types and have fairly wide applications, others are very specific. Figure 8.8 shows a typical line layout for a bottling line. The principles of this layout are described here together with some of the more common variations. Bottles are supplied oriented on pallets and are unloaded on to the empty bottle conveyor system. Alternative layouts can be constructed for plastic bottles arriving in large bulk bins and being unscrambled (oriented) on to the line or (especially for PET) blown direct on to the line from preforms. Returnable bottles go through a decrater, which unloads them from crates, then to a bottle washer which cleans them, and finally to a bottle sorting area to remove those that do not belong. De-cappers and de-labellers may also be required. The bottles are then transferred in bulk (many bottles across a conveyor) towards the filler. Bottles are usually rinsed (the rinse water quality is important and it is often treated with ultraviolet light to disinfect it), filled and capped on one machine (really three machines monoblocked together with a common mechanical drive). This provides better bottle handling and control at high speeds, leading to better efficiency. In some cases, a wet glue labeller is also blocked with the filler.

Fillers are often equipped with two cappers for two different capping duties (different types of cap) and can handle a range of bottles by changing parts.

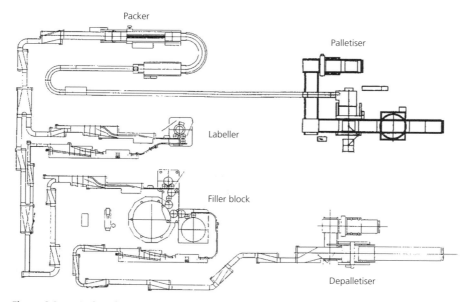

Figure 8.8 Typical packaging line layout. Scale 1: 200.

Filling speed is dependent on the number of valves and the physical properties of the liquid (different manufacturers' valves will perform differently and also affect speed); some fillers are vacuum assisted to speed the filling rate. Filling machines are now very sophisticated in terms of operation and can be equipped to steam a bottle to cleanse it or pre-evacuate it to reduce oxygen levels, fill from the bottom of the bottle to reduce fobbing using a long tube valve, take a bottle and fill a precise weight, fill to a level, fill a precise volume, etc. (Another benefit of the long tube filling valve is that it eliminates the risk of the bulk product in the filler bowl being affected by contaminated return air from the bottles.) The choice of filler will depend on the image required by marketing, the cost of the product in terms of 'give aways' or overfilling, and the demands of the package or product.

Another factor affecting the choice of filler is its ability to be thoroughly cleaned. The simpler the valve the easier it will be to clean, and the more small passages, springs or complicated seal arrangements there are the more difficult cleaning will be. Most filler manufacturers will test fill and test clean products in their laboratories to demonstrate the cleanability and filling performance to be expected from their machinery.

One popular misconception concerns the counter-pressure filler. It is often stated that the overpressure of gas in the filler bowl pushes the product into the container, but this is not the case. Once the valve is open the pressures in the bottle and the filler bowl equilibrate and product is filled by gravity. (Note: a counter-pressure filler is usually used for carbonated beverages but it can successfully fill still products with an inert overpressure.)

Ultimately, the speed of a filling machine is a function of the engineering limits of metal machinability. The valves are mounted on a ring that has to be machined to very tight tolerances and mounted such that it rotates evenly, hence the overall diameter of a filler has a maximum value and so the number of valves that can be mounted on the pitch circle has a maximum.

There are many decoration options for bottles but the most common is the wet-glue labeller, which is described here, although self-adhesive labellers, sleeving and even a form of bottle printing using heat to transfer the print from a backing paper to the bottle are all available, together with machinery of application.

In wet-glue labelling, labels are stacked in a magazine and removed by pre-glued pallets that apply a controlled quantity of glue to the reverse of the label. A thin glue film is applied to the pallet, which in turn removes the labels from the magazine. The glued label is then removed from the pallet by a transfer cylinder usually using gripper fingers to hold the leading edge of the label and a vacuum pad to control the label's orientation on the cylinder. The label is now held glue-outwards, which is suitable for application to the bottle. Once firmly transferred to the bottle the dress is completed by brushes that smooth the label on to the bottle. Some simpler machines will transfer a simple label directly from the glue applicator to the bottle using a comb or similar device. Systems exist for cold-glue or hot-melt labelling, wrap-around labelling (where glue is applied only at the overwrap area), over-the-cap labelling (to provide tamper evidence), application of decorative foils (to give a premium image) and a host of other operations. A quick look along a supermarket shelf will show the many options available, some of which are very complicated in that a sequence of successive operational steps is required to achieve the final result.

Intricate label shapes can be applied but generally labels with an aspect ratio greater than 1:5 present problems with consistency.

The speed of labelling is generally a function of the bottle handling; the labelling stations described above can handle labels far faster than the technology of bottle handling can present bottles to be labelled.

The next machine on the line is the packer, which collates the bottles into the required sales unit (often 12 or 24 bottles). Sales units are generally shrink-wrapped trays, cardboard cartons or crates for returnable bottles. These are stacked on pallets for ease of distribution and storage.

Bottles are normally coded with a 'best before' date and often a closed code giving the batch or production record for tracing purposes. These are applied by laser printer or ink jet onto the label, cap or neck of the bottle. There is a variety of on-line inspection devices available for ensuring the quality of the product, for example, empty bottle inspectors, hydrocarbon sniffers (to check that returnable bottles have not been used for storage of paraffin, etc.), fill height detectors, label application detectors and date code verifies. All of these can be found on the modern bottling line.

To gain maximum efficiency from a production line it is normal for the machinery upstream and downstream of the filler to run faster than the filler and to provide accumulation between machines so that minor stoppages (for example changing the shrink-wrap reel) can be accommodated without stopping the filler. Figure 8.9 illustrates this for the factory layout shown in Figure 8.8.

Canning lines are very similar to bottling lines except that the container is sealed with a can end, which is rolled on by a seaming machine. There are also many specific machines for special filling operations in use in beverage packing, for example, laminated cartons, foil pouches, etc.

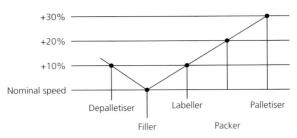

Figure 8.9 Line equipment reference speeds (relative to filter nominal speed).

8.11 Conclusion

To ensure that good quality, consistent, wholesome product is delivered to the consumer is a complex operation consisting of a number of sequential steps. It is essential that these steps are well understood. The best way of achieving quality is to design it into an operation from the beginning and maintain it through the adoption of the principles of good manufacturing practice.

To summarise, a plant will only be effective and the product safe if all the following aspects of design are taken into account:
- the nature and cleanliness of the raw materials
- the selection of materials of construction
- the nature and cleanliness of the packaging
- the process design of the plant
- the functional specification of the plant
- the correct operation of the plant
- the good management of the plant.

CHAPTER 9
Packaging materials

David Rose
A Pkg Prf. Packaging Development Manager, Britvic Soft Drinks, Hemel Hempstead, UK

9.1 Introduction

The importance of packaging in the soft drinks and fruit juices market, be it to preserve, protect or promote the product, cannot be underestimated. Indeed, the market would not exist without packaging, as every product will need to be packaged in some form throughout the supply chain. The current focus on 'healthy' lifestyles, and the consumer pressures for natural products without artificial additives, will require the packaging to do more than simply contain. Preserving and protecting the product will become ever more important.

Research has shown that the decision to purchase or not is made in a fraction of a second, during which time the package has to shout for attention, to look and feel appealing, and be right for the moment, so the importance of the packaging cannot be overemphasised. However, it is no use packing a cheap product in an expensive package, as the consumer will see though this and, if the product does not match their expectations, the chance of a repeat purchase is low. The pack and product must be balanced and developed together.

In the marketplace, competition is very strong, and brand equity is used to establish the position of a product. The type and size of a drinks container, its shape and its pack graphics, are all used to help differentiate one product from another. Those involved in the development of drinks products need to be aware of the marketing requirements and commercial opportunities, in order to ensure that their development efforts align with the brand strategy, costs and lead times.

It is not possible to cover all aspects of such a significant topic as packaging in a single chapter, so the purpose here is to give readers a basic insight into the options that exist. They should explore packaging requirements further with their own in-house packaging specialists (if they have them), or with their packaging suppliers, national packaging institutions or other trade associations relevant to the beverage industry. Some potential contacts are given at the end of this chapter.

Chemistry and Technology of Soft Drinks and Fruit Juices, Third Edition. Edited by Philip R. Ashurst.
© 2016 John Wiley & Sons, Ltd. Published 2016 by John Wiley & Sons, Ltd.

9.2 Commercial and technical considerations

9.2.1 General considerations

Soft drinks and fruit juices compete with many other beverages, such as water, milk (natural or flavoured), tea, coffee, and so on, for our attention. Consumers are more price-conscious than ever so, whether the products are still, carbonated, sweetened, unsweetened, energy boosting or complement diets, all will claim to be needed in our hectic lifestyles.

Packaging formats and materials will all play their part in appealing to the consumer. It is important to consider whether the product is to be in a multi-serve or single-serve pack, is to be purchased in a supermarket as part of a weekly shop, or as an impulse purchase in a garage forecourt or corner shop – or sold in both outlets, maybe in different formats. This will influence the pack size and material format and, hence, the product processing conditions available to the drink development technologists. The pack must also fit into the increasing range of shelf displays, chillers and vending machines used in the various outlets.

Another consideration is how long the consumer will interact with the package. If the pack is a multi-serve, stored in a refrigerator, and the product is poured into a glass or cup, then the interaction time is minimal. However, if the pack is a single-serve size, with a 'sports closure', then the interaction time will be longer and the tactile properties of the package may have more importance. The functionality of the package must last for the life of the pack, which may be several hours, and the pack must be fit for its intended purpose. For example, water in a bottle with a sports closure, intended for use during exercise, must be capable of being held in one hand while running, so an ultra-light PET bottle which crushes and spills is unlikely to get a repeat purchase.

Often the closure is last issue to be considered. We call it a 'closure', as it is one of the last primary pack processes on the filling line, yet removing it is one of the first actions taken by the consumer. Thus, it might equally well be termed an 'opening', for the very significant role it plays for the consumer. Easy access can enhance the product experience, often without the consumer even thinking about it. If the consumer cannot unscrew a closure, or breaks a nail tearing open a closure, the whole experience is ruined and a future customer may be lost.

Whether the annual sales volume will justify a customised pack, or a standard format can be used. Relying on the decoration to provide product differentiation should be considered during the development phase. The timing to introduce the product in a stock pack will be significantly faster than if a new pack has to be developed, new tooling produced and the pack evaluated, as well as affecting the level of capital investment required and the return on that investment.

When considering the introduction of a new product formulation, the first point to take into account is whether the packaging format already exists. If a new packaging design is being requested for a re-launch, it likely to be handled on the existing filling line(s), so must be compatible with the same filling processes.

A decision is needed as to whether changes in the packaging size, shape or closure would prove prohibitive, as there may be only certain sizes with which the existing lines can cope. The existing equipment may be a glass or plastic bottle filling system, or may have been purpose-built to deal with both, although it is best to have one material running on one line and not have a dual-purpose line. Fillers may be designed for still or carbonated products, and may not be able to handle both cold and hot-fill products.

Often, a new juice product will need a higher-performance package (e.g. if it is to be hot-filled or in-pack pasteurised). However, with new developments, it may be possible to reformulate the product to be cold-filled without special package treatment. The use of aseptically filled packs, particularly bottles, is capital-intensive, and would have to be balanced against other process and pack configurations in the context of market opportunity. The use of aseptic filling brings the ability to pack products without preservatives, which can command a higher retail price, thus potentially recouping the higher capital outlay. The use of barrier materials in the bottle can further enhance the product, by allowing the use of high juice content, or natural colours and reductions in anti-oxidant additives.

Another opportunity for change might be decorative. If labels are used and need to be retained, then the method of printing should be discussed with the label manufacturer and examined in terms of print effect versus cost. The label substrate can help to create an impression of quality through the use of plastic, either in conjunction with paper, in laminate form, or in monolayer form. Whether the labels are pre-cut or reel-fed will also impact on whether the label can be a shaped patch label or is wrapped around the package.

Shrink-sleeving is a very popular decoration process that is used to increase the graphic area of the package and to convert a standard material pack (e.g. a high-density polyethylene (HDPE) bottle) into an eye-catching pack. This process allows three-dimensional decoration, which changes a 'labelled' bottle into a 'decorated' package. It is then possible to change the graphics, colour and so forth on one base bottle design, to create packaging for more than one brand. Even bottles made in quite bland and low-valued materials, such as HDPE, can appear to be of premium quality in full-length shrink-sleeve decoration. The flavoured milk market has made use of this technology to great effect. This form of structural package standardisation allows for more than one product or brand to be launched quickly, and with minimal extra cost. The money saved can be used either to enhance the quality of print decoration, or to increase the level of advertising spend (an example of the packaging process supporting the commercial process).

Pre-decorated packaging has been available for many years, particularly on metal cans. However, as two-piece beverage can technology is geared to high volume, the minimum run lengths required are high and can be prohibitive for niche or small-volume products. While sleeving can also be used on metal cans, there is very limited capability to carry this out, except at specialised co-packers. Digital printing, directly onto the can or bottle, is being developed, but it is still in its infancy in terms of speed and availability.

In summary, deciding on a packaging process requires consideration of: the commercial opportunities; the product processing requirements and the packaging options available; the flexibility of the filling line; and the packaging materials and their ability to withstand the processing requirements and shelf life demands. It is also important to consider the decoration formats, and their ability to match the container type, carry the necessary information and withstand (if pre-decorated) the processing itself, while maintaining the brand quality. If all this is put together successfully, then the brand will benefit from the added value created by packaging.

Another consideration is potential future developments for a particular brand or pack. Once a new production line is installed, or an existing line modified for a particular range of formats, adding a new format, such as a larger or smaller primary pack, or a different size multipack or traded unit, can be very costly. Building the capability in during the initial design, even if the specific handling parts are not purchased at the time, can result in a significant saving if the additional format is then used.

9.2.2 Packaging materials

The range of packaging materials covered in this chapter includes those for still and carbonated products, some of which can be cold-filled, and others that need to be heat-treated. Heat treatment can either be within the package, or by pre-treating and filling, hot or aseptically, into a pre-sterilised package, to ensure an acceptably long shelf life to suit market requirements. The role of packaging is to contain, protect and preserve the product from producer to the end user – that is, to present the product in the condition intended by the producer for the enjoyment of the consumer. The packaging materials must, therefore, meet the processing requirements and be capable of running through the production lines efficiently, as well as meeting the necessary economic constraints so vital in a competitive market.

The primary pack must be capable of being collated into suitable formats, and must withstand both the distribution requirements and those at the point of sale. It must also perform a marketing role in promoting brand positioning through package shape, material and decoration, enabling identification and providing selected information (nutritional and statutory). The packaging must remain secure in terms of providing tamper evidence and microbiological hermetic seals. In performing all these functions, it must additionally be capable of meeting ever more stringent environmental legislation and requirements for packaging waste, all at an economical cost.

Polyethylene Terephthalate (PET) bottles have significantly changed the face of soft drinks packaging since their introduction into the world market in the late 1970s. However, glass and metal cans have continued as important packaging materials, as they offer excellent process conditions, high-speed filling and very good barrier properties. Glass bottles are now ultra-light single-trip bottles, and are 50% lighter than they were 30 years ago, while cans are still a major player in the convenience sector.

PET packaging has allowed drinks companies to move into in-house bottle production, and to operate more independently from packaging manufacturers. In turn, the packaging manufacturers have had to offer a more comprehensive service, from managing in-house operations to training customers' operating personnel for fully owned operations, allowing drinks companies to drive their own packaging developments and product portfolios.

Consumers still perceive glass as 'quality', and it imparts a good shelf life. In addition, glass bottles can give a much greater impression of overall size, compared to plastic bottles of the same capacity; a 750ml glass bottle will look the same size as a 1 litre PET bottle. However, the use of modern, ultra-light glass bottles requires the correct handling of empty and filled bottles on the production lines. Simply switching to a new lightweight bottle on an old production line may not be possible and, often, significant capital expenditure is required in order to modify or upgrade conveyors and equipment to handle such bottles.

Cans, whether made from either aluminium or steel, are highly developed to deliver a drink to the consumer efficiently and at a comparatively low cost. The machines which make, fill and seal the can are equally well-developed and understood, running at high line speeds and operating efficiencies. These packages are robust, and offer excellent product protection, including retention of carbonation and low oxygen ingress. The cost of cans is highly competitive, and one major consideration has to be whether there is enough business to justify a line which typically runs at up to 1500 units per minute or around 300 million per year, with a minimum can-manufacturing run of 0.5 million units or more. Contract packing is often an option for a canned product.

Laminated paperboard drinks cartons for non-carbonated still/juice products provide a high technology packaging format that is well understood, and accepted, by the consumer – although they are often considered a 'commodity' pack with minimal differentiation between them. They typically have a square or rectangular cross-section, although some polyhedral shapes are available, with easy open and re-close systems, developed in an attempt to increase differentiation and add a 'premium' look. However, due to the manufacturing process, this is minimal, and it does not present the options of, for example, a PET bottle. Cartons are manufactured by a number of well-developed systems, with high output and efficiency, so unit costs are low. They are thus not only ideal for the home market, but are also capable of global distribution for brands with a wider appeal. Most countries around the world have drink carton-filling capabilities, which is important for global branding and distribution.

9.3 Processing

The processing of a product plays an important part in the choice of packaging material and format. Since most fruit drinks require heat processing to ensure microbiological stability, consideration must be given to the packaging material and its tolerance to elevated temperatures. If the product is also carbonated, then the choice of packaging material and closure system becomes even more critical because of the required heat processing; as the heat builds up, so does the internal pressure. This section looks at the following processes and their impact on packaging:
- Cold-filling.
- In-pack pasteurising.
- Hot-filling.
- Aseptic filling.

9.3.1 Cold-filling

Many soft drink products are formulated with ingredients that either do not need to be heat-treated, or can be 'flash' pasteurised and cooled, in order to be microbiologically stable over the required shelf life. Still products can be filled into a wide range of packaging materials, with the limiting factor being the barrier properties (i.e. oxygen transfer rate) of the more flexible materials. Carbonated products need to be packaged in more rigid materials, or in structural designs that can withstand internal pressure, with containers being circular in cross-section to resist the internal pressure. Specific base designs and smooth shoulder contours enable PET bottles to withstand the required pressures and impact if dropped. Any embossed features, particularly in a PET container, will require careful design, as the features may be distorted by the effect of the internal pressure.

9.3.2 In-pack pasteurising

Some soft drinks and juices need to be heat-treated to provide a microbiologically stable product. One method that is considered very reliable and robust is in-pack pasteurising (IPP), which is particularly suitable for treating carbonated products, as these cannot be hot-filled. Whether still or carbonated, it is normal to fill the products cold, although some still juices are heat-treated first, filled hot and then pasteurised in-pack, as this allows for a different temperature regime in the IPP cycle, as well as giving a better balance of internal/external pressures during and after the IPP treatment. This method is suitable for glass, certain plastic containers and metal cans.

Once filled and capped, the bottles (or packs) are taken through an IPP unit, which heats them to the pasteurising temperature (in about 20 minutes), holds them at that temperature for the required period (normally 10 minutes at 65–75°C), then cools them down. The whole cycle takes about 50 minutes, with the actual pasteurising temperature and time cycle being determined by the product specialists. The packaging materials (including heat-shrink sleeves, if pre-decorated, and the printing on cans) are all subjected to continual dowsing in water – warm, then hot, then cold – over a long period of time. Care needs to be taken when selecting the appropriate packaging format, which must be resistant to heat and water. For carbonated products, the package must also be resistant to the build-up of internal pressure (as high as 100 psi/690 kPa) as the contents are heated to the correct pasteurisation temperature.

The choice of bottle closure is also critical, since leakage of both product or gas out, or pasteuriser water in, is possible during the cycle, because of the internal pressure and differential shrinkage of the closure and bottle materials. For bottles with wide-mouth neck finishes (35 mm and above), it is standard to use metal roll-on (RO) closures as they are more stable at high temperatures and can resist the internal pressure better than plastic closures. Care should be taken in selecting the appropriate sealing system and neck finish profile, as these will have a part to play in giving consistent performance. When selecting these components, it is essential to discuss performance requirements with the packaging suppliers.

9.3.3 Hot-filling

Still products can be hot-filled into a wide range of packaging materials (provided they are heat-resistant). The bottles are normally hot-filled at 85°C and are then either inverted immediately after capping, or passed through a short 'temperature holder' to sterilise the whole of the internal surface of the package. After this time, the packages pass through a cooling system until they are cool enough to prevent the product from 'cooking', thereby affecting the desired flavour profile.

Care must be taken that the packaging is completely dry when packing in-pack pasteurised or hot-filled products for distribution. If they remain damp, there is a severe risk of stress corrosion of can ends. If the secondary packaging is corrugated board, the board may soften and the stacked products collapse.

9.3.4 Aseptic filling of bottles

There are several systems available for aseptically filling glass and plastic bottles for still juices (the aseptic filling of drink cartons is covered in section 9.7) and, while it is possible, carbonated drinks are generally not aseptically filled. The most commonly used process sterilises the pre-made bottle, fills and seals it; the second takes a sealed, pre-cleaned bottle, removes the seal in a sterile environment, fills and re-seals the container. The third system is known as 'blow fill seal' (BFS) or 'form fill seal' (FFS), where a container is formed, filled, and sealed in a continuous process, in a sterile enclosed area.

In the first system, the inner surface of a pre-made, open-necked bottle is sterilised, traditionally by wet sterilisation using peracetic acid ($C_2H_4O_3$) or, increasingly, by dry sterilisation using hydrogen peroxide (H_2O_2) in gaseous form. The sterility of the bottle is maintained while it is transferred to the aseptic filler and then the capper. The bottles used in this process are usually wide-mouthed, as the sterilant has to be driven out of the bottle before filling, which becomes more difficult as bottles become bigger and taller and their necks narrower as shown in figure 9.2. Since heat is used to flush out the hydrogen peroxide, plastic bottles have to be heat-stabilised, even though the fill itself is at ambient temperature or under chilled conditions. Care should be taken with the specification for the bottle and its shape, in order to counter distortion and shrinkage when it is subjected to heat. The layout of this type of filler is shown in Figure 9.1.

A more recent development of this process is towards pre-form rather than bottle sterilisation, which has the advantage of lower media use and allowing lighter bottles to be blown, as shrinkage is less of an issue. Before deciding on a system, the individual product characteristics (pH, CO_2, etc.) and environment must be looked at to choose the best, most applicable solution.

It is common to fit a foil seal before the main closure or dust cap is applied, because it is easier and more reliable to sterilise a foil using heat or hydrogen peroxide sterilant, than to sterilise a pre-formed threaded closure. Also, if a foil is used as the primary seal, then the plastic closure can be of a simpler design, as it is only used to re-close the bottle once opened.

It is possible to use pre-sterilised (irradiated) closures, but these need specialised processing, distribution and handling to maintain sterility prior to application to the bottle. Heat-resistant materials can be steam-sterilised, but plastic materials are usually sterilised with hydrogen peroxide, which is then flushed off using hot air. Care must be taken to ensure that all the sterilant has been flushed away from the thread form or sealing system.

A less common aseptic filling system uses an extrusion blow-moulding operation to produce a bottle (which can be multi-layered, to add a gas or light barrier) moulded under sterile conditions. While the bottle is still sterile, and the polyethylene is soft, the trim material above the neck finish (moil) is crimped closed to seal the bottle. The sealed, sterile, empty bottle can then be packed and transported to a filling line, which may be in another location from the blowing operation. On the filling line, the external surfaces of the bottle are sterilised, which is easier than sterilising the internal surfaces of an open-necked bottle. The closed, sterile bottle is then passed to a trimming station, where the moil is removed and the bottle is filled aseptically. Again, a foil seal is used to close the filled bottle, and a closure of a suitable design is applied to protect the foil during transit and to provide re-closure during use. This manufacturing process requires a material that impact welds while it is still hot, so is only suitable for HDPE or PP.

In the third system, 'blow fill seal', a blow-moulding machine is integrated with an aseptic filler and capper. Its principle is to sterilise the preform, not the formed bottle, with hydrogen peroxide at the exit from the blower heater oven, then blow the preforms with sterile air in a sterile environment and maintain this sterility throughout the filling and capping process. There are pros and cons for this system versus a more conventional open-neck aseptic filler. These centre

Chapter 9

Basic design for PET-Aseptik L

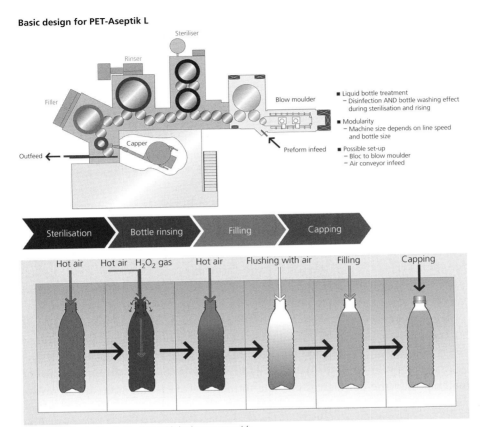

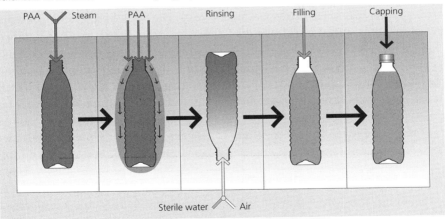

Figure 9.1 Schematic of Aseptic Process (courtesy of Krones AG.).
Schematics for bottle sterilisation with hydrogen peroxide.
Schematic of bottle sterilisation with peracetic acid.

Figure 9.2 Examples of aseptic packs (courtesy of Britvic Soft Drinks Ltd).

on capital investment, production volume and bottle design flexibility. However, as the preform is smaller and has a more simple shape than the bottle, sterilising this reduces the quantity of sterilising agent used. Also, as the preform is thicker material than the bottle, it is possible to increase the temperature of the treatment without risk of shrinkage that would affect the shape of the bottle. This allows the bottle weight to be reduced, compared to traditional aseptic systems.

The alternative blow fill seal format is an extrusion blow-moulding system, normally used for sterile products in the pharmaceutical or medicine markets. The process is similar to the extrusion blow moulding process described below (section 9.4.3: High-density polyethylene). Following the formation of the container, however, the mandrel used to blow the bottle is also used to fill the container with liquid. Following filling, the mandrels are retracted, and a secondary top mould seals the container. All actions take place inside a sterile, shrouded chamber inside the machine. These bottles are typified by the use of an integrally moulded, twist-off, winged 'nib' to open the bottle. However, they do not seem to have taken off in the mainstream drinks market – possibly because of the pack image, or because the system is relatively slow for volume use and is more suited to added value markets.

Another alternative is the 'form fill seal' system, which produces a thermoformed tapered container with a foil seal, very much like a yoghurt pot. Containers are thermoformed, starting from a roll of multi-layer plastic film, then filled with the sterile product, closed by a thermos-sealed peel-able lid and, finally, cut to the desired configuration. One of the benefits of this system is that it has been used for many years, and it is well established for package materials, such as toughened polystyrene and high-contact clarity polypropylene. If barrier properties are an issue, then a multi-layer specification laminate should be used.

9.3.5 Liquid nitrogen injection

As two-piece cans and PET bottles are now made with very thin side walls, their ability to resist vertical top loads is limited. This is not an issue if filled with carbonated product, but it can be a problem for non-carbonated products. However, there are systems available which can inject a precise small volume of liquid nitrogen into a filled pack, just before the end/closure is applied. As the liquid converts to a gaseous state, it expands. This also helps to expel excess oxygen from the headspace, which may otherwise affect shelf life, and it also provides the internal pressure required for side-wall strength and package stability. By adjustment of the injection time and duration, the desired pressure can be accurately controlled and the nitrogen gas does not go into solution with the liquid, changing it to be like a 'carbonated' drink.

9.4 Bottles

When choosing a bottle and the material, it is important to consider not only the intended market and package size, but also the product to be contained, the shelf life and the filling requirements. An important factor is the volume of unfilled headspace designed into the bottle for pasteurised or carbonated products. Typically, this is not less than 5% of the total volume and, ideally, it is more like 7%. This is because, as the product expands during pasteurising, the headspace becomes squeezed and, the smaller the headspace volume, the higher the internal pressure becomes. The larger the diameter of the closure, the more force will be exerted on the closure during heat processing, which could lead to product or gas leakage or, more seriously, to closure detachment during processing.

The profile of the bottle shoulder and the neck diameter are also important factors. The smaller the diameter of the closure, and the more tapered the neck/shoulder, the more exaggerated the change in fill level becomes. Therefore, in order to create a 5–7% headspace, the fill level height will be much lower, which can give the appearance of a short fill or deceptive volume.

9.4.1 Glass

Glass bottles offer an excellent shelf life, as the material is impermeable to gases, preventing both oxygen ingress and the loss of carbon dioxide from sparkling drinks. Thus, while PET now dominates the multi-serve carbonated drinks market for products requiring only cold-filling, glass has continued to enjoy success in the single-serve still and carbonated juice markets. For small, single-serve bottles (250–400 ml), the shelf life of carbonated products is an issue, because of the surface-to-volume ratio, particularly in the case of PET. The use of barrier materials can be used to extend the shelf life, but this does add cost to the materials. However, glass is not always the best material for beverage product bottles in sensitive locations, such as sports/event stadiums and swimming areas, or the beach, where broken glass can be a significant hazard.

One reason why metal closures, mainly roll-on pilfer-proof, (ROPP) are used on glass is because the greater dimensional tolerances of glass are such that it is vital to form each closure to fit each individual bottle neck (the ultimate form of customisation), and this can be done with ductile metal.

During in-pack pasteurising, the metal closure is likely to expand more than the glass neck and, although the reverse is true on the cooling cycle, the pressures will still be higher and the compound lining softer.

In recent years, the use of plastic closures on glass bottles has grown. Consumers usually find the tactile properties, ease of grip and safety (from metal edges) of plastic closures superior to ROPP metal closures. However, care has to be taken over how the sealing system works to retain the carbonated gas, since the internal pressure has to be released before the closure releases itself from the thread finish. Otherwise, the closure can release with some force, potentially causing injury or damage. Most plastic closure systems used on glass and other bottle materials now incorporate a liner seal that releases the internal pressure soon after the consumer begins to open the closure, thereby ensuring that all the excess pressure has been released by the time the closure detaches from the bottle. The liner also creates a more flexible seal, accommodating a broader range of glass finishes and tolerances from more than one glass manufacture, even if they are all at the same nominal diameter.

In order to create a satisfactory seal on a glass neck, some closure suppliers have developed single-piece plastic screw closures with bore seals, as the neck bore has a more controlled section of the neck, due to the glass-blowing process. As the molten glass shrinks on cooling, the thread detail can shrink away from the thread-forming part of the mould cavity, but a blowing pin is positioned in the neck bore section during the blowing process. Thus, the glass can shrink only on to the fixed-diameter pin, which helps to control the bore.

9.4.2 Polyethylene terephthalate

PET represents one of the most significant changes in terms of the packaging materials available to the soft drinks and fruit juice markets. Introduced to the UK market in commercial volumes in the late 1970s, PET has driven the most important market opportunity since the introduction of aseptically filled juice drinks cartons. As mentioned above, PET has allowed the drinks industry to develop into more flexible and larger supply chains. PET offers a package with superb clarity, unbreakability and consistent neck finish. In some countries, there is the option of refilling, or recycling as either polyester fibres or as resin, to be reused in the production of more bottles. PET's unbreakability means that only minimal secondary packaging, such as shrink film, is required for distribution.

In considering PET for a particular application, thought needs to be given to the process and shelf life requirements, in terms of the bottle wall thickness, rigidity and barrier properties. With carbonated products, the internal pressure holds the bottle rigid, making it ideal for palletisation and distribution. For still products, nitrogen can be used to create internal (headspace) pressure, again giving the bottle a degree of rigidity for distribution, with the advantage that this headspace pressure option can be used with sports closures. The alternative for still products is to increase the wall thickness of the bottle, but this adds a significant degree of weight and cost.

PET bottles are typically produced in two stages. First, a preform, similar in appearance to a test tube with a threaded top, is produced by injection moulding. This preform includes the final neck finish, and it is only the body of the preform which is stretched into the final bottle shape a typical stretch blow moulding machine is shown in figure 9.3. The stretching process imparts the vital properties to the finished bottle by creating strain-induced crystallinity, mainly into the

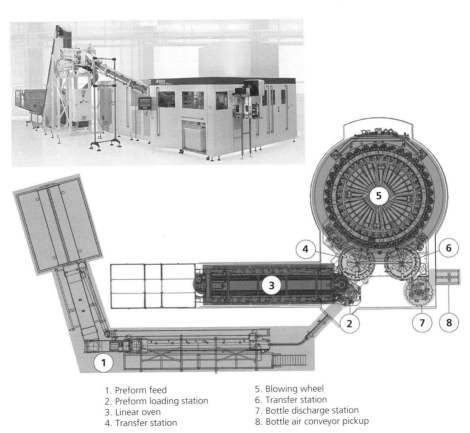

1. Preform feed
2. Preform loading station
3. Linear oven
4. Transfer station
5. Blowing wheel
6. Transfer station
7. Bottle discharge station
8. Bottle air conveyor pickup

Figure 9.3 Stretch blow moulder diagram (courtesy of KHS Corpoplast GmbH). 1. Preform feed; 2. Preform loading station; 3. Linear oven; 4. Transfer station; 5. Blowing wheel; 6. Transfer station; 7. Bottle discharge station; 8. Bottle Air Conveyor pickup.

side walls of the bottle, which have a thickness of around 0.2–0.3 mm for a 2 litre bottle. The un-stretched portions contain amorphous PET which, although thicker than the sidewalls, has a lower barrier performance per unit thickness than the stretched, crystallised material ('crystallinity' refers to material in which the molecular structure is aligned in a regular crystalline fashion; material in which the molecular structure remains random is termed 'amorphous').

The stretch ratios of preform diameter to bottle diameter, and preform length to bottle length, are used to calculate the stretch ratio, which must be optimised to ensure maximum barrier performance. Preform and bottle design, therefore, play a vital role in determining bottle performance, as does the surface-to-volume ratio. Small bottles will have a shorter shelf life than their larger equivalents. As designers have come to understand better the importance of material stretching and its impact on bottle performance, so bottle weight has decreased.

As with all plastics, PET is permeable to gases, letting carbon dioxide out and oxygen in. The shelf life requirements need to take into account the sensitivity of the liquid to oxygen, the ambient temperature and storage conditions, both in distribution and at the point of sale. Also, the level of carbonation at the time of consumption is also likely to be important to the mouth-feel and flavour profile of the product. Therefore, measuring carbonation levels during a real-time storage trial, or even a shortened trial made at elevated temperatures (i.e. 35–40°C), will be an important part of the package validation protocol.

A general rule of thumb for PET is that the shelf life performance is halved for each 10°C rise in ambient storage conditions. Improvements in the barrier performance of PET have addressed this, and improvements in supply chain efficiency have reduced the time between filling and consumption, so the CO_2 loss is reduced. However, it is still common practice to increase the carbonation level at the time of filling, compared with the equivalent product packed in glass or cans.

For sensitive products, the barrier performance of PET can be improved by the addition of other materials to the PET or, probably better, by adding discrete layers of high-barrier material, such as ethylene vinyl alcohol (EVOH) or nylon during injection moulding of the PET preform, or by coating the bottle after it has been blown.

Co-injection of a barrier within the bottle preform requires a more specialised co-injection process, but this is well proven technology and, while the preform cost will be higher, no additional equipment is required on the filling line. A number of systems for coating the inside of freshly blown bottles are available, using either carbon or silicon oxide. These are based on a plasma coating process, and require an additional machine coupled to the bottle blowing machine. Therefore, while the initial capital investment is higher, the unit cost will be lower for high volume production.

One factor to consider is the effect on the recyclability of the barrier component. Additives in the PET can sometimes downgrade the recyclability of the recovered resin, whereas most coatings allow 100% bottle-to-bottle recovery.

The level of crystallinity within the PET plays an important part in the level of heat stability of a bottle. Again, the evolution of preform designs and grades of material has increased the level of heat stability. In order to extend this stability to the temperatures used for hot-filling (typically 85°C), additional methods have been required to increase the level of crystallinity in PET bottles. Success has been achieved by effectively 'heat-setting' the bottles during the normal blowing process, by heating the moulds to temperatures high enough to provide the required hot-fill tolerance. Since the heat-setting process increases moulding time, there are commercial implications that limit application of this process to those products for which heat treatment is critical for microbiological stability, and to those markets that require an unbreakable bottle, such as a sport drink made with a high juice content.

When the product cools, it contracts, creating a vacuum which distorts the normal shape of the PET bottle panel. Hot-fill bottles are traditionally made with panels in the body section and/or in the bottle base that distort in a consistent way, so that the overall bottle does not distort as it cools. However, a recent development is the patented 'reheat base' technology. After hot filling, the bottles are cooled as normal, where they deform under the vacuum created. The bottles are then fed into the base ThermoShape™ re-shaping machine, where the base is heated and re-shaped back to its original form as shown in figure 9.4. ThermoShape allows for hot-fill bottle shapes and weights similar to those used for cold aseptic fill.

Figure 9.4 Base re-heat diagram (courtesy of Plastipak).

The vacuum that creates the bottle panel also has to be contained by the closure, so hot-filled bottles require specific closure design and application that are different from standard carbonated closure designs (see section 9.5). However, ThermoShape™ also allows for the use of a lightweight neck finish and single-piece closure, due to a lower fill level and reduced vacuum during the product cooling, compared with traditional hot-fill technology.

9.4.3 High-density polyethylene

High-density polyethylene (HDPE) bottles are widely used for fresh milk, and can be a lower cost alternative for PET. They are often used for juice drinks and flavoured milk drinks; however, HDPE has relatively low oxygen and gas barrier properties, so it is often used for short shelf life distribution products that appear mainly in the chill cabinets.

HDPE bottles are made using an extrusion blow process, where the plastic is melted and extruded into a hollow tube, which is then captured between two halves of a cooled metal mould. Air is then blown into the parison, inflating it into the shape of the bottle. After the plastic has cooled sufficiently, the mould is opened and the part is ejected. The action of trapping the tube creates excess material at the top and bottom of the bottle, which are cut off in a separate trimming process. This trapping also creates the facility to form cut-through handles, something which PET bottles, made by the injection-stretch blow-moulding process, have not yet been able to replicate in a one-piece design.

9.4.4 Polypropylene

Polypropylene (PP) bottles have the advantage that they are inherently hot-fillable (at 85°C) and retortable to 120°C. The bottles are normally extrusion blow-moulded (as HDPE), and can also be made with multi-layering, to include barriers against oxygen permeability. This technology is commonly used for sauce bottles and pure juice products. Since most of the bottles are of an oval shape, product contraction, resulting in volume reduction, is countered by allowing the bottle to become more oval under the effects of vacuum. Polypropylene also has good deformation recovery properties, making it ideal for squeezable dispensing packs as shown in figure 9.5.

Polypropylene usually has a milky appearance (particularly when empty). However, the contact clarity is really quite remarkable, making a filled bottle look attractive. Full shrink-sleeve decoration changes the whole visual impact if total clarity is an issue and PET proves not to be suitable (e.g. for oxygen-sensitive products).

9.4.5 Polyvinyl chloride

As PET technology has advanced, and in-house operations have grown, the use of polyvinyl chloride (PVC) has discontinued in the wake of health scares and environmental concerns in recent years. Much of this has been linked to public and media misconception, and confusion with other PVC additives and uses, which are not covered in this chapter.

Figure 9.5 Example of polypropylene pack (courtesy of Britvic Soft Drinks Ltd).

Table 9.1 Properties of plastics.

	Polyethylene terephthalate	High-density polyethylene	Polypropylene	Polyvinyl chloride
Density gcm^3	1.37	0.91–0.925	0.9	1.32
Oxygen transmission cm^3.mm/m^2.day.bar	2.4	74	96	3.1
	Medium	High	High	Medium
Gas barrier	Good	Poor	Poor	Fair
Natural colour	v. Clear	Opalescent	Opalescent	Clear (hint of yellow)
Clarity	Excellent	Poor	Some clear grades available	Good
Toughness	Excellent	Excellent	Good	Poor/good
Impact resistance	Good	Good	Good	Good
Low temperature embrittlement	Good	Good	Poor	Good
Carbonated or still products	Carbonated and still	Still	Still	Still
Hot-fill	Heat set form only	No	Yes	No

9.4.6 Plastic properties

The properties of the common plastics used for packaging depends on many factors, including:

- *Temperature*: gas permeability increases with increasing ambient/storage temperature.
- *Humidity*: atmospheric gas permeability increases with humidity in polymers that absorb moisture. This is an issue when EVOH is used as an oxygen barrier layer in a liquid drinks bottle.

- *Crystallinity*: gas permeability decreases with increasing crystallinity.
- *Orientation*: in-plane biaxial orientation decreases gas permeability.

Table 9.1 outlines the properties of plastics, but details of gas permeability data for all common packaging materials are widely available via the internet.

9.5 Closures

The choice of closure materials is generally dependent on the required processing conditions and the nature of the contents and closure system (i.e. the application and sealing method). There is a wide range of closure systems, designed to cope with internal pressure caused by carbonation, the pressure created during in-pack pasteurising or the vacuum resulting when the product cools after hot-filling. Functional closures are now used on a significant number of soft drinks, as part of the overall package to improve the usability for the consumer.

Just as important as the closure design is the neck finish to which the closure is applied. The closure and container neck must match exactly, to ensure that the performance meets the required standards. The majority of container neck finishes used with the closure systems mentioned below are defined by either industry or national/international standards, such as ISBT for PET and BS/EN/ISO for glass.

If there is sufficient volume demand, special sizes could be produced, based on a standard sealing system, but a balance would have to be achieved between the required investment, development, the validation programme and the commercial opportunity. Even a 1 mm change in the overall closure size can considerably affect the dynamics of the closures. It requires considerable expense (up to £200 000) and time (nine months or more) to produce pilot samples and complete robust performance testing, to ensure that the closure is 'right first time' in the market place. Thus, in practice, the options for a customised closure are limited, other than for colouring, embossing or printing, except for a few multinational brands with significant funds and sales volume.

9.5.1 Metal roll-on or roll-on pilfer-proof closures

Metal roll-on (RO) closures are used mainly on glass bottles, where the neck finish tolerance makes it very difficult to produce a pre-threaded closure that will ensure a reliable seal every time. The capping machine drops a plain shell (made from aluminium and lined with a sealing compound) over the (filled) bottle neck. The capping head applies a downward force sufficient to create the primary seal, and then roller wheels deform the side walls of the closure shell to the shape of the neck thread and follow the thread form over the length of available neck thread. Each closure is, therefore, effectively custom-made for its bottle.

This places a great deal of dependence for closure reliability on the bottler and the capping machine. For those drinks which are sugar-based, any spillage or breakage will affect the capping heads, which require ongoing maintenance and quality checks. The advent of more accurate necks on PET bottles has allowed the increased use of pre-formed plastic closures and simplified capping heads. In addition, as plastic closure technology has improved, it has been possible to fit plastic closures on to a wider range of glass bottles, and the use of metal RO closures has reduced over recent years.

RO metal closures come in two forms: tamper evident (TE) and non-TE. There are two TE formats: detachable TE ring and retained TE ring. If the bottles are to be returned and re-filled, it is important that the TE ring is removed at the time the bottle is opened. However, the retained type of ring has sharp edges, which may be an issue for some markets. The ring normally splits into six sections, which look like wings when opened. It is these sharp edges that could be of concern. It is also possible to push the wings back into place and pass the bottle off as being unopened.

The retained TE ring is normally used for one-trip bottles. If the design intends the consumer to drink directly from the bottle, then importance will be placed on the height of the neck finish from the TE bead. This is to ensure that the consumer's lips do not touch the sharp edge of the TE ring. If the product is carbonated and needs to be in-pack pasteurised, then consideration will have to be given to the thread form on the bottle (usually glass) neck.

There are also combination RO closures that have the metal shell, and then a plastic TE band, fitted into a holding collar shaped into the shell. This arrangement offers the advantage that the plastic band can be a different colour from the shell, and so can make tampering more evident.

9.5.2 Vacuum seal closures

Vacuum seal closures are generally of a lug seal design (sometimes called Whitecap or Twist-Off®) and made in tinplate. The closures require an internal vacuum within the package to maintain the seal and microbiological integrity, and are found in a wide range of applications, such as jams and preserves, juice drinks, sauces, condiments and baby foods. They are suitable for hot, cold and aseptic filling, pasteurisation or closed retort processes, and they come in a range of standard sizes, from around 27 mm to 110 mm. In the filling process, the product is hot-filled (typically at 85°C), and the headspace flushed with steam, in order to create a vacuum seal under the closure once the product cools and the steam condenses. In the application process, a hermetic seal is formed when the closure's lugs engage the bottle threads, and the sealing compound is compressed against the bottle sealing surface. As the pack cools, the vacuum-indicating button is drawn down when sufficient vacuum is achieved.

The closure is lined with a sealing compound which can be adapted for different product applications, including stacking and tray packing, where the top load exerted on the container may compress the seal. The closure requires a unique rigid neck finish, with non-continuous threads and specialised capping equipment. The closure typically has four points (lugs) of contacts, which means that a quarter turn of the closure removes it and reapplies it, offering the consumer significant convenience of use. There are also closures with three or six lugs, depending on the diameter. In some markets, these closures have been used on plastic bottles. However, the lug seal can distort the flexible neck, and the closure can be twisted to override the neck threads. Thus, care must be taken if using these types of closure for this application.

Tamper evidence is provided by either a circular portion in the top panel in the closure, which pops up on opening, by a secondary plastic fitment, or the use of a shrink sleeve over the closure. In the first method, a distinctive 'pop' is heard when the consumer releases the vacuum held by the closure and the vacuum-indicating button pops up, providing both an audible and a visual indication of opening. For the secondary plastic fitment, an injection-moulded plastic fitment is inserted into the closure by the closure manufacturer. This composite (plastic and metal) provides the same vacuum-indicating button, and can be applied with the normal application speed for a twist-off closure. The third format is the application of a heat-shrink sleeve over the sealed container neck, which the consumer has to remove before opening the closure in the normal way.

If a vacuum seal is required on a more flexible material (e.g. PET), then a multi-start thread should be used. This allows a 360° grip on the neck. A number of juice and sports drinks bottles are made in heat-set PET for hot-filling. Neck sizes are typically in the 38–43 mm size range, although a 48 mm size is also made. The closures for these bottles are made in plastic (normally PP), and combine the capabilities of vacuum retention and tamper evidence. The seal is designed not to increase the torque retention on removal. The design of these closures enables them to maintain the same imagery in PET, because they incorporate the same neck finish on glass.

9.5.3 Plastic closures

Plastic closures have become major components in the closures market as PET has replaced glass bottles. These closures are almost universally made with an integral tamper-evident ring. The first-generation plastic closures used for carbonated products were made mainly from high-density polyethylene (HDPE) in a one-piece design with a spigot (bore) seal (note: the most common expression is bore seal). The quality of the seal and the superior concentricity of the neck of PET bottles made it possible to remove the closure with ease. Typically, these closures have nearly two-turn thread engagement with the bottle neck. Since the spigot (bore-seal) creates its seal by entering the bore into the bottle neck, it requires a certain amount of closure rotation to remove the bore-seal from the neck and start venting the carbonation gases in the headspace of the product. By the time the closure disengages from the bottle neck, pressure in the headspace must be reduced enough not to 'blow' the closure off the bottle (sometimes referred to as missiling). This is achieved by the use of vent slots in the threads of the closure, and also in the neck of the bottle, which, together allow the gas to escape before the threads disengage.

Closures made with a liner (or wad) type sealing system break the seal faster, thereby allowing more time to vent the carbonation gases. The technology of liners has developed significantly over recent years, giving improved reliability of the seal across a wide range of bottles, both glass and plastic, and offering the potential for an improved barrier in the closure. However, the additional cost of the two-piece closures and the performance of the single-piece closures means that single-piece closures are now the most commonly used type in the marketplace.

One alternative for producing flat plastic closures is injection moulding, where the mould has two parts: a core, which has the inside features of the closure; and a cavity, which has the features of the outside of the closure. Molten plastic is injected into the water-cooled mould, the plastic solidifies, the mould opens and the completed closure is ejected. If the closure is a two-piece design (i.e. fitted with a loose or flowed-in liner), then this is added later.

Another technology for moulding plastic closures for the beverage market is compression moulding. A hot pellet of PP is positioned in a mould, similar to the cavity of an injection mould. As the mould closes, the pellet is squashed and deformed into the shape of the closure by the core of the mould. An advantage of this system is that a liner can be formed as part of the same operation; as the closure leaves the mould, a hot pellet of sealing compound enters the newly formed closure, and another forming mould shapes the liner, which is then firmly attached to the closure.

As the plastic closure market has grown, designs have been developed that are also suitable for glass bottlenecks. This increased flexibility of closures has led to a further decrease in the market share of metal RO closures in the soft drinks sector.

Plastic 'push-pull' and 'flip-top' sports closures as shown in figure 9.6 improve the multi-serve function by making opening and re-closing the closure a simple, and often single-handed, action. In some juice and concentrated drinks, pour control and/or anti-drip functionality has been transferred from the edible oil market in order to improve the consumer experience and use. Dosing closures can contain a sensitive ingredient in powder format which, when activated, is dosed into the drink at the time of consumption.

Figure 9.6 A range of 'push-pull' and 'flip-top' sports closures (courtesy of GCS and Bericap Ltd).

9.5.4 Crown corks

The crown cork has been in existence for over 120 years and, though losing popularity against re-sealable and functional closures, it is still widely used for single-serve and returnable bottles. While only suitable for smaller single-serve bottles, it is a cheap, simple-to-apply closure that is well understood by consumers.

9.6 Cans

The two-piece beverage can is used where high-volume throughput is assured, with more than nine billion two-piece drinks cans made in the United Kingdom each year. The real growth in beverage cans came with the development of the easy-open end in the 1960s, which made access to the product much more convenient. The first cans had the familiar ring-pull ends but, as the technology developed and as a response to market demand, the retained tab has become almost universally used.

In terms of market positioning, the beverage can is the epitome of an effective, standardised pack for carbonated and still drinks products. The format can be handled at very high speeds on filling lines, and its inherent strength and stackability when filled make cans ideal for high-volume products sold either in multi-packs or through vending machines. The size for the soft drinks market is 330 ml, but other sizes used are the 250 ml and 150 ml 'slim can' sizes.

The larger 440 ml and 500 ml cans, which were originally used for beers, are also used for some products – most recently, stimulant/energy drinks. Manufacturers of some premium drink products have found the 250 ml format size attractive from a retail price-break point of view, and the tall, slim size is seen as an attractive profile, particularly if targeted at the female market. Recently, however, a 66 mm diameter 250 ml 'dumpy' can has begun to gain popularity as a point of size/price difference. The 150 ml can is almost exclusively used in the travel (airline) market.

Can sizes are based around two core diameters and neck (end) sizes (see Figure 9.7 for an explanation of the terms used for can dimensions). The core dimension for 330–568 ml cans is a main body diameter of 66 mm, with either a '206'(57 mm) or (predominantly) '202' (52 mm) diameter end. The slim can utilises a 52 mm diameter body, with a 200 (50 mm) end. To explain the end size designations: 206 refers to a diameter of $2\frac{6}{16}$ inch, and 202 to a diameter of $2\frac{2}{16}$ inch.

Early cans used 209 diameter ends, but improved manufacturing has allowed reduction in the end size to 206, and now almost universally 202 diameter. This reduction in the neck diameter reduces the size of the can end, thereby saving around 17–20% on the weight of metal used in making the end.

Can bodies are made either from steel or aluminium, requiring the use of the advanced engineering and sophisticated technology of two-piece can-manufacturing equipment. The quality of the materials used is critical for the forming of metal

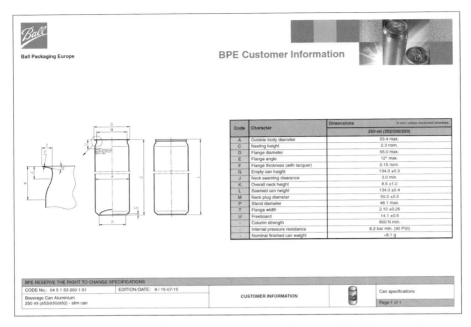

Figure 9.7 Explanation of can dimensions nomenclature (Courtesy of Ball Packaging Europe and Beverage Can Makers Europe).

beverage cans. For steel cans, a special grade of low-carbon steel is used that is coated on each side with a very thin layer of tin. The tin protects the surface against corrosion, and acts as a lubricant while the can is being formed. For aluminium cans, the metal is alloyed with manganese and magnesium to give greater strength and ductility. Aluminium alloys of different strengths and thickness are used for the can body and end (i.e. the can 'lid' or closure). Considerable development resource has gone into reducing the amount of metal that goes into making a can. Body thicknesses have been reduced by wall-ironing techniques, to the point where they are extremely thin (the wall thickness of a 330 ml steel can is less than 0.065 mm).

In the can manufacturing process, coils of sheet metal are fed into a press to form a shallow cup which, using a punch, is then forced through a series of carbide rings (the wall-ironing process), which re-draw the cup, thinning the side wall and increasing the height until the final can shape is developed. The top of the can, which is irregular at this point, is trimmed to a precise height specification, and the surplus material recycled. The can body is then cleaned, coated and decorated. After this, the can body has its open end reduced in diameter, or 'necked in'. The top is then flanged outwards, to accept the end or lid, to close the can after filling.

Two-piece cans are decorated 'in the round', in one pass, after they have been formed. Traditionally, the cans have been coated externally with a clear or pigmented (normally white) base coat, which forms a good surface for the printing inks. After drying, the cans are typically printed with up to eight colours. Later developments print the inks directly onto the can, then apply a protective over-varnish.

The base of the can body is also protected by the application of a coat of varnish. Further protection is added by spraying the internal surfaces of the can with a lacquer to protect the can from corrosion and its contents from any possibility of interaction with the metal.

The can ends are made from aluminium, because it can be formed into complex shapes and offers good opening quality; the strength of aluminium can maintain pack integrity up to the point of consumption. Steel easy-open ends have been tried for the beverage market several times, in order to have a single material on steel cans, but the poor formability and functionality of steel has prevented the commercialisation of any functional steel can ends.

The manufacturing of the can end is an even more complex metal forming process. Figure 9.8 shows a can end illustration and typical dimension nomenclature. The material is supplied in a coil of special alloy aluminium sheet that is fed through a high-speed press, which stamps out end 'shells'. The edges of each shell is curled, and a compound sealant is applied into the curl. In the next step, the shells are fed into a sophisticated conversion index press, which carries out two processes with each stroke – the formation of the tab, and the formation of the finished shell. The pull tab is formed from a separate, narrow coil of aluminium. The strip is first pierced and cut, with the tab being formed in two further stages before being joined to the shell to form the finished end. The shells themselves move through a number of stations to create the rivet to hold the tab, mark an identification on the end, form the panel and mark and score the opening or mouthpiece of the end. At the final position, the shell and tab are aligned, and the rivet swaged to join the two together.

In smaller, specialised markets, cans may be made using three-piece can technology, comprising a can body and two ends. The body is made from tinplate, as is the (plain) base end. The easy-open top end is still made from aluminium. Since three-piece can technology uses flat-sheet tinplate, a range of high-quality printing options are available to the designers. This means, for example, that they can use photographic quality reproductions, create surface effects by mixing areas of gloss and matt, or use special effect varnishes that produce textured finishes. Close registration of the graphics is possible which, combined with surface effects, can produce a striking visual result. The sheet can also be printed in more than one pass.

However, developments in digital printing and improvements in the decorators have resulted in significant improvements in the decorative quality of two-piece cans, to the point where they can almost match three-piece can printing. Also, developments in inks have allowed the introduction of textures on two-piece cans.

It is not unusual to see shaped (two-piece or three-piece) cans in certain markets, such as the beer or syrup markets as shown in figure 9.9. Cans are shaped either mechanically by die shaping, or by hydraulic or air pressure,

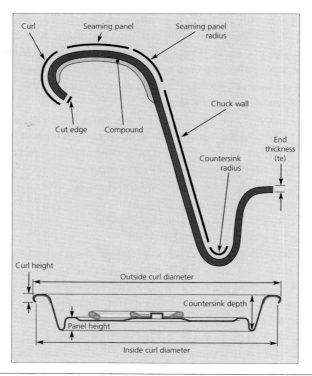

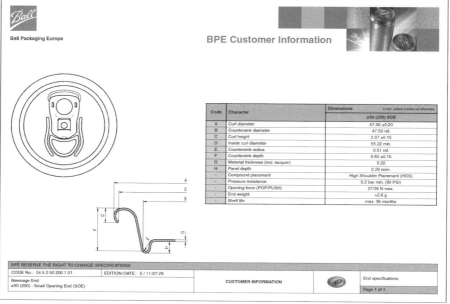

Figure 9.8 Can end dimension nomenclature (Courtesy of Ball Packaging Europe and Beverage Can Makers Europe).

Packaging materials 217

Figure 9.9 Teisseire shaped cans (courtesy of Britvic Soft Drinks Ltd).

expanding the can internally against an exterior moulding shape. Mechanical shaping is the oldest and most widely used shaping technology, where a forming tool inside the can body opens and shapes the can body according to the tool geometry. This system has advantages of lower investment costs and higher production speeds, but the tooling can leave marks and cause damage to the internal lacquer. The hydraulic or air pressure systems are similar to bottle blowing, and the 'blank' finished printed can is blown into a mould. During this blow moulding process, no tooling is required inside the can, so that the integrity of the decoration and internal coating of the can is fully kept intact, but the process is slower and more capital investment is required.

Protecting the can from the effects of corrosion is very important, in order to achieve the very long shelf life that metal cans offer. Discussion with the can manufacturer will ensure the use of the appropriate grades of lacquers on the internal surfaces to prevent primary corrosion. The external surfaces must also be considered, in order to prevent secondary corrosion, which will result in leakage. Handling on the filling line must ensure the smooth flow of cans and eliminate any sharp objects that may scratch or pierce them. A lot of moisture is present on the line, because of the use of conveyor lubricants or from the pasteuriser, and it is important that cans are dried before being packed, especially if they are to be shrink-wrapped. Cold cans, below the dew point, must be warmed to ambient temperature, or else they will become wet when packed.

It is not only the can body that has to be kept free from corrosion; the easy-open end must also be protected. This end is scored through to create a locally thinned line of metal – the tear line when the can is opened. If this area of the can is subject to stress corrosion, then rapid leakage can occur. Naturally, storage conditions such as daily temperature fluctuations and high humidity are also important.

9.6.1 Metal bottles

Aluminium bottles have recently emerged for specialist markets; however, they are still relatively expensive. The manufacturing process is almost identical to cans, with the bottles requiring internal lacquer to prevent interaction between the product and the metal of the bottle.

Some bottles use a standard crown cork but, where a screw closure is used, the closures for these packs is often specialised, having different thread geometry and requirements from PET or glass closures. Thus, supply can be limited and costs higher than a standard closure for PET or glass bottles.

9.6.2 Plastic cans

With the significant development of PET in the beverage market, even the can format has been challenged. Some years ago a significant effort was made to develop a PET two-piece can, using similar technologies to metal cans, to produce high-output plastic cans. Unfortunately, the shelf life of such a small container, with a high surface-to-volume ratio and high product costs, proved to be a major drawback, and the format was never widely adopted.

9.7 Cartons

Typical brick-shaped laminated board cartons (often named 'briks') have been available for 70 or more years for liquid products. Initially developed for the dairy market, they are still a major packaging format for still drinks and, particularly, fruit juices, either in small single-serve or larger multi-serve versions. A significant factor is that carton systems are available and well understood by consumers worldwide, making global production of new products much easier.

The development of aseptic technology has radically opened up new possibilities for efficient and economical handling of perishable liquid products. Aseptically processed and packaged juices and liquid milk will keep for months at room temperature in sealed packs. The key advantages of aseptic cartons are: they can be stored and distributed without refrigeration; they keep for a long time with no added preservatives; and the quality of their contents remains outstandingly high. The rectangular brick format makes very effective use of materials and economical use of bulk volume in palletisation, distribution and shelf utilisation in retail units. All surfaces of the carton are also available for branding images, as well as for the ever-increasing regulatory labelling requirements.

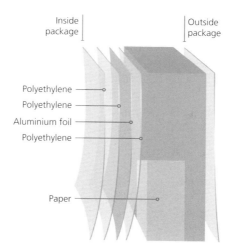

Figure 9.10 Typical carton composition (courtesy of Tetra Pak).

The packaging material is a laminate made from layers of paperboard, plastic (polyethylene) and aluminium foil. The paperboard is the main material in a carton, providing stability, strength and the decorated surface. The polyethylene protects against outside moisture, and bonds the paperboard to the aluminium foil, which protects against oxygen and light which would cause deterioration of product quality a typical carton composition is shown in figure 9.10.

One limitation of cartons is their lack of integral structural strength, which requires a greater degree of secondary packaging for protection, to ensure effective supply chain distribution due to the limitations.

Two basic systems are available, but the finished packs are similar in appearance.

In the first process, developed by Tetra Pak, a web of laminated material is fed through the filling machine, which sterilises the material in a bath of hydrogen peroxide. It is then dried with sterile hot air before the filling zone. The web is kept in a sterile area as it passes through the machine, where it is formed into a continuous tube (Figure 9.11a). The sterilised liquid to be filled flows into the tube, which is then sealed through and below the level of the liquid, using heated bars. This produces a continuous flow of sealed pouches that are completely full, with no headspace. The machine then uses formers to shape the brik container along predefined creases in the laminate. Finally, the top and base 'ears' are folded down and stuck to hold the brik in shape. For products that need to be shaken before opening, the pouch cannot be filled completely, with a headspace left in the pouch. This is achieved by sealing the package through the product flow before the pouch is completely filled.

The other process, developed by SIG Combibloc, uses pre-cut and creased laminate, welded along the vertical seal (to form a box open at each end), which is folded flat for delivery to the filling line as shown in figure 9.11b. In the filling

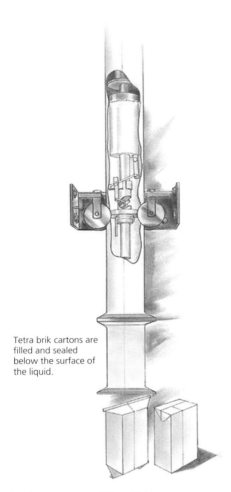

Figure 9.11a Carton production (courtesy of Tetra Pak).

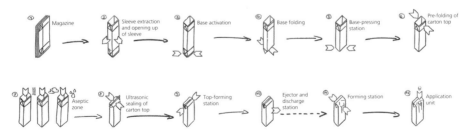

Figure 9.11b Carton production (courtesy of SIG Combibloc).

machine, the flat carton blanks are automatically decanted from the shipping containers, then opened and placed on a forming mandrel, where the base is sealed. The open-topped carton is then transported to an aseptic chamber, sterilised with hydrogen peroxide, dried with sterile hot air and filled with sterilised liquid, before the top is folded and ultrasonically sealed above the level of the product. Steam and or nitrogen may be injected into the headspace before closing, depending on product requirements.

The sterilised liquid, filled in this way, may range from plain milks or juices, up to highly particulate products such as chunky soups and chopped tomatoes. A recent innovation has seen fruit drinks containing real fruit pieces, or cereal to add fibre, being made possible.

In both systems, a product can be given its own distinctive image by choosing a particular combination of package volume, shape and printing techniques. Traditionally, only a limited number of pack variations were possible on one filling machine, but recent developments have allowed an increased range of pack sizes and options to be available from one filling line. Thus, careful consideration of the pack sizes chosen can negate high capital investment.

Pack openability was an issue for cartons in the early years, which used various forms of perforated corners to open and produce a spout for pouring and form a rudimentary closure for storage. However, this system was not easy for people with limited hand mobility, so many resorted to scissors or knives, with resulting injuries. This led to the development of a range of pull tabs and plastic closures, to give ease of access, as well as improved re-sealability. Single action, re-sealable screw closures are now a common place addition to many cartons.

The usual opening system for the smaller single-serve pack sizes is a drinking straw. During the manufacturing process, a small hole is made in the paperboard before lamination to reveal the PE/foil barrier layer, and a plastic drinking straw is attached to each package. The consumer removes the straw from the carton and its own sealed wrapping, and then punctures the package using the angled end of the straw, and pushes the straw into the package. The size of the hole in the carton is such that it holds the straw tightly and minimises leakage. This is a simple and effective package, although it does require the consumer to be aware of how to handle a less rigid package while attempting to gain grip, to press home the drinking straw.

9.8 Flexible pouches

Flexible pouches offer new opportunities for many still juices, and are increasing in popularity for sports and children's drinks. Some consumers perceive them as modern, and differentiating the product in a general beverage market. Others will see the unbreakable, collapsible format as being suitably 'sports' oriented.

Initially these were simple flat packs, and the product was consumed by piercing the pouch with a straw. However, recent developments include the inclusion of plastic closures to give a simple re-closable system, and machine systems which produce shaped packs that are able to be free-standing.

Either a cold or hot-fill system, pouches can be made in a variety of sizes without a large capital cost, because most pouch fillers support the pouch by the neck and therefore one machine can be used for all sizes. Since the film laminate is decorated while flat, most printing systems are available to the designer, as are the lamination formats widely used for other pouch or sachet types. The main material is metallised PET, laminated to aluminium foil. Low density polyethylene (LDPE) is used as the sealing medium. The products need to be packed aseptically, or hot-filled to permit ambient storage for fruit based or unpreserved drinks. For high-temperature retort processing, a polypropylene-laminated film is also available.

The main advantage of pouches is that they are lightweight. However, they are susceptible to puncture damage and, having no inherent stacking strength, are not self-supporting. This requires that the distribution packaging must provide all the support and protection required. Thus, the secondary packaging will be more substantial than for a bottle, as it must support the full pack weight, which may be considerable on the lower layers of a palletised load.

9.9 Multipacks

Recent years have seen a significant increase in the number and size range of consumer multipacks. Sold mainly in the supermarkets and discount stores, they form a popular pack for convenience and value, and are almost mandatory for a mainstream brand. This can bring financial challenges where consumers expect a multipack to be cheaper than the equivalent number of single packs, with the additional cost of the multipack to be included. One option to mitigate this is to use a smaller primary pack for a multipack, compared with the single sale pack.

There are two popular formats – carton board or shrink film – with carton board packs being either a wrap or full enclosure style. A range of multi-pack styles is shown in figure 9.12. The machines to form these packs are dedicated to one style/material, and one machine can be capable of producing a range of sizes – although, generally, machines capable of producing the smaller sizes (4pk) are not capable of producing the largest sizes (18pk and larger). Therefore, if a wide range of sizes is required, two machines may be required.

Shrink film packs use printed polyethylene shrink film, applied in register to a collation of packs to form a simple, low cost pack. However, once opened, the pack loses its integrity, so consumers may not find them convenient where they will be stored for a long period. They are more suitable where simplicity and customer shopping convenience is a priority, and the primary packs will be consumed over a short time. For larger packs, a self-adhesive tape handle can be applied to the finished pack, using a dedicated handle applicator machine.

The wrap style of board pack has similar consumer limitations, although it is possible to remove individual packs by careful tearing the board. Full enclosure board packs are effectively a small case, and can include features such as integral handles, or opening and dispensing flaps, to allow the individual primary packs to be removed easily while maintain the integrity of the multipack.

Figure 9.12 Range of multipack styles (courtesy of Britvic Soft Drinks Ltd).

9.10 Secondary packaging

Secondary packaging plays an important, but often underestimated, role in collating, protecting and providing identification for and about the primary packaging and product that it supports. In wholesale outlets, such as a cash and carry, the secondary package becomes the primary package and, therefore it must work at that level – that is, to be eye-catching, functional and representing the added value of the consumer units it carries. The secondary packaging may need to be a sales enhancement feature in its own right, or it may have to be unobtrusive enough not to distract from the decoration and marketing message on the primary package. It is disappointing to see a well-decorated primary package half hidden by a deep, plain corrugated tray. On the other hand, a well-selected and highly decorated tray can provide a visually attractive point-of-sale item.

The trade outer is often forgotten about at the time the primary package is designed. In fact, the reverse practice should be followed and, before finalising the primary pack design, the fit to the transit case and fit to pallet is checked because, if the packs do not fit the case/pallet, any changes will be extremely costly and time-consuming. Ensuring efficient case and pallet loading through the supply chain will give significant economic benefits and will minimise transit damage.

For many plants that supply a European market, pallet sizes are often a source of complexity, because collated primary packs are not modular for the whole pallet range. It is interesting that many drinks carton dimensions have been chosen to be modular, and this should also be done for other pack formats. By taking secondary packaging into account early in the packaging design and material selection process, a speedier and more economical result will be achieved.

In many multiple-store distribution networks, pallet loads are broken down into smaller, mixed-product units or cages, to be distributed to individual stores. This means that trays or cartons of primary packs may be stacked (possibly unevenly) above or below heavier items. Thus, care must be taken in selecting the secondary format. If the primary pack is unbreakable, then the role of its corrugated tray should be closely examined. Is it there just to collate packs prior to shrink-wrapping, or does it provide a valuable feature at the point of sale? If only the former, can it be dispensed with? Filling lines, however, usually have case or tray erectors that will handle only certain sizes or formats. It is important to understand which of these are available when undertaking the feasibility and capability phases in the product development cycle.

Modern automated warehouses have quite critical requirements concerning the loading of the pallets, and will not tolerate any significant overhang. A target of ± 10 mm to the nominal pallet base size is a good starting point. There are a number of computer programs that will calculate pallet configuration from the unit pack dimensions, and which can facilitate the rapid analysis of a number of options. Before committing to a selected palletising plan, it is advisable to check the capability of any existing palletising equipment to actually reproduce the pattern at operating speeds.

9.11 Pack decoration

This section covers package decoration formats, rather than the detail of their printing. Packaging that is delivered pre-printed, such as cans, drinks cartons and pouches, are covered in sections 9.6 to 9.8 above. Decoration formats include labels applied to a pack, and heat-shrink plastic sleeves, which may be applied through labelling technology or as a pre-made sleeve that is shrunk into place.

Labels can be made of paper or plastic, and are usually applied to the filled package by a labelling machine, either before or (more commonly) after filling. Labels must be applied to a flat or singularly curved surface. Applying a label to a compound curve will result in the label creasing and looking unsightly. The use of clear plastic labels can give the appearance of the print being directly onto the primary pack – a 'no label' look.

Labels can either be patch, a small label only partially around a container, or full wrap, where the label ends overlap. For patch labels, more than one can be applied to a container, such as a front, back and neck label. Patch labels can be square, rectangular or shaped (punched) into a particular profile. Complex shapes are not recommended, due to the difficulties in applying at high speeds. Wrap labels are generally rectangular, but some profiling of the upper and lower edges is possible. If the package has been filled below the dew point, then the package is likely to be wet by the time the label is applied. The use of an air knife to remove excess water and/or special adhesives should be used.

Label application is by one of three systems:
- The first is 'cut and stack', where pre-cut labels are fed from a magazine, then a labelling machine applies the adhesive and applies the label to the package. For small patch labels, one machine applies all labels in one operation. Patch labels will typically use a traditional 'wet glue', with wrap-around labels using hot melt. Cut and stack labellers are generally used for paper labels, but plastic full wrap labels can be applied with this system.

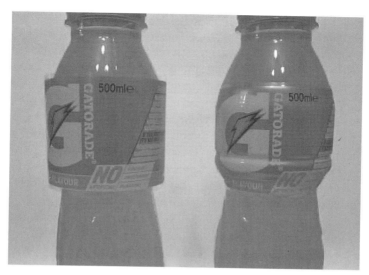

Figure 9.13 Shrinkable label, as applied and shrunk to a bottle (courtesy of Britvic Soft Drinks Ltd).

- The second system is reel-fed, where the labels are supplied to the labelling machine in a continuous reel. The machine cuts the label using a registration mark printed on the label, applies the adhesive and applies the label to the package. A variation on this system uses a reel-fed shrinkable plastic label, applied as a normal label, then shrunk to the bottle contour (see Figure 9.13).
- The third system uses self-adhesive or pressure-sensitive labels, supplied pre-cut on a reel of backing material. The machine removes the label from the backing material and applies the label to the package.

Labels must be applied to a flat (or conical) section of the bottle. To apply a 'label' to bottles with compound curves (i.e. curved in two dimensions), a shrink sleeve is required. Pre-made heat-shrink sleeves are widely used on beverage bottles for all-over decoration that can make a plain bottle look attractive. This system also allows one basic bottle, to be used for more than one product or more than one flavour – a potential logistical saving.

Manufactured from heat-shrinkable materials, the plastic shrink sleeve is supplied either as a flat tube, either as pre-cut sleeves or in a continuous reel similar to reel fed labels. The sleeve is placed over the bottle and shrunk using either hot air or steam. The sleeve design is dictated by the primary pack, with the print controlled or even pre-distorted when printed, so that it appears normal when shrunk to the pack contours.

9.12 Environmental considerations

The environmental impact of packaging is an emotive and important area for discussion. Waste and 'green' pressures are increasing significantly, with continuous requirements for material reductions, recycling and/or renewable materials.

Most general litter seen on the streets is packaging and, due to the bright colours used, it is usually highly visible. This is further fuelled by the regular media coverage about packaging, which drives general public perception that most waste is packaging, and the opinion that packaging is generally excessive and wasteful, rather than fulfilling a very beneficial function in getting products to the point of sale in good condition, preventing spoilage, damage and pilferage. The mantra of 'Reduce, Reuse, Recycle' is becoming more critical in the world of beverage packaging. The main focus in the UK beverage market over the last few years, however, has been on material reduction and recycling, as returnable/refillable packaging has largely died out, driven by market forces and supply chain changes.

Returnable bottles for beverages can have applications where there is a closed-loop supply chain, to ensure an effective (<80%) rate of return of empty containers and energy use in returning and cleaning the containers ready for reuse. Such containers are usually much heavier than single-trip containers, and there have to be sufficient numbers to service the total loop. Within Europe, there are a large number of returnable/refillable bottles (mainly glass, although PET has made inroads into this market). Unlike glass, PET is permeable, and rigorous checks have to be undertaken on returned bottles to ensure that the bottles have not been filled with noxious products, or other drinks with strong flavourings that will be retained by the PET. Filling lines have automatic 'sniffers' that detect noxious or strong odours and, if found, the bottle is rejected.

For recycling, glass is a very good material, and a wide range of local collection and recycling schemes ensure a high recovery rate – in some cases, too good, resulting in an excess of green glass recyclate, which is re-used as road in-fill. Aluminium and steel are also recognised as good materials for recycling, again with many schemes supporting the recycling of beverage cans.

Plastic bottle recycling has seen a number of schemes and processors emerging, where plastics are recovered and re-processed into food-grade materials. However there are still issues to overcome regarding separation and contamination, as beverage products are susceptible to flavour taints and migration from the pack. Further research and trials are required to develop this area.

The multi-layer structure of cartons used to be perceived as an environmental issue. However, much has been done to develop recycling technology for cartons, with a number of recovery systems now in place. The valuable virgin fibre paperboard component can be easily separated for recovery and recycling, while the aluminium and plastic can be used for generating energy or a number of other applications, including aggregation to produce feed stock pellets for garden furniture and roofing materials. Life cycle assessments have shown cartons to have one of the lowest environmental impacts of any beverage package.

Whatever the pack material, when designing new packaging, the ability for the pack to be sorted, separated and recycled must be included, as well as using materials that promote recycling and that are easily recycled in the mainstreams. WRAP (Waste & Resources Action Programme), INCPEN (Industry Council for Packaging and the Environment) and RECOUP (Recycling of Used Plastics) have a number of documents regarding recycling and recyclability which give guidelines for recycling, sustainability, and so on.

Weight reduction of the packs is often seen as the easiest way to increase the environmental credentials of a pack. However, simply light-weighting the primary pack can affect the performance of the transit packaging through the supply chain, which may then require an increase in weight, leading to a net increase in materials (and cost). It is imperative to investigate and evaluate the lightweight pack fully through the supply chain, including any potential changes in secondary or tertiary packaging. It is the overall supply chain cost/benefit that is important.

One area of weight reduction for plastic is the closures and the neck finish, where both have been reduced in height. There are a number of recent developments where reducing material in both the closure and the neck finish saves around 2 g of plastic per bottle. Figure 9.14 shows one example – two neck finishes used for carbonated drinks, where the previous (1810) finish is being replaced by the lighter (1881 finish).

Half-litre bottles for still water which, a few years ago, weighed 20 g, are now typically around 15–18 g, with some as low as 10 g being found, although the functionality in holding and drinking from these ultra-light bottles is a subject for debate.

Another recent development are bioplastics, derived from renewable biomass sources, such as vegetable fats and oils or starches, rather than fossil-fuel plastics derived from petroleum. However, there is still considerable debate and discussions

Figure 9.14 1810 and 1881 neck finish heights.

regarding their benefits over traditional plastics. They are still reliant upon petroleum to power farm machinery and irrigate growing crops, to produce fertilisers and pesticides, to transport crops and crop products to processing plants, to process raw materials and, ultimately, to produce the bioplastic. There is also the question regarding diverting agricultural output from food to fuel and plastics.

Whatever one's point of view, packaging must be used effectively to preserve and protect the product. The energy used to produce packaging materials is much less than that used to produce the product they contain, and that used to distribute the product. If the product were to become unfit for sale due to the packaging, it would be a far greater waste of energy.

9.13 Conclusions

Packaging for liquid beverages is a large and complex area, which continues to develop at an ever-increasing pace to maintain and increase market share. This chapter covers the basic topics in an informational way, and is not intended to be a comprehensive document. For more details, it is recommended that the reader contacts the various manufacturers, or one of the relevant institutes or trade associations. Details of some of these are given below. Almost all have comprehensive websites, which give a considerable amount of information as well as contact details.

The Packaging Society (formerly the
   Institute of Packaging)
A division of The Institute of Materials,
   Minerals and Mining.
1 Carlton House Terrace
   London
SW1Y 5DB
Tel: +44 (0) 20 7451 7300
www.iom3.org/content/packaging

RECOUP (Recycling Of Used Plastics)
1 Metro Centre
Welbeck Way
Woodston
Peterborough
PE2 7UH
Tel: +44 (0) 1733 390021
www.recoup.org

INCPEN (Industry Council for
   Packaging and the Environment)
Soane Point
6-8 Market Place
Reading, Berkshire
RG1 2EG
Tel: +44 (0) 118 925 5991
www.incpen.org/

WRAP (Waste & Resources Action
   Programme)
2nd Floor
Blenheim Court
19 George Street
Banbury
Oxfordshire
OX16 5BH
Tel: +44 (0) 1295 819900
www.wrap.org.uk

British Soft Drinks Association
20-22 Bedford Row
London
WC1R 4EB
Tel: +44 (0) 20 7405 0300
www.britishsoftdrinks.com

The Can Makers
Westminster House
Kew Road
Richmond
Surrey TW9 2ND
Tel: +44 (0) 020 7437 0227
www.canmakers.co.uk

British Glass
9 Churchill Way
Chapeltown
Sheffield
South Yorkshire
S35 2PY
Tel: +44 (0) 114 290 1850
www.britglass.org.uk

British Plastics Federation
BPF House
6 Bath Place
Rivington Street
EC2A 3JE
Tel: +44 (0)20 74575000
www.bpf.co.uk

CETIE (International Technical Centre
  for Bottling and related Packaging)
112–114, rue La Boétie
75008 Paris
France
Tel: +33 1 42 65 26 45
www.@cetie.org

British Bottlers Institute
53 Basepoint
Caxton Close
Andover
Hampshire
SP10 3FG
Tel: +44 (0) 1264 326478
www.bbi.org.uk

MPMA (Metal Packaging
  Manufacturers Association)
The Stables
Tintagel Farm
Sandhurst Road
Wokingham
Berkshire
RG40 3JD
Tel: +44 (0) 1189 788433
www.mpma.org.uk

The Institute of Packaging
  Professionals
1833 Centre Point Circle, Suite 123
Naperville
Illinois 60563, U.S.
Tel +1 (630) 544-5050
www.iopp.org

PAFA (Packaging and Films
  Association)
2nd Floor, Gothic House
Barker Gate
Nottingham
NG1 1JU
Tel: +44 (0) 115 959 8389
www.pafa.org.uk

ISBT (International Society of
  Beverage Technologists)
14070 Proton Rd
Suite 100, LB 9
Dallas TX 75244-3601
USA
Tel: +1 972 233 9107 xt208
http://www.bevtech.org

Acknowledgements

The author would like to thank the following companies for their help in supplying information and pictures for this chapter:

Krones AG	www.krones.com
KHS Corpoplast GmbH	www.khscorpoplast.de
Plastipak	www.plastipak.com
Ball Packaging Europe	www.ball-europe.com
Tetra Pak	www.tetrapak.com
SIG Combibloc	www.sig.biz
Britvic Soft Drinks Limited	www.britvic.co.uk
Global Closure Systems	www.gcs.com
Bericap Closures	www.bericap.com

CHAPTER 10

Analysis of soft drinks and fruit juices

David A. Hammond

Fruit Juice and Authenticity Expert, Wolverhampton, UK

10.1 Introduction

Although marketing and 'image' have a lot to do with a consumer's desire to try a product, if they do not like the taste they, will not purchase it again. In addition, if the quality of the product is variable, it will not live up to the consumer's expectations and, once more, it is unlikely that they will purchase it regularly. This means that it is critical to control the quality of the product. This can be done in a number of ways, including sensory assessment and analysis of certain key ingredients, such as sweeteners, acidity, colour and, in a carbonated soft drink, the level of dissolved carbon dioxide. If a formulation uses high-intensity sweeteners, preservatives or colours, there are legal limits set for most of these materials, so it is critical to ensure that their concentration(s) do not exceed the legal maxima.

The fortification of some food products, such as cereals, has been common practice for many years, but the addition of vitamins to soft drinks has been less usual. However, with the drive for a healthier lifestyle, it has become more common. This was also partly driven by the phenomenally successful launch of the Sunny Delight brand in the UK. This product was marketed on its benefit of containing three vitamins – A, C and E – and this led to the launch of a number of 'me, too' products. However, the sales of Sunny Delight were not sustained in the long run, and the product had to be re-launched with a much higher juice content in order to meet more closely the consumers' expectations that the product had previously offered. With fortified products, there is an analytical need to ensure that the vitamin levels claimed on the label are met at the end its shelf life. These are only a few examples of why the analytical control of ingredients in a product is critical.

There is a growing interest in beverages with associated special benefits. Such products will often contain sugars for energy, caffeine for stimulation, vitamins, amino acids and, possibly, herb extracts to impart some 'unique selling property'

Chemistry and Technology of Soft Drinks and Fruit Juices, Third Edition. Edited by Philip R. Ashurst.
© 2016 John Wiley & Sons, Ltd. Published 2016 by John Wiley & Sons, Ltd.

(USP) or 'health' benefit. A very good example of this type of product would be Red Bull. According to its website:

> 'Red Bull Energy drink is a functional beverage. Thanks to a unique combination of high quality ingredients, Red Bull Energy Drink vitalizes body and mind.'

There is also an increasing number of beverages that make wonderful claims for themselves due to the 'superfruits' they contain, for example pomegranate, acai and mangosteen and/or herbs. These complex formulations can pose a real analytical challenge, to prove that the special ingredient/s is/are actually present. At the end of 2010, regulators on both sides of the Atlantic took action to limit some of the rather outlandish claims that were being made for some of these beverages, as the regulators felt that they were misleading the consumer. It will be interesting to see if Red Bull and other products will still be able to maintain their claims in the EU in the longer term, in light of the 2006 Nutrition and Health Claims Regulations 1246 (Anon, 2006b), which is discussed below.

With the widespread use of computer-controlled production and packaging lines within factories, there is now little chance of a mistake being made in a formulation, so a limited set of equipment is now usually found in the quality assurance (QA) laboratory. This would probably consist of: a refractometer, for checking soluble solids content (as °Brix); an autotitrator for acidity; an instrument to check carbonation levels, if dealing with 'fizzy' drinks; and some equipment for checking the microbiological integrity of the product. There may even be less testing equipment in some factories. However, if there is a problem with a 'bottling syrup' in one of the modern factories which is not detected before packing, this is a very expensive mistake to rectify!

In a juice-producing factory, additional equipment, such as a gas chromatograph with mass spectrometric detection (GC-MS) or high performance liquid chromatography (HPLC) may be found, to ensure that the company's products conform to 'normal' standards. Depending on the factory's size and the company's culture, these checks may be handled internally, or externally by a third-party laboratory that specialises in this type of analysis. The choice of internal/external analysis will depend on many factors, which will include the firm's culture and the economics of running its own laboratory.

There have been many important developments in the chemical analysis area over the last 15 years, since the first edition of this book was published. HPLC systems have developed extensively, and will now typically be computer-controlled, thus requiring less-skilled operators to use them. They are also probably cheaper than they were, even allowing for inflation, so they are much more commonplace. One major development has been in the use of HPLC linked to a mass spectrometric detector (HPLC-MS). Although GC-MS was common in 1998, HPLC-MS was still in its infancy; however, over the last few years, there has been much activity in this area by the instrument manufacturers to make such equipment easier to use. They now can be supplied with search libraries and the like, allowing for easier interpretation of data, and it is not uncommon to see, in some 'official' analysis, for regulatory purposes, that the method prescribes the use of MS for confirmatory purposes. This is very common in veterinary residue analysis, especially if dealing with prohibited substances such as chloramphenicol.

There have also been developments to reduce the analysis times and solvent usage, by using specialist liquid chromatography (LC) columns with very small particle sizes, and pumping systems with small delivery volumes, such as the Waters ultra performance (UPLC) or Dionex UHPLC systems. This type of enhanced HPLC offers much shorter analysis times, 25–50% of that for a 'normal' HPLC procedure, and smaller solvent usage per analysis. This also offers an advantage when interfacing the HPLC system with a mass spectrometer, as there is less solvent to remove before introduction into the high-vacuum area. Some manufactures offer software to assist analysts in converting their existing regular HPLC procedures to one suitable for use on an ultra-performance LC system.

Since publication of the first edition of this book, a number of validated methods have been published for the analysis of soft drinks ingredients. When the first edition was published in 1998, there were only a handful of methods for the analysis of soft drinks ingredients that had been collaboratively tested in the AOAC official methods manual, and only two of these used an HPLC approach.

One of the advantages of membership of the AOAC is online access to the Official Methods Book (OMB) over the internet, which can be extremely useful. In 1998, no methods were found in the British Standard catalogue. Inspection of the British Standards website (http://shop.bsigroup.com) now shows three standardised approaches which use HPLC for the analysis of high intensity sweeteners in soft drinks. The most recently published procedure, from 2010 (BS EN-15911), uses a light-scattering detector to determine nine high-intensity sweeteners. This is a very useful, highly sensitive, non-specific detector that has been added to the analyst's armoury. It can be used for the analysis of many compounds that do not contain a useful chromophore in the UV or visible regions, but which need to be detected at the mg/l level.

The general lack of method standardisation in the soft drinks area is probably because the testing matrix is relatively straightforward, and does not pose a lot of the associated problems that are found in other areas of food analysis. Thus, the industry and regulators have not felt the need to standardise the test methods. However, with an increased emphasis on results from accredited laboratories, this may well change in the future.

Although there are only a limited number of methods that have been specifically validated for soft drinks, a large number (≈40) of validated methods are available for the analysis of fruit juices, most of which will work equally well for soft drinks. These methods are published in the International Fruit Juice Union (IFU) handbook of analytical procedures, which offers the best reference collection of methods for the analysis of fruit and vegetable juices in the world, with new methods added on a regular basis (Anon, 2010a). The IFU's collection covers all the main procedures required to assess the quality and authenticity of fruit and vegetable juices and nectars. There is also a collection of microbiological methods that are specifically designed for fruit juices. The methods are listed on the IFU's website (http://www.ifu-fruitjuice.com) from which they can now be purchased directly.

Twenty-eight of the IFU procedures provided the basis of the European standardisation organisation (CEN) methods that were published for fruit and vegetable juices in the 1990s. These have been adopted by the EU member states as national standard methods (e.g. BS, DIN etc.). However, five other procedures have also been adopted as British Standards, so that there are now 33 procedures listed on their website (http://shop.bsigroup.com) for fruit juices and soft drinks. These methods can be purchased from the website, and should also be available from other standardisation organisations around the world such as DIN, AFNOR, ANSI, and so on. During the finalisation of the Codex standard for fruit juices and nectars (2005), a number of IFU, AOAC and EN methods were adopted as suitable procedures for the '*assessment of the quality and authenticity fruit juices*', and these are listed in the standard. A number of IFU methods have also been proposed by the Russian Fruit Juice Association, for adoption by their Government when it drafts its new fruit juice regulation. Thus, it is clear that these procedures have international recognition.

Most of the IFU procedures are applicable for the analysis of soft drinks, perhaps with some modification to recommended dilutions, for example. However, one that will require modification for use in a carbonated soft drink is the method for titratable acidity, which has to include an additional degassing step in order to remove dissolved carbon dioxide and to ensure that reliable acidity results are obtained.

This chapter will cover the methods of analysis for different key elements of a soft drink formulation or a fruit/vegetable juice (e.g. sweeteners, acids and preservatives, etc.). When ingredients or components are found in both soft drinks and juices, their analysis is discussed together. In other areas, where there is no overlap (such as for preservatives which are not permitted in fruit juices by the Codex standard and other regulations), these components are covered separately.

Although it is beyond the scope of this chapter to discuss, in any great detail, the methods used to assess the authenticity of fruit juices, a brief summary of this area is given at the end, with some recent developments that have been made since the previous edition.

10.2 Laboratory accreditation

Since the initial publication of the book, there has been a significant growth in the number of laboratories that are accredited for the analysis they undertake. This has largely been brought about by external pressure from customers for the laboratories to be able to demonstrate their competence in the analyses they offer. Adoption of the ISO 17025 standard, or its predecessor, the ISO guide 25, is an intensive and costly exercise to get an accreditation for any laboratory. As it hopefully improves the reliability of results, it should be considered as 'one of the costs of doing business' for a third-party laboratory. For an internal laboratory, within a company, it is a rather onerous standard to comply with, but there are at least two less demanding standards that can be adopted if required such, as CLAS, offered by Campden BRI Research Association, or LABCRED, is offered by Exova (Anon h and i, no date). At the time of writing, AIJN's Code of Practice Expert Group is looking at drafting some simple guidance on minimum requirements that should be expected within a Quality Assurance laboratory in a fruit juice factory.

The ISO 9000 quality standard, first published in 1987, was aimed at manufacturing sites, as a way to enhance and maintain the quality of their finished products. In 1990, ISO published their guide 25, which detailed the requirements for a laboratory to gain accreditation for their analyses. This guide, and the subsequent standard (17025), introduced and maintained the critical concept of 'competence' in the analysis, and this is the feature that distinguishes 17025 from the 9000 standard series. When the latter is checked for conformance, the system's paperwork, and how it is implemented throughout the organisation, is inspected. Guide 25 laid down similar requirements for a detailed and documented quality system, which included a copy of the method, training records for staff, maintenance records for instruments, and so on. However, it also required the laboratory to demonstrate competence in the method and data, showing the long-term performance of the procedure by use of a shewhart (control) chart. It also required, where possible, participation in proficiency schemes, such as FAPAS for external verification of the laboratory's performance.

In 1999, the ISO guide was replaced by a full standard (17025), which did a very similar job, but was slightly less prescriptive, so giving the laboratory more flexibility within the framework for its implementation. Verification of compliance with this standard requires an inspection of the quality system and the paperwork associated with it, and it also assesses the laboratory's competence in carrying out the analysis. This covers suitability of the method selected, training, and demonstrated competence of the actual laboratory staff for the particular method. It also involves a regular assessment, by onsite inspection of the system by the accreditation body, at least every other year.

Although the accreditation of laboratories seems to be more developed in Europe than it is in the USA, this aspect of laboratory quality will develop quickly in the US over the next two or three years, because of the introduction in 2011 of the new US Food Safety Modernization Act (public law 111-353, 2011). Due to the scares seen in the USA over the last few years, caused by microbiological and chemical contamination of foods, the Food and Drug Administration (FDA) decided to update their rather old food safety legislation. The new regulation incorporates a number of the existing regulations, like HACCP, across the whole food arena. Unlike Europe, HACCP had previously only been mandated in specific areas in the US, such as for meats and juices. Another aspect that the new regulation mandates is the use of an accredited laboratory for food analysis. This will force third-party and internal laboratories to adopt some type of laboratory accreditation system quickly.

Most countries have their own accreditation bodies, such as UKAS in the UK. More information about the requirements for accreditation is available on their website (www.ukas.com). One group that can assist with this topic in America is the American Association for Laboratory Accreditation, which is a non-profit, non-governmental public service membership society. Further information about their offer can be found at their website (http://www.a2la.org).

The 17025 standard is available from a national standards institute (e.g. BSI, DIN AFNOR etc.). However, a number of guidelines have been published by various bodies to assist in its interpretation, and these are often easier to read and comprehend. One such publication is the AOAC guidelines published in 2006 (Anon, 2006a).

When contracting work to outside laboratories, their accreditation scope, which can often be examined on their website, should be checked, and enquiry made about their performance in proficiency trials in which they participate. This will allow better assessment of their suitability to undertake the work for an organisation.

10.3 Sensory evaluation

The sensory assessment of soft drinks and fruit juices is discussed briefly here for completeness.

As the flavour and odour of a soft drink or fruit juice are essential elements of a product, they should be closely controlled. This is generally carried out by trained panellists, who have been screened to ensure that they have an aptitude for this type of assessment. They are often quality assurance personnel or workers from the factory. Sensory assessments should be carried out in a surrounding where the panellists can concentrate on their work without distractions.

At least every batch of finished product should be checked to ensure that it is free from off-tastes, and tastes 'normal'. For continuous production, appropriate checking should be instituted. Although the tasters should be familiar with the product's flavour, odour and mouthfeel, they should always be provided with a control for reference purposes. This will generally be the previous satisfactory production batch of product. Incoming raw materials, such as juices or juice concentrates, sugar and water, should also be assessed to ensure that they taste normal, and will not impart any off-tastes to the finished product.

10.4 Water

In a soft drink, or a reconstituted fruit juice from concentrate, the quality of water used is an essential element. Checking the water quality involves assessment of its odour and taste, to ensure that it does not contain any off-tastes or odours. It also involves checking that any water treatment processes have been effective, and that they have not introduced any defects into the water. The water should also be assessed to ensure that it does not contain materials that are likely to give a precipitate or floc in the product on storage.

The most common cause for floc formation in a clear soft drink is microbiological contamination by yeasts. If this occurs, the yeasts can be identified by allowing the floc to settle in the container, or precipitating it by centrifugation, and very gently decanting off the liquid from the precipitate. The precipitate can then be examined under a microscope, and the yeast cells will be seen as about 5 μm ellipsoids.

Another cause for a floc in a clear soft drink, particularly in the autumn, can be the presence of algal polysaccharides in the water. Although these polysaccharides are soluble in water at neutral pH, they will precipitate from solution when the water is made acidic. To check for the presence of these polysaccharides in the water, the following test should be carried out:

To a sample of raw water (200 ml), concentrated hydrochloric acid (2 ml) is added to ensure that the pH is below 2. The water is then held at 60°C for 12 hours, and examined against a dark background under strong illumination, to detect any floating particulate materials. The heating period can be extended for a few days, to ensure there is no precipitate.

On a less frequent basis, and as part of good manufacturing practice, samples of water should also be analysed more thoroughly for a wider range of components, including toxic heavy metals (e.g. arsenic, copper, chromium, cadmium, lead, mercury, selenium and zinc), polyaromatic hydrocarbons (PAHs), some of which are known to be carcinogens, and pesticide residues. If using potable water from a commercial supplier, there is less need for checks on the water, as the water company will be undertaking the testing. However, if operating in a developing country, it is worthwhile being more proactive. The levels of fluoride and nitrate ions in the incoming water should also be checked on an infrequent (6–12-monthly) basis.

The analysis for heavy metals, PAHs and pesticides residues is very specialised, and is better left to a specialist laboratory. However, analysis for fluoride and nitrate levels can readily be determined by ion exchange chromatography, linked with detection by conductivity. A typical example of this is given in the Dionex application note 25 (Anon (a), no date) and a chromatogram of a standard for this type of separation is given in Figure 10.1.

Ozonolysis is a process which can be used for water treatment to remove any microbiological contamination. However, if this approach is used it is critical

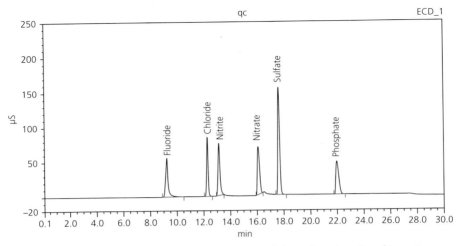

Figure 10.1 Ion exchange chromatographic separation of the main anions found in water, using a Dionex HPLC and AS-11HC column. Conditions: Column: Dionex AS-11HC 250 mm × 4 mm; solvent Milli-Q water and 3 mM NaOH for 6 minutes, then to 30 mM NaOH over 15 minutes; flow rate 1.5 ml/min; suppressed conductivity detection.

that the water be checked for bromide ions (Br^-). If bromide is present, during ozonolysis, trace levels of bromate (BrO_4^-) will be formed. This is considered harmful, and has a strict maximum level (10 µg/l) set down in The Drinking Water Directive (DWD), Council Directive 98/83/EC. Although bromate can be determined by ion chromatography (EN ISO 15061:2001), it is again best left to a specialist laboratory at these low levels. Ozone can, in some circumstances, oxidise soluble iron in water, giving rise to traces of precipitates.

It should be remember that the Fruit Juice Directive (2012/12/EC) is very prescriptive as to what additives are permitted in fruit juices and nectars. Although it is allowed to add sugar or lemon juice, or concentrated lemon juice, to most juices, this is strictly controlled, and adding both to the same juice is prohibited. Under the additives directive (95/2/EC), citric acid may be added to most juices, or malic acid to pineapple juice. It is not permitted to add both sugar and acids to the same juice. No preservatives are allowed to be added to juices or nectars, with the exception of sulphur dioxide to grape juice. If this preservative is added to grape juice, it has to be removed by physical means to below 10 mg/l before the product can be sold to the consumer. The addition of sugars, lemon juice or its concentrate, citric acid or lactic acids, are allowed to be made to nectars and, in low-calorie nectars, high-intensity sweeteners are also permitted. No colours, natural or synthetic, may be added to juices or nectars. Pectin is allowed in two juices and nectars – pineapple and passion fruit – which, sometimes, show unstable clouds. It should also be mentioned that sorbic and benzoic acids can actually be added to grape juice, but this is only for sacramental use.

The list of additives permitted in soft drinks is much broader than is allowed in fruit juices, and these are controlled by three main directives and their respective amendments: the Sweeteners for Use in Foodstuffs Directive (94/35/EC), the Colours for Use in Foodstuffs Directive (94/36/EC), and the Food Additives other than Colours and Sweeteners Directive (95/2/EC).

10.5 Sweeteners

After water quality, sweetness is probably the most important feature of a soft drink. Until 1995 in the UK, it was essential that a soft drink contained a minimum level of sugar. This level was set at 45 g/l, unless the product was listed as a 'low calorie' soft drink (Anon, 1964, UK Soft Drinks Regulations 1964). These regulations were revoked in 1995 (Anon, 1995a), and it is now possible to make a soft drink with or without added sugars, as required, provided the product is clearly labelled with its ingredients. In most fruit juices and many soft drinks, except diet varieties, sugars are a major component of the product.

The sweeteners used in soft drinks can be divided into two main categories. There are the calorific sweeteners, such as sucrose, invert syrups, glucose syrups, high fructose corn syrups and honey, and the high-intensity sweeteners, which can also be split into two classes: artificial sweeteners, such as aspartame, acesulphame-K, sucralose, saccharin and so on, and a natural high intensity sweetener, Stevia.

There has been a lot of interest in this natural high-intensity sweetener, which is extracted from the leaves of a South/Central American plant, *Stevia rebaudiana*. The use of Stevia has been permitted in Japan for many years, and has gradually been gaining regulatory approval for use in foods and beverages around the world. This sweetener is now approved for use in foods in the EU.

All fruit juices naturally contain glucose and fructose, and many also contain sucrose. The proportion of the individual sugars seen in a juice depends on the actual fruit concerned, and the processing/storage conditions to which it is subjected. Some juices, such as orange and pineapple, are rich in sucrose, whereas others, such as raspberry, only contain low levels. The typical levels of the main sugars seen in a range of juices are given in Table 10.1.

In soft drinks the type of sugar used in a formulation will depend on where the product is being produced. In EU countries, where the price of sugar had previously been adjusted to support locally produced materials, the sugar used to prepare soft drinks would often be sucrose, or an invert sugar syrup prepared from beet or cane. In the USA, most soft drinks are prepared from syrups derived from starch, such as a high-fructose syrup prepared from corn (HFCS), as these syrups are cheaper than sugar derived from cane or beet. However, as there has been a lot of bad publicity around HFCS and its purported link to increased incidences of obesity, there has been a move away from this type of sweetener in some drinks.

Table 10.1 Typical levels of the main sugars seen in a range of fruit and vegetable juices.

Juice	Sucrose (g/l)	Glucose (g/l)	Fructose (g/l)	Glucose/fructose ratio
Apple	15	30	60	0.3–0.5
Apricot	30	50	20	1.0–3.0
Blackberry	Trace	40	40	0.95–1.05
Blackcurrant	Trace	27	36	0.6–0.9
Carrot	40	10	8	0.9–1.5
Cherry (sour)	5	40	35	1.0–1.4
Cherry (sweet)	Trace	65	60	0.9–1.3
Cranberry (European)	<1	18	14	1.5–2.5
Cranberry (North American)	<1	28	8	3.0–5.0
Grape	<1	80	80	0.9–1.03
Grapefruit	25	30	30	0.9–1.02
Lemon	2	8	8	0.95–1.3
Mandarin	45	20	20	0.95–1.0
Orange	40	22	22	0.9–1.0
Passion fruit	25	35	35	0.95–1.2
Pear	10	25	70	0.2–0.4
Pineapple	45	23	23	0.9–1.1
Pomegranate	Trace	60	65	0.8–1.0
Raspberry	<1*	25	28	0.6–0.9
Strawberry	6	25	28	0.8–0.95
Tomato	<1	12	15	0.8–1.0

* Can be seen at higher levels in juices/purees that are rapidly heated after pressing by elimination of the natural invertase.

High-intensity sweeteners are used in diet formulations, but may also appear in regular soft drinks. Their use depends on the requirements of the manufacturer, and also on the relative price of sugar and the high-intensity sweetener(s) involved. This substitution of sweetener for sugar was particularly common in the UK in the 1970s, where part of the sweetness of a product was often provided by the high-intensity sweetener saccharin. This was used to control the cost of the formulation, as sugar was a major part of the product's formulation cost.

10.5.1 Analysis of natural sweeteners

The natural sweeteners used in soft drinks formulations are generally the same as those found in fruit juices (i.e. sucrose, glucose and fructose). As sugars are generally the second largest component in a soft drink or fruit juice after water, one of the quickest and simplest ways of assessing whether a product is within specification is to measure its refractive index using, say, IFU 8. This can be achieved using a simple hand-held refractometer, costing about £100, or a sophisticated temperature-controlled refractometer, which costs several thousands. The soluble solids content can also be determined by measurement of the density of the product using a densitometer, such as that manufactured by Anton Paar, and a procedure for this is detailed in IFU method number 1A (Anon, 2005a).

The assessment of the refractometric solids content, generally referred to as Brix, is one of the basic tests that is carried out in today's soft drinks and fruit juice factories to assess the soluble solids content of an incoming sugar syrup, a soft drinks bottling syrup before dilution, fruit juice/fruit juice concentrate, or a diluted finished product. This measurement of soluble solids content is only an estimate, as it assumes that everything in solution has the same refractive index as sucrose. As the refractive index of citric acid solution is lower than that of sucrose, it is normal practice to adjust the refractometric solids content of citrus juices for its acidity by application of a correction factor. This correction factor can be calculated using the following equation:

$$\text{Acid correction to }°\text{Brix} = 0.012 + 0.193\,m - 0.0004\,m^2$$

where m is the total acidity obtained by titration to a pH of 8.1, expressed as anhydrous citric acid in g/100 g.

In the case of soft drinks, this is frequently the limit of analytical testing for sugars. However, it is common to check the levels of the individual sugars in a fruit juice, to assess its authenticity and quality. Today, most practitioners in this field use either HPLC or enzymatic methods to achieve this.

A wide range of HPLC methods have been used to detect sugars, and a quick look in the literature shows a long list of publications on the separation and quantification of sugars by HPLC. As the sugars used in soft drinks formulations, except for fructose, do not have a strong absorbance in the ultraviolet region, they are generally detected using a differential refractometer, which is a non-specific detector with a relatively low sensitivity (typically in the region of 0.01–0.05%). This technique is, however, ideally suited for the analysis of sucrose, glucose and fructose at the levels detected in soft drinks and fruit juices, which are usually in the region of 0.5–5%. The separation of the sugars can be performed on a range of different columns, three of which are discussed below, and any one of these would be appropriate for soft drinks analysis.

The first method employs the use of an amino-bonded silica column. On this column the free hydroxyl groups on the silica particles have been reacted and replaced by amino groups. This type of column, sold by most column manufacturers, will separate sucrose, glucose and fructose in about 10 minutes, as illustrated in Figure 10.2.

As with all methods, this approach has some limitations; it uses acetonitrile, which is toxic, and the separation of glucose from fructose can sometimes be problematical after extended use of the column. However, sample preparation is easy, and only requires dilution to the required level (often 1 : 10) and filtration, prior to analysis, to remove particulate materials. This protects and extends the column's useful life. The degradation of the resolution between glucose and

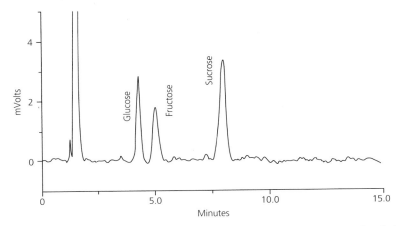

Figure 10.2 Separation of the main sugars found in orange juice using an amino-bonded silica column. Conditions: column: Highchrom 5 μm 250 mm × 4.5 mm amino-bonded silica; solvent = acetonitrile/water (8 : 2); flow rate of 2 ml/min; refractive index detection.

fructose is caused by the partial inactivation of the column by materials in the matrix, but this resolution can be recovered by reducing the acetonitrile concentration in the solvent. This column can also be used to assay the level of ascorbic acid (vitamin C) in a soft drink or fruit juice, although different solvent and detection systems are used.

The second method, using refractive index (RI) detection, is carried out with a resin-based polymer column. Sucrose elutes first from this column followed by glucose, fructose and then sorbitol. This type of column is generally more robust than the amino-bonded column and if handled well will last much longer; however, it is much more (*ca* three times) expensive. The method has been collaboratively tested for the analysis of sugars and sorbitol in fruit juices by the IFU (No 67). The HPLC conditions are given below:

HPLC conditions: column: benzene divinylstyrene sulphonated resin in the calcium form 300 mm × 7.5 mm; solvent = 0.1 mM Ca EDTA in HPLC-grade water at 0.6 ml/min; at 80°C using RI detection.

The third HPLC method for analysing sugars uses a very different detector system, a pulsed amperometric detector (PAD) and column. If sugars are analysed using a conventional electrochemical detector, they very soon 'poison' the electrode and reduce its sensitivity. This problem is overcome with the PAD system, where the electrical potential applied to the cell is varied, to automatically clean and recondition the electrode during the analysis. The Dionex Corporation pioneered this type of detector in the 1980s for the quantification of sugars and other oxidisable materials.

Figure 10.3 shows a chromatogram of an adulterated apple juice using this approach. The four major peaks that are visible are, in order of elution: sorbitol, glucose, fructose and sucrose (Anon, 1992). The column used for this analysis is

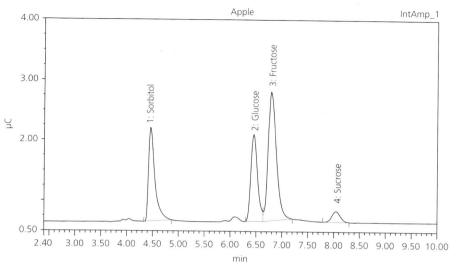

Figure 10.3 HPLC trace of a suspect apple juice run on Dionex HPLC, using PA-10 column and PAD detection. Conditions: Column: Dionex PA-10 250 mm × 4 mm; solvent = 100 mM sodium hydroxide in HPLC water; PAD detection; sample diluted 1 : 100 in water prior to filtration and analysis.

a mixed-bed resin column containing both sulphonated and amino-bonded resin beads (PA-10). From the size of the sorbitol peak, it is clear that this apple juice contains undeclared pear juice.

More recently, another type of detector has been introduced for sugar analysis, which is an evaporative light scattering detector (ELSD). This relies on the separation of the sugars on an amino-bonded or resin column, as for the methods using RI detection above. However, in this case, the eluent passes into the detector, is nebulised, the solvent evaporates and the remaining materials are carried into a laser beam. As the materials pass through the beam, they scatter the light, and the scatter is proportional to the analyte concentration. It should be noted that this type of detector does not always show a linear response with concentration.

A typical chromatogram for a mixed sugar standard using an ELSD is given in Figure 10.4. One of the advantages of this type of detector is that it has a much greater sensitivity than an RI detector. However, as with the electrochemical detection, due to the detector's high sensitivity, these two methods require a high dilution to be used, which sometimes can introduce a larger analytical error if it is not carried out carefully.

Sucrose (Bergmeyer and Brent, 1974a), fructose (Bergmeyer and Brent, 1974b) and glucose (Bergmeyer *et al.*, 1974c) can also be quantified by the use of an enzyme-linked assay. This type of approach is most commonly used in some areas of Europe (e.g. Germany), for the analysis of sugars in fruit juices, or in laboratories that have a spectrophotometer but no HPLC equipment. Kits

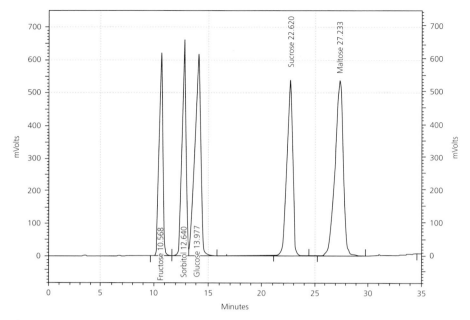

Figure 10.4 HPLC trace of mixed sugar standard using ELSD. Conditions: column: Phenomenex Luna 5 µm, 100A NH2, 250 × 4.6 mm; solvent water/acetonitrile (80 : 20); ELS detection. Note on other NH_2 columns sorbitol and glucose co-elute.

for this type of enzymatic analysis are available from a number of different suppliers, including r-Biopharm. It is also possible to buy the individual chemicals and enzymes to carry out the assay (Anon, 1987a). These methods have been collaboratively tested for the analysis of fruit juices (IFU No 55 and 56; see Anon, 1985, 1998a, respectively).

Generally, in fruit juices, the enzymatic methods are slightly more reproducible for glucose and fructose than the HPLC procedure (results from IFU analytical commission ring tests). However, the HPLC methodology, using RI detection, gave slightly better quantification of sucrose, because the enzymatic procedure uses a two-stage process, which introduced more analytical variation. Unless an automatic enzyme analyser is available, it is quicker to use HPLC to quantify the sugars in fruit juices and soft drinks. If glucose, maltodextrin or inulin syrups are being used in a soft drinks formulation, the total sugars may not add up to a value close to the soluble solids content. This is because the higher oligosaccharides that these syrups contain will not be quantified in these procedures. If this is an important feature, then Dionex HPLC or HPLC-ELSD can be used to quantify these higher oligosaccharides. On the Dionex system, a sodium acetate gradient is used to elute the higher sugars from the column and a procedure is detailed in their application note 67 (Anon (b), no date).

A typical example of a soft drink containing a maltodextrin syrup is given in Figure 10.5. Maltodextrin syrups are often added to soft drinks to control the

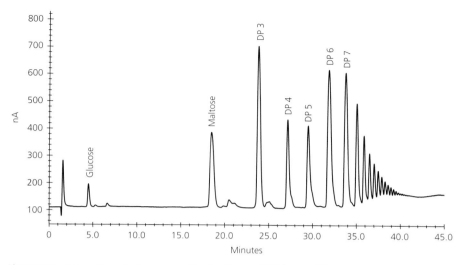

Figure 10.5 Separation of oligosaccharides by Dionex HPLC. Conditions: column: PA-100 5 µm 250 mm × 4 mm; solvent = 0.002 M sodium acetate in 0.1 M sodium hydroxide and 1 M sodium acetate in 0.1 M sodium hydroxide; flow rate 1 ml/min; using pulsed amperometric detection, time 0 = 100% A then a gradient to 46% B over 50 min using curve 9.

osmolality of the product, such as in 'sports' drinks. As the proportion of the different oligosaccharides vary from one manufacturer's product to another, it is impossible to determine the actual level of addition of the maltodextrin syrup, unless the actual material used in the formulation is available to the analyst.

10.5.2 Analysis of high-intensity sweeteners

Although sugars are found in juices, the high-intensity sweeteners are not permitted, but they may be used in low-calorie drinks and nectars. Here, it is important to ensure that there is resolution of the sweeteners of interest from the compounds naturally present in the fruit juice(s). This can be a particular problem in drinks with a high juice content, as the polyphenolic components can interfere with the analysis.

It is not uncommon to use more than one high-intensity sweetener, together with sugar, in a soft drink formulation. If some or all of the sugar is replaced with high-intensity sweetener(s), it becomes important to check if it/they have been added at the correct level(s). This is not only important to produce a consistent product for the consumer, but also because, in most countries, legal maxima are set for these materials. This not only applies to the high intensity sweeteners, but also to the preservatives, bittering agents and other additives (which will be discussed later). In the UK, these limits are either given in the Sweeteners in Foods Regulations 1995 (Anon, 1995b), or the Miscellaneous Food Additives Regulations (Anon, 1995c) and their revisions. References to the relevant EU law are contained within the UK Statutory Instruments, which may be viewed at http://www.uklaws.org/statutory/instruments.

Saccharin is a particular high-intensity sweetener that has been commercially available for about 100 years, and is still widely used as a low-calorie sweetener around the world. The traditional approach for quantifying saccharin involves its extraction into diethyl ether from a strongly acidic solution, removal of the solvent, and quantification of the saccharin by titration against sodium hydroxide, using bromothymol blue as an indicator (Egan *et al.*, 1990a). Although this is a reliable procedure, it is rather time-consuming, and is unlikely to be used extensively today.

With the wider availability of HPLC, it is now much more common to assay most sweeteners, and some preservatives, by HPLC. A wide range of procedures has been published for the analysis of saccharin in soft drinks by this technique. There is one validated method for the analysis of saccharin in soft drinks in the AOAC manual, and two in the British Standards catalogue. The AOAC procedure separates saccharin, caffeine and benzoic acid using a reverse phase column, and was published as AOAC method number 978.08 (Woodward *et al.*, 1979). A reverse-phase C_{18} column is used with acetic acid/water/propan-2-ol as the mobile phase, with detection at 254 nm. The level of propan-2-ol is selected between 0% and 2% (v/v), to ensure that there is an appropriate resolution between saccharin, caffeine, benzoic acid and any other compounds in the formulation, such as sorbic acid or colours.

The older of the two British standard methods used a phosphate buffer and acetonitrile mixture as the solvent with a C_{18} column (Anon, 1999a). This method actually allows the separation of a range of other components in soft drinks and foods (e.g. preservatives), plus the degradation products of aspartame, although the chromatographic resolution of the method between acesulfame-K and phenylalanine, which elute very early in the run, is not baseline. The other BS method (BS EN 15911) (Anon, 2010b) is a modern approach that uses a light-scattering detector for quantification. As with the other methods, the chromatographic resolution is carried out on a C_{18} column, but the ELSD provides a good non-specific detector that allows nine high-intensity sweeteners to be assayed in one run of 36 minutes. Although the procedure was not assessed directly on a juice product, it did have canned fruit and an energy drink as part of the validation study.

There are a large number of other procedures that have been published for the separation of a number of sweeteners and preservatives at one time; these are all based on reverse-phase HPLC. Perhaps one of the most startling is still the method published by Williams (1986). This used a small particle size (3 µm) C_8 column, and allowed the separation of a range of colours, sweeteners and preservatives in less than 5 minutes! The materials separated were: amaranth, quinoline yellow, quinine sulphate, sunset yellow, caffeine, aspartame, saccharin, vanillin, sorbic acid, benzoic acid and green S. This is similar to the results that can now be obtained using ultra-performance liquid chromatography, but preceded it by 25 years!

HPLC conditions: column: 3 µm Spherisorb RP_8 100 mm × 4.6 mm, solvent = 17.5 % acetonitrile, 12.5 % methanol, 70% buffer (0.85% sulphuric acid in HPLC grade water containing 17.5 mM KH_2PO_4 at pH 1.8); flow rate 1.35 ml/min; UV detection at 220 nm.

Two methods have been published which were designed to analyse a range of sweeteners and preservatives in one run. The first method, published in German by Hagenauer-Hener *et al.* (1990), describes the analysis of aspartame, acesulfame-K, saccharin, caffeine, sorbic acid and benzoic acid in soft drinks and foods. The method relies on a similar system to that given above, but with a less complex solvent system (Figure 10.6). The solvent system has been modified to include a gradient portion to elute the preservatives more quickly.

The final method was specifically developed for the analysis of diet soft drinks and food products by Lawrence and Charbonneau (1988) at Health Canada in Ottawa. The procedure allows the separation of a broad range of sweeteners, including cyclamate, alitame, sucralose and dulcin. This technique again uses reverse-phase HPLC with an acetonitrile/phosphate buffer system but, in this case, as the compounds have a wider range of polarities, a gradient elution is used. Aspartame, saccharin, alitame, acesulfame-K and dulcin were detected using ultraviolet absorbance at 210 nm or 200 nm. Sucralose, which has become a very popular high-intensity sweetener, was measured by refractive index detection. BS method 15911, as mentioned above, offers better sensitivity, as the evaporative light scattering detector (ELSD) is a better detector than using end absorbance in the UV region.

HPLC conditions: Column: C_{18} 250 mm × 4.6 mm; solvent = 0.02 M phosphate buffer/acetonitrile from 97 : 3 at pH 5.0 to 80 : 20 at pH 3.5 using a linear gradient.

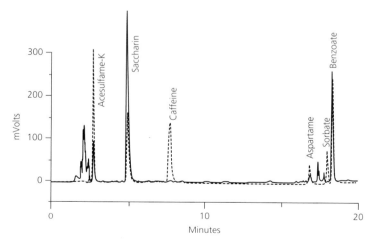

Figure 10.6 Separation of soft drinks ingredients, using the method of Hagenauer-Haner *et al.* Solid line is orange squash; dotted line is mixed standard. Conditions: column: PR8 5 µm 150 mm × 4.6 mm; solvent = 0.02 M phosphate buffer/acetonitrile (90 : 10); flow rate 1 ml/min; using UV detection 220 nm, time 10: 10–70 % acetonitrile over 15 min.

Aspartame was introduced as a high-intensity sweetener in the mid-1980s, for use in foods, and has been widely adopted for use in soft drinks. It has one limitation: in an acidic environment, as found in a soft drink, it is gradually lost by hydrolysis. However, its analysis and that of its decomposition products is quite straightforward by HPLC. Tsang *et al.* (1985) used a reverse-phase separation for the analysis of aspartame, whereas Argoudelis (1984) used a strong cation exchanger. As stated above the EN/BS methods (EN12856:1999 and 15911:2010) will also separate aspartame from its decomposition products (Anon, 1999a, 2010b).

Acesulfame-K was introduced as a high-intensity sweetener at a similar time to aspartame. It is also much sweeter than sucrose, and is stable under the low pH conditions of soft drinks. Its analysis in a soft drink is relatively straightforward by HPLC. It is included in BS method 15911, and in an older method given by Grosspietsch and Hachenburg (1980).

Although capillary electrophoresis is not as common as HPLC, there has been interest in its use in some areas. It offers one major advantage, in that it uses extremely small quantities of solvents, and so may become more popular in the future if solvent prices and their disposal costs increase rapidly. There is a growing number of methods published for this type of method, and an example of one of these describes the separation of high-intensity sweeteners, preservatives and colours in one run of 15 minutes (Frazier *et al.*, 2000).

Although the use of cyclamate was banned in the UK in the late 1960s, after concerns about its safety, it was re-approved for use in foods in the EU in 1996 after re-examination of toxicological data. A number of the traditional approaches used for its analysis are given in *Pearson's Analysis of Foods* (Egan *et al.*, 1990b), and a number of these involve the oxidation of the molecule to give sulphate ions, which are then measured gravimetrically or colorimetrically.

When cyclamate was banned, this was before HPLC techniques were developed, so there have been fewer methods published for its analysis using modern procedures. The substance offers a challenge to the analyst, as it does not have a useful chromophore in the ultraviolet region, and its detection by a change in refractive index would be difficult at the levels used in soft drinks (a maximum of 400 ppm). There is a European standard method (12857) for its analysis using HPLC (Anon, 1999b). In this case, the cyclamate is reacted to form a derivative, N, N-dichlorocyclohexylamine, prior to its analysis by HPLC using UV detection at 314 nm, so this is a rather cumbersome method. The separation is conducted on a C_{18} column, and the analysis run takes about 12 minutes. The more recent EN/BS method (15911) using ELSD again, is very useful here, but does require the rather specialised detector, whereas the older method would be a possibility in most laboratories. A further paper described the analysis of cyclamate using capillary electrophoresis (Zache and Gruending, 1987).

Sucralose is a very popular sweetener that has been introduced since the last addition of this book. It is now used widely in both foods and soft drinks, and is stable in acid solution. Again, its analysis offers a challenge, as it has only a low absorbance in the UV region. It can be assayed using RI detection, but a modern, highly stable detector is required for this analysis, as the usage levels are rather low (300 mg/l) for this type of detection. There is also a method using electrochemical detection, but the method uses a complex waveform on the PAD detector to enhance the sensitivity of the assay and so gives a complex looking chromatogram (Anon (f), no date). Again, this material can be analysed using EN BS method 15911 if ELSD is available.

Stevia is now approved for use in the EU. Again, the EN BS method (15911) can help, as it will separate Stevia materials from a range of other sweeteners, however, the method will not resolve two of the important materials within the Stevia sweetener – stevioside and rebaudioside A – from one another, using the method's defined conditions. However, there is a method published by Dionex using either UV at 210 nm or ELSD for its quantification in a range of preparations (Anon (g), no date). It should also be possible to separate these two compounds by adjustment of the elution gradient given in the EN method.

10.6 Preservatives

There are three main preservatives used in soft drinks in the UK and Europe, benzoic and sorbic acids, and sulphur dioxide. In some countries, para-hydroxybenzoates are also allowed in beverages. Although benzoate is a well tried and tested preservative, there have been problems over the past 15 years when it is present in a beverage with Vitamin C. This combination can produce low (parts per billion), but detectable, levels of benzene, which arises from the decarboxylation of the benzoic acid, catalysed by ascorbic acid. This has led most manufacturers to move away from this preservative. Even when ascorbic acid is absent, many retailers will not accept products containing benzoates, because of negative consumer attitudes to the ingredient

The analysis of these ingredients will be covered in two sections: the first will examine the analysis of benzoic and sorbic acids and the para-hydroxybenzoates; and the second will discuss the analysis of sulphur dioxide.

10.6.1 Benzoic and sorbic acids

One of the older methods used to detect the presence of preservatives in a soft drink and juice is thin layer chromatography (Woidich *et al.*, 1967), which provides a useful method to detect benzoic and sorbic acids, as well as the substituted benzoic acids. The first stage of this method involves the extraction of the preservatives with diethyl ether, prior to their chromatographic separation on polyamide plates. Although it is difficult to use this procedure to quantify the level of these preservatives in a sample, it is not impossible. If a laboratory does not have access to HPLC, this can still be of use today.

Extraction using diethyl ether has also been used in another traditional method to determine the level of preservatives in a sample. In this case, benzoic acid can be extracted from a product at low pH using diethyl ether. By adjusting the pH of the product and, hence, the ionisation of the acids themselves, it is possible to quantify benzoic acid in a soft drink in the presence of saccharin. After extraction, the benzoic acid is assayed spectrophotometrically or by titration (Egan *et al.*, 1990c). There is also a BS/ISO method (5518), which was published in 2007. However, this uses a spectrophotometric assay after sample pre-treatment. If access to HPLC is available, this method would not normally be recommended today. However, it offers another option if HPLC is not available.

Benzoic and sorbic acids are now normally assayed using HPLC. As discussed in the section on the analysis of sweeteners, some of the HPLC methods developed for soft drinks actually allow the separation of both sweeteners and preservatives in one run (e.g. Williams (1986), Hagenauer-Hener *et al.* (1990) and the EU method for sweeteners (Anon, 1999a)). However, the preservatives were not included as analytes in the collaborative trial of the method. The separation of benzoic and sorbic acids can be difficult, so care should be taken that the system will actually resolve these two preservatives if they are present, otherwise spurious results can be obtained. The pH of the solvent is the critical feature that allows the separation of these two preservatives.

A validated method (No. 63) for the analysis of benzoic, sorbic and parahydroxybenzoic acids in fruit juices has been published in the IFU Handbook (Anon, 1995e). In this method, the resolution takes place on a C_8 reverse-phase column, with a methanolic/ammonium acetate buffer (pH = 4.55) as the solvent. Here, sorbic and benzoic acids elute quickly, and the less polar parahydroxybenzoate esters elute much later in the run. The method stipulates that caution has to be taken with orange juice, due to possible interferences with natural materials in the juice that elute close to benzoic acid. Here, the use of diode array detection would help solve this issue. Due to their short retention times, this method would probably not be very useful for soft drinks containing some of the high-intensity sweeteners, as they may co-elute with these two preservatives.

HPLC conditions: column: 5 μm 250 mm × 4.6 mm RP_8; UV detection at 235 nm; solvent = 0.01 M ammonium acetate and methanol (50:40) adjusted to pH 4.55 with acetic acid; flow rate 1.2 ml/min.

There is a method in the AOAC manual for the analysis of benzoic acid in orange juice (994.11). However, this method was not designed to detect benzoic acid as a preservative. In Florida, when pulpwash (PW) concentrate is prepared, benzoic acid is added to it as a marker, so that, if the PW is added back to orange juice, later on it can be detected from this tracer.

10.6.2 Sulphur dioxide

Sulphur dioxide is a widely used preservative for foods and soft drinks, particularly squashes and cordials. In principle, it is quite easy to detect and quantify. Due to the importance of sulphur dioxide as a preservative, an extensive review of its chemistry in foods was undertaken in the mid-1980s (Wedzicha, 1984). It should be noted that, where this is used in soft drinks or in one of the ingredients, if the concentration is above 10 mg/l in the finished product, it must carry a warning on the label, as it can cause an allergenic reaction in some people. Some people find that it imparts an unpleasant taste to the product, even at the low levels permitted in beverages.

There is a very simple and quick method that can be used to detect the reducing power of sulphur dioxide, developed in the last century and often called the Ripper titration (Ough, 1988). In this method, the sulphur dioxide is titrated against iodine or potassium iodate/potassium iodide solution, in the presence of starch. When all of the sulphur dioxide has been oxidised, a blue colour is produced by the reaction of free iodine with the starch indicator. This is a very quick method, but will only give an estimate of the level of sulphur dioxide, because other reducing substances, such as ascorbic acid, will interfere. Consequently, this method is not particularly appropriate for juices or soft drinks that contain a high concentration of ascorbic acid.

The concentration of sulphur dioxide in a soft drink is generally determined by a steam distillation procedure. Here, the sulphur dioxide is driven out of the acidified solution by heating. The acidification displaces any sulphur dioxide that was bound to other food materials, and the sulphur dioxide is carried over by the steam into a trap(s) containing a hydrogen peroxide solution. The sulphur dioxide is oxidised to sulphuric acid, and this can then be measured by titration against a standardised sodium hydroxide solution.

There have been many modifications published to the original method published by Monier-Williams (1927) for different applications. However, for grape juices, the Tanner modification (Tanner, 1963), which uses phosphoric acid rather than hydrochloric acid for the acidification step, and nitrogen instead of air to flush the system out, should be used, as other procedures can give false positive results at the very low levels required to determine if the product is acceptable for sale to the consumer. This method is less liable to these interferences than the other modifications (IFU method No. 7a (Anon, 2000)). There are a number of critical steps that need to be taken to ensure that the method works correctly, and these are detailed in the IFU procedure.

Although some chromatographic methods have been published for the analysis of sulphur dioxide in solution (the sulphite anion), these have often involved steam distillation to ensure that all the sulphur dioxide is liberated. However, as the distillation step is the time-consuming stage, there is little point in using these procedures in systems that do not liberate large amounts of other acidic or sulphurous materials, such as garlic and cabbage, which interfere with the titration method.

Some newer methods have also been developed which will assay both free and total sulphur dioxide. These procedures rely on the separation of the sulphite anion from the other materials in the food, by ion-exchange chromatography and measurement by electrochemical detection. There is an AOAC-approved method using direct current (DC) amperometric detection (Anon, 1995f), which uses an alkaline extraction medium (20 mM phosphate buffer containing 10 mM mannitol at pH 9), so that free and total sulphur dioxide can be detected. After extraction, the sample is injected directly onto an HPLC, with an ICE-AS 1 column and DC electrochemical detection. In the AOAC method, it suggests cleaning the electrode at the end of each run to maintain optimal detector sensitivity. However, Dionex has improved the original AOAC method by using pulsed amperometric detection (Anon (c), no date). Using this approach, the detector response is much more stable, as the electrode is constantly cleaned during the analysis, rather than at the end of each run. However, this modification still awaits formal verification using the AOAC procedure.

HPLC conditions: column: 250 mm × 4 mm Ion Pac ICE-AS1; solvent = 20 mM sulphuric acid; PAD detection using a platinum electrode.

10.6.3 Dimethyldicarbonate

This preservative (marketed as Velcorin) is, when used, normally injected into a stream of product just prior to bottling. The preservative is rapidly hydrolysed into methanol and carbon dioxide, and may be detected by measuring the methanol content of products, typically using GC-MS. Products containing fruit juices where some degradation of the pectin may have occurred can also contain methanol as a breakdown product. In such situations, a baseline level of methanol should be established.

10.7 Acidulants

After sweetness, the second most important feature of a fruit juice or soft drink is its acidity. In fact, it is actually the balance of the sweetness to sourness (acidity), commonly called the Brix to acid ratio (B/A), which is the important parameter. An acceptable range for the B/A ratio of a fruit juice or fruit juices concentrate is often laid down in purchase specification; the higher the B/A ratio, the sweeter a product tastes. For European tastes, an orange juice might typically have a B/A ratio set around 15. However, this does vary from country to country, and between consumers. It should be noted that, in Europe, B/A ratios for apple juice are typically around 20, which seems to appeal to European tastes, whereas in the USA, apple juice is typically consumed with a B/A ratio around 40. This is one reason why far more Chinese apple juice is imported into the USA than is used in Europe, as this material generally has a ratio around 35–45, so is more suited to US tastes.

The normal method used to assess the acidity of a soft drink or fruit juice is by titration using sodium hydroxide. Traditionally, phenolphthalein was used as an indicator to detect the end-point of the titration. However, a pH meter is now the instrument of choice, and the end-point is taken as 8.1 (IFU No. 6). This has a major advantage, as it allows the titration to be carried out automatically and avoids any problems in highly coloured products, such as blackcurrant juices, where the indicator end point (colourless to pink) can be difficult or impossible to detect. The ability to carry out the titrations automatically is also much more time- and cost-effective.

In a carbonated product, the carbon dioxide has first to be removed from the product by boiling prior to analysis. Applying the energy of sound (sonication) alone is not enough to ensure that all the carbon dioxide has been liberated from the soft drink. It should be noted that, if dealing with 'cola products', which contain phosphoric acid, if the titration is taken to an end-point of pH of 8.1, this will not affect the final ionisation of the hydrogen phosphate ion (HPO_4^{2-}) to the phosphate ion (PO_4^{3-}), as the pKa of HPO_4^{2-} is 12.67. Under these conditions, this final dissociation will not take place, so this has to be allowed for when the acid content of cola products is being determined (i.e. the acid only 'appears' to be divalent).

In the analysis of a fruit juice, it is important to determine the concentrations of the individual acids, to assess if it is authentic and of a good quality. A range of organic acids can be determined using enzyme-linked assays, and these procedures have been collaboratively tested and published in the IFU compendium of methods (citric No. 22, isocitric No. 54, d-malic No. 64, l-malic acid No. 21 and d- and l-lactic acids No. 53). Kits for these tests are available from r-Biopharm and a range of other suppliers.

Most acids can also be assessed by HPLC, using either ultraviolet or conductivity detection. The following procedure was developed to assess the organic acid profile of cranberry and apple juices, but has been found to work well for other organic acids (Coppola and Starr, 1986) and has been published as an AOAC 986.13. A typical trace for cranberry juice run using this method is given in Figure 10.7. The method recommends the separation of the acidic materials from the sugars by prior treatment with an ion-exchange resin.

Separation of the organic acids is also possible using an ion-exclusion column, such as a Bio-Rad HPX-87H. However, in this case, it is essential that the acids are separated from the sugars, otherwise a number of peaks co-elute, thus distorting the quantification. This method is particularly good for the analysis of fumaric acid in apple juice, because no pre-treatment is required here as the acid elutes late in the chromatogram, well separated from other components (IFU 72). A chromatogram for a mixed organic acid standard is given in Figure 10.8.

Fumaric acid was found to be a good marker to detect the addition of d, l-malic acid to apple juice (Junge and Spandinger, 1982). With the advent of the specific enzymic assay procedure for d-malic acid, the method fell out of use.

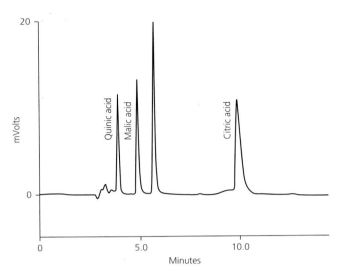

Figure 10.7 HPLC separation of the organic acids in cranberry juice using Coppola's method. Conditions: column: 5 μm ODS2 250 mm × 4.6 mm; UV detection at 210 nm; solvent = 0.05 M phosphate buffer at pH 2.5; flow rate 1 ml/min.

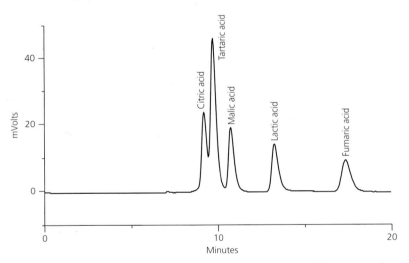

Figure 10.8 HPLC separation of organic acid standard by ion exclusion chromatography with UV detection. Conditions: column Phenomonex Rezex monos 30 cm × 8 mm; UV detection at 210 nm; solvents 4 mM H_2SO_4 in Milli-Q water; flow rate 0.6 ml/min.

However, in 1995, a number of samples of apple juice in Germany were found to contain elevated levels of fumaric acid, which was attributed to the addition of l-malic acid, with high levels of fumaric acid. Those who want to cheat now have access to l-malic acid with little or no fumaric acid, making this method of detection less sensitive. It should be remembered that fumaric acid can also

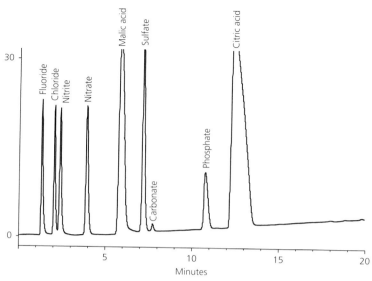

Figure 10.9 HPLC separation of organic acids and anions by ion exchange chromatography and conductivity detection. Conditions: column Dionex AS-11 25cm × 4 mm; suppressed conductivity detection; solvents 18 MΩ water and 0.3 M NaOH; flow rate 1.5 ml/min 2 mM NaOH to 33 mM NaOH over 15 minutes.

be formed via microbiological action in the juice, so the presence of an elevated level, above 20 mg/l, is not instant proof of adulteration – it may just be a quality issue, or mean that the analytical sample is rather old and has started to ferment!

Organic acids can also be quantified using HPLC linked to a conductivity detector. This has one advantage over ultraviolet detection, in that only charged species are measured, which means the method is liable to less interferences. Depending on the actual approach chosen, it is sometimes possible to detect other anions, such as Cl^-, SO_4^{2-} and PO_4^{3-}, in the same run. If the conditions given in the Dionex application No 21 are used, this allows the organic acids to be separated without inferences from the fully ionised anions, such as Cl^-, SO_4^{2-} (Anon (d), no date).

HPLC conditions: column: 2xICE-AS1 300 mm × 7.9 mm; conductivity detection; solvent = 2 mM octanesulphonic acid in 2% propan-2-ol; flow rate 0.5 ml/min.

However, if a method similar to Dionex application No 123 is used, both organic acids and anions can be separated on the AS-11 column (see Figure 10.9; Anon (e), no date). Care has to be taken with this method, as nitrate and malate ions elute very close to each other using these conditions. This can be a particular issue if the column is not operating well, and there is a need to determine the low levels of nitrate ions (0–10 mg/l) seen in juices in a product containing a high level of malic acid (2000 mg/l).

10.8 Carbonation

For carbonated soft drinks, the level of carbonation is a key parameter and, if an incorrect level is used, this can disrupt the overall flavour balance of the product. The level of carbonation can be measured in a number of ways, and four are discussed here. The first is a manometric method published by AOAC for beer (940.17), while the second is also a manometric procedure, but is somewhat less complicated than the AOAC procedure. The final two methods use specialist equipment, with the first employing the electrical conductance of carbon dioxide gas to do the determination and the final method using ion chromatography.

In the AOAC procedure, the container's cap is punctured with a strong needle. The gas is then shaken out of the product and collected in a gas burette. The pressure is adjusted to atmospheric before the volume is read off the burette. The carbon dioxide can also be removed by dissolving it in sodium hydroxide solution; any other gases in the container can be measured in this way.

The second method is a similar procedure. The top of the product's container, plastic, glass or can, is punctured with a needle. The carbon dioxide is expelled from solution, by vigorous shaking, and the headspace pressure increase is recorded. From the pressure rise, the volume of carbon dioxide dissolved in the product can be determined. This method is probably one of the most commonly used procedures for measuring carbonation, although the exact method used varies from one manufacturer to another (the number of times the product is 'sniffed' (pressure released and re-shaken) before the headspace pressure is recorded). One supplier of this type of carbonation tester is Stevenson and Reeves, of Edinburgh, who also sell a slide rule to convert the pressure rise measurements into carbonation volumes. Information of their products can be found on their website (http://www.stevenson-reeves.co.uk).

> *Caution*: *care has to be taken when measuring the carbonation level of products using this method in glass containers, as there is a risk that the container might fracture. To remove this risk, special carbonation testers can be purchased, in which the glass bottle is contained inside a shield so that, if it does fracture, no glass fragments can escape.*

The first of the more sophisticated approaches requires a specialised instrument – the Corning 965 carbon dioxide analyser.

> *In this methodology, the soft drink is initially cooled to 4°C in a fridge. The container is then opened and an aliquot of concentrated sodium hydroxide (40%) is added to 'quench' the carbon dioxide in the product (typically 10 ml of NaOH is added to 284 ml of product). The carbon dioxide is quenched by reaction with the sodium hydroxide to form bicarbonate and carbonate ions. An aliquot (50 μl) of the quenched product is removed and pipetted into the instrument's cell. The cell is closed and the solution acidified to release the carbon dioxide; this is then detected by the change in the thermal conductance of the vapour phase.*

This is a rapid procedure, which gives very reproducible results, similar to the pressure rise method, and is extensively used in the brewing and soft drinks industries. However, this type of detector is now no longer made by Corning.

The final method uses ion chromatography and showed comparable results to the Corning method (Harms *et al.*, 2000). Here, the carbon dioxide is again converted to carbonate by the addition of sodium hydroxide. The treated sample is then analysed, using HPLC linked with conductivity detection. Harms *et al.* (2000) found that the method worked well, both for beers and soft drinks, and that there were no significant differences between the results produced by this procedure and the data obtained using the normal Corning method discussed above.

HPLC conditions: column: ICE-AS1 150 mm × 4 mm; conductivity detection; solvent = 18MΩ HPLC grade water; flow rate 0.4 ml/min.

During the carbonation of water, the levels of oxygen and nitrogen need to be reduced to a minimum. This not only means that the over-pressure of the carbon dioxide applied to the carbonator can be lower, it also means that fewer problems are seen with fobbing during filling, which is rapid out-gassing after filling, which causes loss of product when the container leaves the filler head. These permanent gases are often typically driven out of solution by rapid agitation of the water as it is fed into the carbonator, and during carbon dioxide sparging to ensure that this problem is kept to a minimum.

In non-carbonated products, oxygen can also be a problem, as it can cause oxidative changes to the flavour and also reduce vitamin retention, particularly vitamin C. Although headspace oxygen measurements can be made by gas liquid chromatography (GLC), this is not a straightforward analysis, and it requires skilled staff and specialised equipment (Anon, 1997a).

10.9 Miscellaneous additives

Two compounds that are used in some soft drinks formulations for specific purposes are caffeine, employed in a range of beverages including colas for its stimulant properties, and quinine, used for its bitter taste and to meet a legal requirement in Tonic water, and sometimes in other products. Traditional methods for the analysis of these two compounds have often involved their extraction from aqueous solution into an organic solvent, and then quantification by one of a range of methods.

10.9.1 Caffeine

There are three spectrophotometric procedures given in the AOAC compendium of methods (960.22, 962.13 and 967.11) for the analysis of caffeine, all of which have an extraction stage followed by a quantification procedure. There is also an HPLC method, discussed earlier, which was designed to measure saccharin, benzoic acid and caffeine at the same time (AOAC 978.08). Again, the HPLC method, EN 12856:1999 (Anon, 1999a), can be used for the analysis of caffeine but, again, this analyte was not specifically included in the collaborative study. There are five methods given in the BS portfolio of methods for the analysis of caffeine, four using HPLC, but these were directed at coffee and tea products, rather than soft drinks.

In *Pearson's Analysis of Foods*, two methods are quoted for the analysis of caffeine. The first is a simple solvent extraction, followed by quantification by ultraviolet absorbance at 273 nm (Egan *et al.*, 1990d), while the other is a GLC method.

Some companies use only natural caffeine in their products. Isotopic methods, offered by Eurofins, are available here to confirm if the caffeine is from a natural source, typically from the decaffeination of coffee or guarana, or has been prepared synthetically.

10.9.2 Quinine

The level of quinine in a soft drink can be measured either by determination of the absorbance of an extract of the soft drink after making the product alkaline with ammonia, or by its detection directly, spectrophotometrically or fluorimetrically (Egan *et al.*, 1990e).

As with the other additives used in soft drinks, caffeine and quinine can be detected using the same HPLC method as other materials, such as in the method published by Williams (1986). This method separates most of the major additives used in soft drinks in a short time (4–5 minutes). Although some of the resolutions are not quite baseline, as would be expected in such a short analysis time, and not all of the synthetic colours are separated from each other, this is still a very impressive method. There is also a method published by Valenti (1985), in JAOAC, that detailed the analysis of quinine, hydroquinone, benzoate and saccharin, using a RP_{18} column, which gave good results.

HPLC conditions: Column μBondapak C_{18} (10 μm) 30 cm × 3.9 mm. UV detection 254 nm, solvent 1% acetic acid in methanol – acetonitrile – water (2 : 1 : 7).

10.9.3 Other additives

Over the last five to ten years, there has been a rapid growth in drinks that are associated with certain attributes. As highlighted earlier, Red Bull has been a great success, based on its energy and stimulating properties. There have been a range of others which have focused on other aspects of health and 'wellbeing', often containing green tea extracts, soy extracts and/or extracts of herbs but, after the recent rulings from EFSA, a number of these claims will probably have to be withdrawn in the short term. If the companies still want to use them, they will have to prepare a dossier that demonstrates the action they are claiming for the product, and this will have to be evaluated by EFSA before they can use the claim.

10.9.3.1 Taurine

Taurine, or 2-aminoethanesulphonic acid, as is it is correctly called using the IUPAC method, is one of the specialised ingredients that is often included in energy drinks or products that claim to stimulate the mind. It is similar to an amino acid, except that the carboxylic acid group is replaced by a sulphonic acid residue. It is involved in a number of functions in the body, such as digestion, and is thought to be involved in brain development.

It can be analysed using the liquid chromatographic method used for the other amino acids. This involves separation using an ion exchange column and detection in the visible region, after post-column addition of ninhydrin. There is a standardised method for the analysis of amino acids in fruit and vegetable juices that would be applicable for this type of analysis (Anon, 1999c). However, this approach needs a dedicated machine, and is costly to set up if not run on a regular basis. Dionex published a method for amino acids in general that are detected using electrochemical detection, which requires no pre- or post-column derivatisation. In this, the amino acids are separated on an ion exchange column (PA-10), which happens also to be used for sugars analysis, using a hydroxide ion or hydroxide/acetate gradient (Anon, 2003a).

There are other pre-column derivatisation methods, such the Waters Pico-Tag (Cohan, 1984) or AccQ-Tag approaches (Anon, 1996a,b). Here, the amino acids and taurine are treated with a reagent that gives the derivatised molecules either a UV (Pico-Tag) or fluorescent (AccQ-Tag) chromophore, and these derivatised molecules are separated on special HPLC column designed for their resolution. The method works well, but some of the resolutions required for the amino acids seen in fruit juices are rather tight, so the column needs to be excellent condition for them to work well.

10.9.4 Fibre analysis

Fibre in diet is better describes as non-metabolisable carbohydrate, and there is a growing interest in its inclusion into the diet to help improve gut health. This is extending to drinks, too, with these types of materials being added to milk- and fruit-based products such as 'smoothies'. One source of soluble fibre which has received interest over the last few years is inulin, or oligofructans. These are oligosaccharides that are extracted from chicory or Jerusalem artichokes, and they are claimed to improve colon function and to have 'prebiotic' properties, enhancing the workings of the gut. Inulin is a complex carbohydrate that can be assayed in a number of different ways, and there are two published methods in the AOAC manual for the analysis of soluble fibre derived from inulin. The method chosen for its analysis depends on the product and the type of fibre added (997.08 and 999.03).

In 997.08 (Anon, 1997c), high pH anion exchange chromatography, linked with pulsed amperometric detection (HPAEC-PAD), is used to analyse the product for sugars at various stages of the analysis process. Initially, the concentrations of the free sugars in the product are measured. The sugars are then determined after treatment of an extract with an amylase, which breaks down starch, and then an inulinase, which hydrolyses any inulin. From the levels of the sugars (sucrose, glucose, fructose, and galactose if milk is present) found at each stage, it is possible to calculate the level of inulin in the product.

Method 999.03 (Anon, 1999d) is slightly different, using two enzymes to hydrolyse any sucrose or starch that the product contains. The glucose and fructose liberated in this step are then reduced, using sodium borohydride, to their corresponding alcohols. Finally, the borohydride-treated extract is mixed with a fructanase to degrade the inulin, and the free sugars that are liberated in this step are quantified using a colorimetric assay. The inulin content is calculated from the level of the sugars liberated by the fructanase action.

There are a number of well-documented, validated methods for the analysis of total fibre in food products in the AOAC manual, namely, 985.29, 991.42, 991.43, 993.19, 993.19, 997.08, 999.03, 2001.03 and 2009.01. Most of these are gravimetric methods, where the fibre is precipitated with ethanol and washed. The fibre content is then calculated from this, after allowances are made for any protein and/or ash that the precipitate contains. These methods rely on precipitation of the fibre with alcohol for their quantification but, as inulin is soluble in mixtures of ethanol and water, it cannot be detected in these tests. The more recent methods, 2001.03 and 2009.01, include a step for some HPLC, to allow for the presence of maltodextrins that are poorly digested in the body. Nearly all pure fruit juices contain insufficient fibre to make a nutritional claim.

10.9.5 Herbal drinks

There has been a growing interest in the health benefits that can be derived from taking herbal extracts, and this has led to the inclusion of these types of extracts in some beverages. However, as mentioned above, the Nutrition and Health Claims Regulations 1924:2006 (Anon, 2006b) may have a detrimental effect on this sector, due to the cost necessary to prepare the dossier to gain approval of any existing or proposed claims.

The analysis of herbal extracts is a very complex issue and is well outside the scope of this chapter. However, it is incumbent upon a manufacture to check the authenticity of the materials that they are buying, in order to ensure that they are 'real' and have not been adulterated or extended in any way with cheaper materials. This can be done in a number of ways but, as with the authenticity of fruit juices, it is best left to a specialist laboratory that has a broad range of experience in this field.

Some useful references in this area are given in the German Pharmacopoeia (DAB 10) that has a large section on the analysis of herbal extracts, a lot of which methods use high-performance thin layer chromatography (HPTLC). Another useful reference is the book edited by Grainger-Bisset and Wichtl (1994), which lists a whole range of herbs, their risk of adulteration, pictures to assist in their authentication and some analytical procedures. Also, a small monograph that details a number of herbal extracts can be found in the British Herbal Pharmacopoeia (Anon, 1996c). There is a growing level of interest in this topic in the USA, and an increasing number of HPLC methods are being published in the US Pharmacopoeia and the AOAC manual from work carried out by interested groups in this area.

If it is a challenge to confirm the authenticity of an herbal extract, it is even more of an analytical challenge in a complex finished product that might contain juice and other ingredients. This means that ingredient traceability is extremely important, and a manufacturer should check the incoming raw materials and ensure that they are delivered with a certificate of analysis from a reputable laboratory that specialises in this area.

10.9.6 Osmolality

Although this is not an additive, it is an important parameter to measure in sports drinks and isotonic products that are being sold for rehydration purposes. These types of products require an osmotic pressure similar to human serum, which is typically 285–295 mOsm/kg. This parameter is measured using one of the colligative properties of solutions that all chemists cover in their first year of physical chemistry at university. These properties are dependent on the numbers, rather than nature, of molecules in solution. Osmometers typically use either the depression of freezing point or the elevation of vapour pressure to measure the osmolality of a product. Two manufacturers who supply these types of instruments are Advanced Instruments Inc. Norwood, Massachusetts, USA (http://www.aitests.com) and Wescor Inc. Logan, Utah USA (http://www.wescor.com). Analysis for osmolality typically requires no sample treatment; all that is done is for the sample to be introduced into the instrument and, a few minutes later, the value is printed out.

10.10 Analysis of colours used in soft drinks

In soft drinks, even those containing juices, the addition of either synthetic or natural colours is permitted, and they are specified by the colours in foodstuffs Directive 94/34/EC and its amendments. For the synthetic colours, and some of the natural substances (lycopene (E160d) and lutein (E161)) there are maximum levels given in the directive, whereas the other natural colours, such as beetroot extract (E162), anthocyanins (E163) and mixed carotenoids or β-carotene (E160a) are permitted at *quantum satis* (i.e. any appropriate level).

The colouring materials used in soft drinks can be split into two classes. The first consists of the synthetic food colours that have been used to enhance the appearance of foods and beverages for many years. They are so called azo-dyes, are generally water-soluble, and include tartrazine, quinoline yellow, sunset yellow and ponceau 4R. However, over the years, these have become far less popular with the consumer, because of bad publicity about their possible side-effects on children. This trend led to the reformulation of a wide range of products – particularly orange squashes – in the UK in the mid-1980s, to remove the colours tartrazine and sunset yellow. They were replaced with natural colours, or synthetic colours based around the carotenoid molecule, such as annatto, β-carotene and β-apo-8'-carotenal. This trend intensified after publication of the UK Food Standards Agency-sponsored study conducted at Southampton University in the UK (McCann et al. 2007), which looked at the possible link between food colours and sodium benzoate with hyperactivity in children. From the results of this study, the FSA concluded that there was a link, so they asked UK manufacturers to adopt a voluntary ban on the following colours – the so-called 'Southampton 6':

> E102 tartrazine, E104 quinoline yellow, E110 sunset yellow, E122 carmoisine, E124 ponceau 4R and E129 allura red.

This has been so successful that almost all soft drinks now on sale in the UK do not contain these colourings based on azo-dyes. The European Food Safety Authority (EFSA) also examined the data from the Southampton study, but they were less sure of the link, so the ban has not been extended to other European countries. However, they did introduce a regulation (2008/1338/EC), which made it obligatory to put a warning on the label of a product that contained any of the colours listed above:

> 'name or E number of the colour(s)': may have an adverse effect on activity and attention in children'.

The initial switch from the water-soluble colours to β-carotene or β-apo-8'-carotenal caused its own problems, as these new formulations, which were prepared to give the same colour levels as seen in the old products, showed a strong tendency to produce a neck ring (i.e. they formed unsightly red/orange rings at the top of the bottle of squash). However, a general reduction in the level of added colour helped reduce the problem, so it is now only a sporadic issue. As a result of the FSA's action, it is now common to see mixed carotenes or anthocyanins in soft drinks/squash formulations in the UK. It should also be noted that the appearance of orange squashes are now yellow, rather than the bright orange they once used to be.

Although the anthocyanin and carotenoid pigments are stable at low pH, and provide a strong colour, this can, in some cases, fade due to oxidation or light action. This is a particular limitation with the carotenoid pigments, but it is generally overcome by the addition of an antioxidant, such as ascorbic acid, which significantly improves their stability.

It should be remembered, if ascorbic acid is present in a formulation, that it is unwise also to use benzoic acid, as the production of benzene can occur. Studies conducted by safety authorities around the world (FDA, FSA, Food safety Australia and New Zealand) have shown that this issue is often worse in low-calorie products.

10.10.1 Assessment of colour

The colour of a soft drink or fruit juice may be assessed in a number of ways. In clear products, the absorbance of the drink at one or more wavelength/s can be measured. The actual wavelengths chosen will depend on the particular colour of the product. For a yellow product, such as apple or grape juice, wavelengths of 465, 430 or 420 nm are often chosen to assess the colour. These values can then be expressed in European brewing convention (EBC) units, by multiplication by a factor of 25. The actual Brix value chosen at which to assess the colour depends on the country; however, levels between 11 and 12 are often taken as the norm. If dealing with a red-coloured product, then the assessment is generally carried out between 520 and 530 nm. Absorbance values are sometimes also taken at 420 nm in red or black juices, to assess the brownness of the product. The two absorbance values (520 nm and 420 nm) are often used to express a colour ratio, which gives an indication of colour verses brownness:

$$\text{Colour ratio} = \text{absorbance at } 520\,\text{nm} / \text{absorbance at } 420\,\text{nm}$$

Sometimes, the absorbance is also assessed at 580 nm, and a ratio calculated for the values at 520 and 580 nm, which is called the blue index. With the red/black juices, the product may have to be diluted so that the intensity of the colour can be assessed within the linear range of the spectrophotometer. The dilution should be done in an acidic buffer, to ensure that the true colour intensity is recorded. If this is not done, the absorbance can be reduced by the shift in pH affecting the apparent colour of the product. In some cases where a dilution of less than 1 : 10 is required, this can be replaced by using a cell with a shorter path length, of 0.5 cm or 0.1 cm. If this approach is adopted, then the colour value has to be adjusted for the revised cell path length. A method is detailed for the analysis of the colour in juices in the IFU handbook method No 80 (Anon, 2010c). Colour values can also be expressed in transmittance units if required, using this method, and it only requires the spectrophotometer to be set up in the appropriate mode before measurements are made, for this to be done.

In cloudy products, colour assessment is more difficult. Removal of the particulate material by filtration/centrifugation can be carried out, but this does not solve the problem, as some colours – particularly the carotenoids and some anthocyanin pigments – are strongly absorbed onto the particulate material/filters, and their removal gives a low 'colour' reading. In addition, the particulate material affects the opacity of the product and, hence, the visual perception of the colour. It has been known for a number of years that the simple approach of measurement of a product's absorbance values at 420 nm or 520 nm does not fully address this issue, because some products can have similar absorbance values at these wavelengths, but will be perceived by the eye as very different.

The way to address this issue is to use a tristimulus colour measurement technique. In this method, the colour is split into three separate primary components, which more closely match the way the eye perceives colour. It can be run in either absorbance or reflectance modes, and values for the product are measured at a range of wavelengths in the red, green and blue areas of the spectrum. The values are taken, and the components X, Y, Z are calculated. These X, Y and Z values are then transformed to split the colour up into three variables that address the colour or hue (red/green, yellow/blue) of a product, the opacity of the colour (how much light it reflects or absorbs) and, finally, the depth or intensity of the colour (deep purple/pale purple).

This three-dimensional space can be represented by either L, a, b or H, C, L values, which are different conventions used to describe a colour measured in a similar manner. In the L, a, b system, used by Hunter Lab, the colour is defined in terms of its lightness (L), which is the extent of black or white it contains (0–100), and in terms of two variables a and b, which define the colour/ hue and intensity. Red colours are situated in the $+a$ direction, while green has $-a$ values. Yellow is in the $+b$ direction, and blue has $-b$ values. The size of the value is indicative of the intensity of the colour (i.e. the larger the value, the more intense is the colour). In the CIELAB 76 convention, L again defines the lightness, as above, whereas H (hue angle) defines what the colour is (red is situated at 0°, yellow at 90°, green at 180° and blue at 270°) and C (chroma) describes the intensity or depth of colour. This area was reviewed by McKaren (1980) some years ago. IFU has recently published a recommendation (No 9) (Anon, 2008) about the analysis of colour using tri-stimulus measurement in fruit juices.

The different instruments (Hunter, Gardner, Instrumental Colour Systems, etc.) process the absorbance or reflectance data in slightly different ways, which means that the values obtained from one instrument manufacturer can differ slightly from that of another. Therefore a product's defined colour has to be qualified with a statement indicating which instrument was used. Notwithstanding this limitation, it is not uncommon to find a tristimulus 'colour meter' in a manufacturer's quality assurance laboratory, so that routine quantitative assessment of the colour of products can be made. This is particularly true for tomato-based products, where their nature makes conventional spectrophotometric assessment meaningless. In the USA, this tristimulus approach is used to determine the colour of Florida orange juice, and is used as part of the USDA screening of the product for grade-setting purposes (which means price).

The discussion above has addressed the assessment of a product's colour or perceived colour in basic terms. In the next two sections, methods to determine what compounds have been added to a product to colour will be addressed. For the purpose of this chapter, the section on synthetic dyes will cover the analysis of the water-soluble dyes (the so-called 'coal tar dyes'), and the section on natural pigments will cover the anthocyanin pigments (such as grape skin extracts etc.) and the carotenoid-based materials, even if they are of synthetic origin.

10.10.2 Synthetic colours

These are materials based around azo-sulphonic acids often referred to as 'azo-dyes'. They are water-soluble and provide strong stable colours. There is an interesting chapter by Wadds (1984) that discusses some of the basic methods (including thin layer, paper and high-performance liquid chromatography) which were used in the past and are still used today to detect these colours.

When colours are added to a food system, their characterisation is sometimes more difficult. This is due to interference from other materials in the food and/or difficulty with their quantitative extraction from the food matrix. This is particularly the case for high-protein foods which bind colours very tightly, and it can make their quantitative analysis very difficult. However, the analysis for azo-dyes in soft drinks is generally straightforward using modern methods. There are generally few problems with recovery here, as the colours are already in solution and not bound to other materials, and this makes the analysis easier/quicker. In some cases, the colours can be analysed without prior concentration while, in others, they can be concentrated on a solid phase extraction C_{18} cartridge, followed by elution with a small volume of methanolic/ammonia prior to analysis.

Due to the varied structures of the food dyes, they can generally be differentiated from one another by their characteristic ultraviolet/visible absorbance spectra. Using HPLC, coupled with a diode array detector (HPLC-DAD), it is possible to collect a compound's absorbance spectrum as it elutes from the HPLC column. This greatly assists in the identification of colours and a wide range of other compounds that absorb in the UV or visible regions. However, this is not possible with everything, and sometimes mass spectrometry has to be used.

A number of different approaches have been proposed for the analysis of water-soluble dyes, but all of these have their shortcomings. The separation of colours has been carried out using ion-exchange resins, reverse-phase HPLC coupled with ion-pair reagents, and reverse-phase HPLC at low pH, where the ionisation of the dyes is suppressed. The latter is the technique used most widely, and is also the approach recommended by Wadds (1984). It offers the simplest approach to this type of analysis, and a typical HPLC profile of some water-soluble pigments is given in Figure 10.10. This type of approach allows the identification of pigments present in a beverage. However, the quantitative analysis of colours can be a little more difficult, because pure standards or pigments of 'known' purity can sometimes be hard to find. This area is further complicated, as some pigments are made up from a mixture of a number of components, rather than a single entity – for instance, in Figure 10.10, there are five peaks attributed to quinoline yellow so, if a quantitative analysis is required, these peaks would have to be summed.

Due to problems with the addition of non-approved dyes to foodstuffs (e.g. sudan dyes to chilli powder, etc.), there has been a lot of work in this area over the last five years or so. Here both diode array and MS/MS detection have been

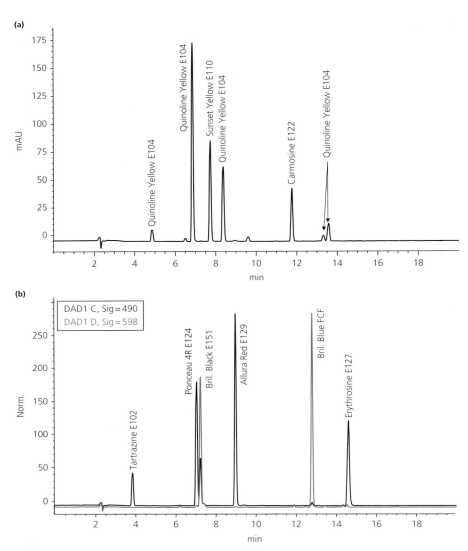

Figure 10.10 HPLC trace of standard pigments using gradient elution system. a) standard mixture (1) monitored at 415 nm; b) standard mixture 2) monitored at 490 nm and 590 nm. Conditions: column: 5 µm C18 150 mm × 4.6 mm, using diode array detection at 450 nm and gradient elution; solvent A = 0.02 M ammonium acetate; solvent B = Acetonitrile; gradient profile 0 min 95% A, 20 min 50% A, 25 min 95% A, 30 min 95% A; flow rate 1.0 ml/min.

used, to assist in conformation of the colour. A number of methods have been developed that allow the detection of a wide range of colours in one run. The UK's Laboratory of the Government Chemist (LGC) did a lot of work in this area, and they have produced a robust LC-MS-MS method that looks at the analysis of thirteen synthetic pigments in foods and. This illustrates the possibilities that exist now with modern LC-MS systems (Walker, 2006).

10.10.3 Natural pigments

This section examines the analysis of two major classes of compounds:

1 anthocyanin-based materials (e.g. grapeskin extracts and highly coloured juices).
2 carotenoid-based materials (e.g. annatto, β-carotene, mixed carotenes and β-apo-8'-carotenal).

The quantitative analysis of synthetic water-soluble pigments is sometimes difficult, but it is often impossible with natural pigments, because of the lack of pure standards, their prohibitive cost (hundreds of US$ per 100 mg) and their poor stability. There can also be a large natural variation seen between one plant extract and another. Quantitative analysis of these natural extracts is further complicated because they are, generally, a complex mixture of compounds. For instance, a concord grapeskin extract will often contain over 30 different monomeric anthocyanin pigments plus some polymeric materials, all of which contribute to the colour, but they will not all be readily quantifiable by HPLC.

Fortunately, quantitative analysis of the anthocyanins is less critical than analysis of the synthetic pigments, as their use is permitted *quantum satis* in beverages (at any appropriate level), whereas there are defined maximum limits for the synthetic colours. Generally, the analysis of anthocyanin pigments is usually carried out to assess the authenticity of red and black fruit juices, rather than to quantify the level of added colour or to show that a particular juice is present in a drink. However, this is changing with increased interest in the health aspects associated with polyphenolic compounds, such as those seen in grapeskin extracts (anthocyanins in this case), where a quantitative result may be required.

There has been a recent review of the methods available for the analysis of all natural colours (Scotter, 2010). This gives an excellent overview of the area and also details where further work is required to improve or develop the available methods. This review is accessible on the UK FSA's website and can be downloaded from there: (http://foodbase.org.uk/results.php?f_category_id=&f_report_id=529).

10.10.3.1 Anthocyanin pigments

Due to the extensive research carried out on anthocyanins over the years, there is a large amount of information available as to which pigments are found in particular fruits. This allows an analyst to assess whether a sample contains the expected pigments, and to determine if there are any added from another source or fruit. A very good reference book which details the anthocyanins found in various plants and also gives details of other phenolic materials found in fruits is Macheix *et al.* (1990).

Most anthocyanin analysis today is carried out using reverse-phase HPLC, often coupled with diode array detection, as this offers the best analytical approach to assess these compounds. The use of reverse-phase HPLC means that predictions can be made about the elution order of compounds from the column, which is based on their polarity and makes the interpretation of the anthocyanin profiles easier. A couple of useful chapters that discuss the analysis of anthocyanin pigments more fully are those of Lea (1988) and Wrolstad *et al.*

(1995). Both of these describe more fully the types of methods used, and give examples of profiles found in different fruit types.

Another method of analysis that is used around Europe was developed by Hofsommer for the analysis of red and black juices, as part of an EU-sponsored project in the early 1990s on methods to determine the authenticity of fruit juices (Hofsommer, 1994a). This HPLC method has been collaboratively tested by the IFU analytical commission, and was found to give acceptable qualitative results (Anon, 1998b). The procedure does not involve any sample clean-up for juices, the product is only filtered and an internal standard is added prior to analysis by reverse phase HPLC, using detection in the visible region of the spectrum at 518 nm. However, for soft drinks, it is often useful to employ a solid phase cartridge (C_{18}), to concentrate the colour prior to analysis, due to the lower levels in these types of products.

Two examples of this method for juices are given in Figures 10.11 and 10.12. Figure 10.11 shows an HPLC trace for an authentic blackcurrant juice, which has four peaks for the juice, plus the late-eluting pelargonidin internal marker.

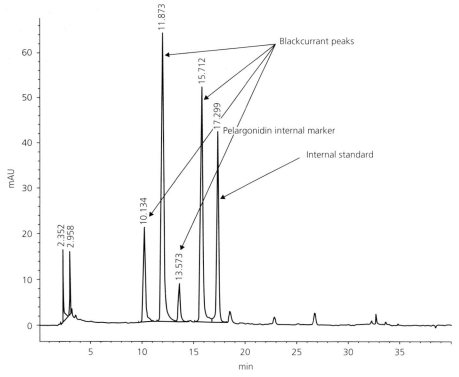

Figure 10.11 HPLC analysis of a blackcurrant juice using the IFU method. Conditions: column: 5 µm ODS2 250 mm × 4.6 mm; flow rate 1 ml/min; solvent A = 10% formic acid; solvent B = 10% formic acid in 50% acetonitrile and 40% water; gradient 0–1 min 12% B, 1–26 min from 12% B to 30% B, 26–35 min 30% to 100 % B, 38 min 100% B, 38–43 min 100 to 12% B; detection at 518 nm with DAD.

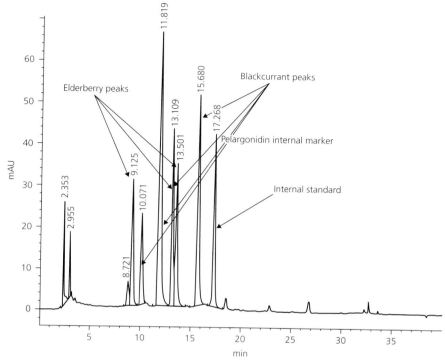

Figure 10.12 HPLC analysis of adulterated blackcurrant juice containing elderberry using the IFU method. Conditions: column: 5 μm ODS2 250 mm × 4.6 mm; flow rate 1 ml/min; solvent A = 10% formic acid, solvent B = 10% formic acid in 50% acetonitrile and 40% water, gradient 0–1 min 12 % B, 1–26 min from 12% B to 30% B, 26–35 min 30% to 100 % B, 38 min 100% B, 38–43 min 100 to 12% B; detection at 518 nm with DAD.

Figure 10.12 shows a chromatogram for an adulterated blackcurrant juice that contains elderberry. In this trace, there are the same five peaks seen in Figure 10.11, but two others are also present (indicated by the arrows), and the intensity of the third peak of the blackcurrant juice (retention time 13.5 min) is enhanced by the co-elution of a pigment from the elderberry (also indicated by an arrow). This demonstrates the importance of this procedure, as the adulteration is instantly detectable using the method. However, juice adulteration is not always as easy as this to detect. Assessment of the anthocyanin profile of a red/black juice should be part of any routine screening procedure designed to assess the authenticity of these products.

The quantitative analysis of anthocyanin pigments by HPLC is more complex, due to the lack of most standards and/or their high costs. For most proposes, it is not essential to measure the absolute levels of pigments, but this could be possible in a semi-quantitative manner, where the concentrations of the individual pigments could be measured relative to a commercially available compound (such as pelargonidin-3-O-glucoside chloride or cyanidin-3-glucoside).

If there is a need to measure the total level of monomeric pigments in a product, this is best done by the spectrophotometric procedure AOAC, 2005.02 (Anon, 2005b). In this method, the solution's absorbance is measured at two wavelengths and at two pH values (pH 1.0 and pH 4.5). From these values, the level of monomeric pigments in solution can then be calculated. The method proposes that the value is expressed as cyanidin-3-glucoside, so this will again only be an, as most fruits contain other anthocyanins. However, if all labs use the same factors in the calculation, consistent results will be obtained.

10.10.3.2 Carotenoid-type materials

The analysis of the natural carotenoid pigments found in orange and mandarin juices is extremely complex, and should be left to an expert laboratory. However, the quantitative analysis of added β-carotene or β-apo-8'-carotenal to an orange drink or squash is much easier, as pure standards are commercially available and these compounds can be readily separated on HPLC. The quantitative analysis of the carotenoid pigments or β-carotene is not critical in soft drinks, as their use is allowed at any level (*quantum satis*).

However, the synthetic materials β-apo-8'-carotenal (E160e) and the ethyl ester of β-apo-8'-carotenic acid (E160f), and the natural materials lycopene (E160d) and lutein (E161b), all have defined limits in beverages and foods, which makes their quantitative analysis more important.

It is very common to find β-carotene, β-apo-8'-carotenal or mixed carotenes in orange-based drinks, flavoured or containing juice, as a colouring material. Although it would sound as if the quantitative analysis of the mixed carotenes could be difficult, this colour is mostly made up of β-carotene, with small levels of the α- and γ-isomers, so it is relatively straightforward. The pigments can be extracted either by solvent (ethyl acetate) or by precipitation of the pigments, using Carraz reagents, followed by extraction of the pellet material with acetone. The extract can then be assessed by spectrophotometer, to give an estimate of the colour usage by determination of the absorbance at 430 nm, or the extract can be assessed by HPLC. Analysis and quantification of these pigments in products with low juice content is relatively straightforward, because there are few sources of interference. Often, a simple gradient or isocratic system, using acetone/water, on a reverse-phase column (RP_{18}), with detection at 430 nm, can be used. A single laboratory validation of an HPLC method for carotenoids was carried out using tomato as a base a few years ago (Dias *et al.*, 2008), and this provides a useful method of the analysis of these materials.

Another pigment that was used widely in soft drinks after the changeover from water-soluble pigments to 'natural pigments' is annatto. This is extracted from *Bixa orellana*, and it contains a range of apo-carotenoid materials based around a C_{24} dicarboxylic acid from the central portion of the carotenoid backbone, as shown in Figure 10.13. This figure also shows the structures of β-carotene and β-apo-8'-carotenal for comparison. After extraction from the plant, the extract is

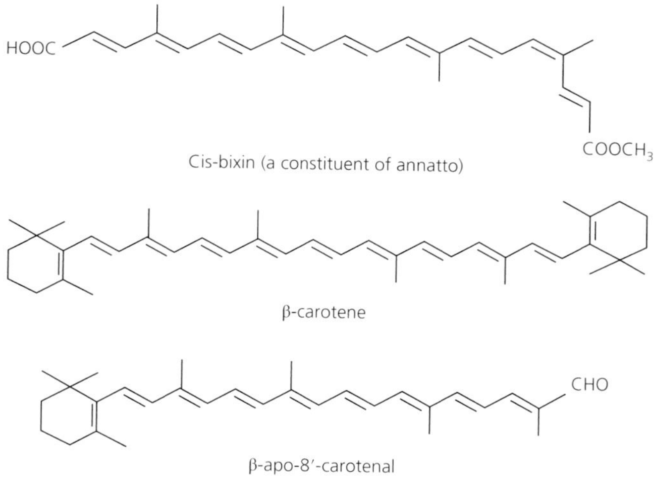

Figure 10.13 Chemical structures for annatto, b-carotene and b-apo-8'-carotenal.

subjected to a basic hydrolysis, to cleave some of the ester groups, giving the water-soluble 'annatto' colour. This is a mixture of free dicarboxylic acids and esterified materials, and its analysis by HPLC gives a number of peaks, because the product is a mixture of a range of closely related compounds.

A recently published method developed for the analysis of bixin and norbixin should be applicable to soft drinks, even though it was developed for meat (Noppe *et al.*, 2009). The chromatographic analysis portion of the method will be applicable, as it uses either diode array or mass spectrometry, but a sample preparation, similar to that described above for other carotenes, should work. It should be noted that annatto is now not permitted as a colour for use in soft drinks within the EU, nor is it permitted in this matrix by the Codex General Standard for Food Additives (GFSA) 1995-192 (Anon, 1995h). However, it is permitted in this matrix in the US and Canada.

A typical example of an HPLC method used to analyse carotenoids is given below, and again comes from work sponsored by the EU in the early 1990s (Hofsommer, 1994b):

Conditions: column: 5 µm ODS2 250 mm 4.6 mm; flow rate 1 ml/min; solvent A = water pH 6.5 (with sulphuric acid), solvent B = methanol, solvent C = acetone, gradient: 0 min 20% A, 25% B and 55% C, 1 min 14% A, 24 % B and 62% C, 22 min 7.2% A, 128% B and 80% C, 42 min 3% A, 7% B and 90% C, 43 min 100% C, 44 min 100% C, 45 min 20% A, 25% B and 55% C; detection at 430 nm.

As β-carotene also has some vitamin A activity, it is sometimes added to products for fortification purposes as well as its use as a colorant. If this is the case, the methods outlined above can be used for this purpose. A similar method was used to look at carotenoid isomers in carrot juices and fortified drinks (Marx *et al.*, 2000).

10.11 Vitamin analysis in soft drinks systems

There is a long history in the UK of selling soft drinks with a high Vitamin C content. Two well-known British companies developed products in the late 1930s to promote the health of children by providing them with beverages rich in vitamin C. Neither of these products were fortified, but relied on the high natural Vitamin C content of orange and blackcurrant juices.

Although the fortification of some foods, like cereals, has been common practice for years in the UK, it was not until the 1990s, with the emphasis on a healthier lifestyle, that the addition of vitamins to soft drinks became more popular. At this time, there were a few fortified products available, such as 'Trink 10' in Germany, which is a mixture of ten juices, fortified with a range of vitamins. However, interest in this type of formulation expanded with the highly successful launch of Sunny Delight in the late 1990s, which encouraged the launch of many similar products on the market.

Vitamin fortification of beverages can offer its own challenges, as the producer has to ensure that the vitamins are stable enough for label claims to be met at the end of its shelf life. Another problem is masking the taste of some of the added vitamins, particularly those from the B group. In a soft drinks system, these can impart a rather 'meaty' aroma/flavour, which would not normally be associated with a soft drink or fruit juice.

The introduction of the revised Fruit Juice Directive in 2003 (Anon, 2003b) also promoted the sale of fortified beverages, as it allowed the addition of vitamins and/or minerals to fruit juices, rather than only allowing their use in fruit juice drinks. If vitamins are added to a product to make a nutritional claim, it is critical that shelf life studies are undertaken to prove that the 'overages' added are sufficient to ensure that the label claim can be met at the end of its shelf life. This is a critical step, as none of the vitamins are fully stable in a soft drink environment, and some are lost very quickly in the presence of oxygen (e.g. Vitamin C). There is no easy way to model this loss using accelerated studies so, if there are plans to launch a fortified product, it is essential that storage trials are carried out early, so that this does do not delay the product's launch.

Trace levels of transition metals can also have a deleterious effect on vitamin loss, so metal scavengers, such as EDTA or phosphate salts, are often added to improve their shelf life. In some countries, there are strict regulations as to how much of an overage may be added to a food/beverage above the RDA whereas, in other countries, the regulations are much more flexible.

The addition of fat-soluble vitamins to soft drinks also offers a formulation challenge, to ensure that they are fully dispersed, and that there are no problems with neck-ringing during storage. Similar problems can also occur in juice drinks fortified with omega-3 fatty acids, which are added to allow a 'promote heart health' claim to be made.

Vitamins are generally assayed in foods and beverages by either a specialist microbiological route or by a chemical procedure. In the former, a specialist organism, whose growth is dependent on the concentration of the particular vitamin of interest, is used. These methods are highly sensitive to interferences that limit the growth of the microorganism, and often take 4–7 days to perform, as they rely on the growth of the organisms under stressed conditions (e.g. limited levels of a critical vitamin). The tests generally offer quite specific results, but require specialist skills.

The second route is the use of HPLC linked with UV/visible, fluorescence or mass spectrometric detection, or a combination. Here, the vitamins are extracted from the product and then assayed by HPLC. Again, this is a specialist area, and it is often best left to an expert laboratory. There can be problems with losses during extraction and/or interferences with materials from the soft drinks matrix.

A few examples of the typical HPLC methods are given below but, if the reader is interested in knowing more about this area, there are good reviews. Ian Lumley, from LGC in the UK (1993), looked at this area in the 1990s, and gives a good background to the area with information on how the methods have developed. The second is a book which covers all the different types of vitamin analyses using modern chromatographic methods, edited by Leenheer, Lambert and Nelis (1992). A further review of methods for water and fat soluble vitamins was published by Wiley and edited by Song *et al.* in 2000. Finally, there is a recent review of methods for water soluble vitamins by Blake (2007).

It is not uncommon to include a hydrolysis step into the extraction procedure before vitamins are analysed. This can either be to hydrolyse ester groups to liberate the free alcohol, in the case of fat-soluble vitamins, or the removal of phosphate groups and so on with some of the B vitamins. There are a wide number of references for the analysis of vitamins in foods, but few of these were designed specifically for juices or soft drinks. Over the last few years, there has been a lot of CEN activity in this area. There now a number of standard methods published in the BSI collection for foodstuffs, and it is likely that there will soon be more methods published for this type of analysis in foods. The present methods cover Vitamins: A (BS/EN 12823-1 and 2), B_1 (BS/EN 14122 2003), B_2 (BS/EN 14152:2003), Niacin (BS/EN 15652:2003), B_6 (BS/EN 14166:2009, 14164:2008 and 14663:2005), C (BS/EN 14130:2003), D (BS/EN 12821:2009), E (BS/EN 12811:2008 and BS/ISO 9936:2006) and K_1 (BS/EN 14148:2003). There are also a number of other standardised methods for vitamin A in milk and milk-based products.

Although there are a number of citations in the AOAC manual, most of these are for vitamin preparations/premixes, infant formula products or fortified milks. There are some methods for foods, rather than beverages, but these should not necessarily preclude their use in fortified drinks. There is a reference to the old titration method for vitamin C and an old microbiological method for

B_6 in foods, and two references for Vitamin A in fortified milk products (2001.13 and 2002.06). There are other references for a range of vitamins in infant formulations, but a number of these are rather old.

10.11.1 Fat-soluble vitamins

If the fat-soluble vitamins (A, D and E) are added to a product, the first step in the analysis will often be that of hydrolysis. In fortified products this converts any esters (e.g. Vitamin A palmitate or Vitamin E acetate) to their free alcohols. After this step, the product is extracted with a hydrophobic solvent such as hexane or diethyl ethyl, prior to HPLC analysis, often using a reverse phase (C_{18}) column.

Some carotenoids also have vitamin A activity, and these can be analysed using the procedures discussed above. Another recent method will separate some of the carotenoids and tocopherol (vitamin E) isomers (Schieber et al., 2002), but this uses a method similar to that of Hofsommer described above for carotenoids in general, rather than specifically for vitamin activity.

10.11.2 Vitamin B class

There are a number of B vitamins; B_1 (thiamine), B_2 (riboflavin), B_3 (niacin), B_5 (pantothenic acid), B_6 (pyridoxine), B_9 (folate) and B_{12} (cyanocobalamine), and their analysis is complex. Column manufactures often show chromatograms demonstrating the resolution of the majority of these vitamins in one run. However, this is generally only for evaluating standards or vitamin premixes. When these compounds are added to a food or beverages at normal levels, the analysis of B_5 in particular is very difficult by HPLC using UV detection, although it is easier using mass spectrometric detection. B_{12} can also be difficult to assay by UV, due to the low levels added in foods although its property of fluorescence may assist. The presence of colours and so on in the beverage can also interfere with the analysis of the vitamins. Diode array detection (DAD) or MS detection is quite critical for this analysis, to be sure the correct material is being evaluated and that the peak is not co-eluting with any interferences (other compounds).

Although Vitamin C is generally recognised as a vitamin of interest in citrus juices, another vitamin which is of interest in this type of fruit juices is folic acid (folate). This substance can be found in low levels in orange juice, but ones significant enough to make a nutritional claim. However, the analysis of natural levels of this vitamin is difficult, due to the many varied forms that the material is seen in nature, so it is often best assayed using a microbiological method. Recent work published by LGC in the UK has provided an HPLC method using either fluorescence or UV detection (Lawrence, 2002) for both natural and added forms of the vitamin. The method uses a folate binding protein in the sample clean-up stage to assist in the clean-up/concentration. The procedure was tested on a range of foods, but not specifically on fruit juices or soft drinks.

10.11.3 Vitamin C

This is one vitamin that most laboratories can measure. There are a number of old-fashioned approaches that use 2,6-Dichloroindophenol (DCPIP) in a titrimetric method such as AOAC 967.21. This works well in some systems, but it can give rise to false positive results if there are other reducing substances present. The method will not detect dehydroascorbic acid (DHA) and so it may well underestimate the actual vitamin C activity in a product if it contains a significant level of DHA. However, even allowing for these shortcomings, it is generally used in production to provide an estimate of the level, using a quick and easily approach. It should be noted that the IFU Analytical Commission is presently evaluating the titrimetric method using both DCPIP and iodine as titrants. Validation data for these two approaches should now be available.

In the AOAC manual, there is a fluorometric method (AOAC 984.26) for Vitamin C. Here, the ascorbic acid is oxidised to dehydroascorbic acid (DHA) with charcoal, and the DHA is then reacted with o-phenylenediamine to give a fluorometric compound that can be detected. This is a robust method that has general applicability and will detect total vitamin C activity (both ascorbic acid and DHA), but it involves a rather long-winded procedure.

Vitamin C can also be effectively detected in juices and soft drinks after treatment with a reducing agent, such as dithiothreitol or cysteine, and then quantified by HPLC on either a reverse phase or an amino bonded column. This type of approach was adopted in EU method 14130:2003 (Anon, 2003c). This is one of the vitamins that should be routinely assayed in orange, grapefruit and blackcurrant juices, as these are all good sources of this compound. In the industry guidelines (AIJN COP) for the quality of juices, a minimum value is set for this vitamin in these products.

10.11.4 Vitamin analysis using immunological procedures

Although microbiological and HPLC methods are the most commonly used methods for vitamin analysis, immunological methods are available for the determination of some of these compounds. These are stand-alone enzyme-linked immunosorbant Assays (ELISA) methods and are offered for B_1, B_2, B_3, B_5, B_6, B_7, B_9, B_{12} and inositol by r-Biopharm, on whose website information can be found (http://www.r-biopharm.de). However, with folate (B_9), there is an issue of evaluating the natural forms compared with added folate, so this type of assay should only be used in fortified products. Some vitamins can also be assayed using an advanced dedicated instrument – the Biacore – a sophisticated instrument that permits immunological assays to be undertaken. Instead of monitoring a change in absorbance of a solution in the UV or visible regions, detection is by changes in the response of the surface of a sensor chip. A method for B_{12} using this approach in infant formulas and vitamin pre-mixes has recently been published in the AOAC compendium (Anon, 2011a).

10.12 Methods used to detect juice adulteration

Over the past 35 years, extensive research has been carried out to find ways to try to stop unscrupulous producers from adulterating fruit juices. Strategies have developed from simple procedures, such as measuring the potassium and nitrogen contents of juices, to the use of highly sophisticated and expensive equipment to detect the most recent approaches that the perpetrators use to extend products. Such adulterations often involve the substitution of some of the fruit juice solids by sugars derived from beet, cane, corn or inulin, or the addition of cheaper juices or second extracts of the fruit. This topic was reviewed a few years ago by Fry *et al.* (1995) and Hammond (1996).

As unscrupulous suppliers adapt their adulteration strategies regularly, it is now common to have to use a battery of tests to confirm that a product is authentic. Although this can be costly, it is the only way to ensure the authenticity of a purchased product in order to protect a company's reputation. The array of tests will often include a number of the procedures described above, such as sugar and acid profiles, and should include stable isotopic and fingerprinting procedures to detect the most sophisticated forms of adulteration.

A number of multi-component approaches have been used to assess fruit juices. Examples include the German RSK system and the French AFNOR system. However, it is now common across Europe to use the AIJN Code of Practice (COP), which can be viewed at their website for a fee (http://www.AIJN.org). In countries that form part of the European Quality Control system (EQCS), it is mandatory to use these guidelines to assess the quality of juices on sale within that market (http://www.eqcs.org). The EQCS is a voluntary industry self-control scheme which evolved from the German system, and it now covers most of the EU states. Its principles are to ensure that products on sale within the countries are all authentic, and that there is fair and even competition between the various companies. In doing so, it also protects the interests of the consumer and the good and healthy image that fruit juices have in the general public's eyes. It offers a number of independent but interlocking schemes that control the quality of the products in their markets. It also offers a network to alert other schemes to any problems that have been detected in their market.

The AIJN COP has also been adopted by the Russian Association as a suitable way to control the quality of juices on sale in their country. A similar system has also been adopted in Australia. However, the values in this guide are a little different from the AIJN's version, as it was set up to cover juices from Australia and the southern hemisphere.

All of the compositional guidelines, RSK, AFNOR and AIJN, define ranges for a number of analytes within which a sample's data should fall. These are valuable approaches, but they should not be used to the exclusion of other, newer procedures, which are sometimes more sensitive to the most sophisticated approaches that adulterators now have to attempt to avoid detection. However, it should not be forgotten that if only a single parameter is outside its accepted range and not covered by the commentary notes, adulteration is not instantly proven. Equally, even if all parameters are within the ranges, adulteration is not necessarily absent. The assessment of whether juices are pure or adulterated is best left to experts who can use their judgement and extensive experience to make a fully rounded assessment. Another important avoidance step involves the availability of a good audit trail from reputable supply sources.

To safeguard their own reputation, individual companies can do some of the analyses, such as most of those laid down in the AIJN COP, but few companies have the time or money necessary to carry out all of these tests or the isotopic procedures. While this should not prohibit them from using these procedures, contract laboratories offering specialist procedures with substantial databases are available around the world.

The risk of buying adulterated or substandard raw materials can be minimised by using a list of inspected and approved suppliers with suitable production methods. This type of QA program can be expensive if an expert needs to visit all of a company's suppliers. However, help is again at hand from SGF, as discussed below.

Although these quality assurance steps are expensive, it could limit a company's exposure to bad publicity or prosecution if adulteration problems occur in the market, such as happened in 1995/1996 with apple juice. A large number of well-known companies were found to be have purchased apple juice adulterated with syrups derived from inulin, which was detected using a newly developed capillary-gas chromatographic fingerprinting method (Low and Hammond, 1996; and IFU recommendation No. 4, Anon, 2004). Recently, there have been issues with the adulteration of pineapple juice and pineapple juice concentrates, where cane and/or corn derived sugars had been added to the product to extend them.

This type of adulteration had been very difficult to prove before the introduction of a new quantitative ^{13}C-NMR method by Eurofins in 2009 (Thomas *et al.*, 2010). This new method allowed this type of adulteration to be more readily detected than it had been before, using conventional or Cap-GC fingerprinting methods. A plot of the carbon-13 contents seen at the methyl (CH_3) and methylene (CH_2) sites of ethanol, prepared from the sugars using the SNIF-NMR® method (AOAC 995.17), is seen in Figure 10.14. As pineapples use the Crassulacean acid metabolism (CAM) to fix carbon dioxide from the atmosphere, it makes it almost impossible to detect, by simple carbon isotopic mass spectrometry, the presence of cane and corn derived sugars added to pineapple juices. However, using the new quantitative ^{13}C-SNIF-NMR method, pineapple sugars are now well separated from those seen from cane or corn, which allows this type of adulteration to be detected. This is illustrated in Figure 10.14, as the sample x is a pineapple juice adulterated with cane sucrose.

Another way of tackling raw material supplier audits is to sub-contract them to an outside body or, to use Germany's Sure Global and Fair (SGF) International

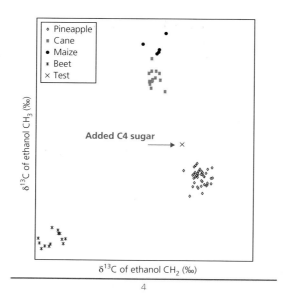

Figure 10.14 ^{13}C-SNIF-NMR plot showing the separation of pineapple, cane, corn and C$_3$ derived sugars.

Raw Material Assurance (IRMA) programme, details of which can be found on their website (www.sgf.org) The IRMA scheme was set up many years ago, and covers a wide range of suppliers who have the same aims as the SGF (i.e. to consistently produce high-quality authentic products). These suppliers are open to unannounced audits by the IRMA team at any time, to ensure that these standards are being met. Although the use of approved suppliers does not exclude the need for a company to test incoming materials, it does limit the risk, and it means that a lower level of testing is required than would otherwise be necessary.

The use of fingerprinting procedures, such as the anthocyanin method (IFU 71) and the Kirksey method for polyphenols (Kirksey *et al.*, 1995) offer two good ways to check for the addition of other fruits in a product. As adulterators have become more sophisticated in the approaches that they use to extend juices, there has been a need to follow them with more complex methods of analysis. This has meant that the use of fingerprinting techniques and isotopic methods to detect adulteration have become more commonplace. The use of internal carbon isotope ratios within the individual sugars of a juice was used to improve the detection limit of C$_4$ (cane/corn) sugars added to the juice (Martin *et al.*, 1997; Hammond *et al.*, 1998; Gonzalez *et al.*, 1999). The work carried out by Hammond's group was developed with financial support from the UK's Food Standards Agency and, using this approach, it was found to halve the detection limit for the addition of C$_4$ derived sugars to orange and apple juices.

In some cases, it is even important to look at the internal isotope ratios seen within a molecule, such as those in malic acid in apple juice. Two groups – Isolab in Germany and Eurofins in France – found it was useful to look at the carbon isotope ratios at the C_1 and C_4 positions of malic acid, relative to the total carbon isotope ratio. This allowed them to detect the addition to apple juice of lower levels of various forms of cheaper synthetic L-malic acid derived from C_3 sources than would be possible in other ways (Jamin et al., 2000).

Sometimes simple methods, such as developed by Low in the early 1990s, using capillary gas chromatography (Cap-GC), can expose major problems of adulteration (Low and Hammond, 1996). Cap-GC is readily available in well-equipped laboratories around the world, and so is easily applied. A well-established technique, the analysis of sugars by capillary gas chromatography, was adapted by a lateral-thinking academic (Prof. Low) to allow the detection of two routes that were known to be used to extend apple juices – the addition of HFCS and invert syrups. This method was also found to be able to demonstrate the adulteration of apple juices with another type of sugar syrup, one derived from inulin, which had only been proposed as a possible adulterant. This method highlighted a major problem of adulteration of apple juices with inulin and corn-derived syrups and invert syrups in 1996. In 2010, using this Cap-GC method, a number of samples of apple juice concentrate were, again, shown to be adulterated with a range of sugar syrups.

Work on the use of internal isotope ratios has continued apace. Eurofins has been able to demonstrate that using the ^{18}O content of the water of a juice, and also the ^{18}O content of the ethanol derived from the sugars after fermentation, allows the detection limit of water added to NFC juices to be improved over the single-parameter approach given in the AIJN's COP (Jamin et al., 2003; Monsallier-Bitea et al., 2006). This technique has recently detected a number of problems in Europe with NFC juices.

In the 1980s and early 1990s, there was a lot of interest in the use of screening methods to control the authenticity of fruit juices, using a range of rapid techniques such as mid-range FT-IR (Smith and Runtz, 1990), NIR (Twomey et al., 1995) and pyrolysis mass spectrometry (Hammond, 1990). However, these techniques fell out of favour as they were found not to be robust enough for normal use. Recently, with the high cost of analysis, there has been a renewed interest in the NIR approach (Leon et al., 2005). Now, although the stabilities of instruments may have improved, similar issues to those seen in the past may recur, so this technique may again fail in the longer term.

However, one rapid fingerprinting approach that has been studied over the last few years, and which looks as if it may offer a significant advance, is ^{1}H-NMR. The results indicate that it may be possible to use this as an initial screening method to detect a wide range of additions to juices. The method is very straightforward; the sample is diluted in buffer, filtered, and then a high field ^{1}H-NMR spectrum of the product is obtained. This data is then subjected to a range of multivariate statistical approaches, which allows samples to be characterised as within or outside the model ('normal' or 'suspect'). The spectra can also be interrogated to determine which region/s is/are causing the discrimination. The method does not preclude the use of other methods for conformation, such as conventional isotopic or DNA approaches. The NMR data can also be used to determine the levels of a number of compounds in the juice, such as the individual sugars and organic acids (Rinke, 2008), from their isolated signals.

FOSS offers a similar system for wines and musts but, in this case, FT-IR is used for the quantification. Here, a wide range of parameters (such as ethanol, pH, volatile acidity, total acidity, levels of tartaric, malic, gluconic, lactic and acetic acids, reducing sugar content, glucose to fructose ratio and density) can all be calculated from one simple infra-red measurement (http://www.foss.dk). This may well be an indication of things to come in the future.

A further fingerprinting method that has come to the fore over the last five years is the use of DNA methods to detect admixtures of juices. For years, mandarin juice was added to early season orange juice, to enhance the poor colour of the product. Now, in the US, it is permitted to add up to 10% mandarin and 5% sour orange juice to orange juice concentrate to bolster the colour, without declaration. The addition of mandarin juice to orange is also permitted by the Codex standard (2005:247) if permitted by local regulation. However, this type of addition is considered an adulteration in Europe. The addition of mandarin to orange affects the carotenoid profile, which can be detected using IFU 59, but the procedure has not been very sensitive in some cases. Surprisingly, it was actually much harder to detect this addition by HPLC, due to the very complex nature of the chromatograms seen for orange and mandarin juices, than it is by the old chromatographic procedure.

To address this issue, work was carried out on behalf of the UK's FSA in the 1990s at Leatherhead Food International, who developed a DNA method to detect mandarin in orange (Knights, 2000). However, this method was not widely used by industry. More recent work in Spain and France has again promoted the use of DNA procedures and, in this case, they have been more widely used. With the method that has been developed at ADNid, in France, it allows the variety/ies of the mandarin/s which have been added to the orange juice to be identified (Moreau and Canivenc, 2008). The method can also be used to look at a range of other citrus juices. This group has gone on to develop a range of other DNA methods for different fruits and mango varieties.

10.13 Methods used to assess the juice or fruit content of soft drinks

The UK Soft Drinks Regulations, 1964 prescribed the minimum juice or fruit contents that certain products had to contain. As examples, a 'crush' had to have a minimum juice content of 5% and, for a 'whole fruit drink', a minimum level of fruit (10%) was required before dilution. When these regulations were revoked in 1995, the various specifications such as drink, crush, and so on, were removed, but the need to define the juice content of a product remained. Under the EU regulations (Anon, 1997b) covering the labelling, presentation and advertising of foodstuffs amendment, the so-called QUID (quantitative ingredient declaration) regulations requires manufacturers to declare, on the label of a food, the levels of characterising ingredient(s). For example, in an orange juice drink, a manufacturer will have to label the orange juice content of the product. The regulations allow the manufacturer to list the content, based on the level of juice in the product formulation, at the so-called 'mixing bowl' stage. Although this means that the producer can put a calculated value on the label, it will still be up to the regulatory authorities to check that this value is appropriate, and they will have to assess the product to determine if the label is correct.

A similar situation exists in the USA with the Nutrition Labelling and Education Act (NLEA). When this was introduced in 1990, it placed the onus on the manufacturer to label the product on the packaging with the individual level of the component juices in, say, a 100% juice blend, or the quantity of juice in a soft drink claiming to contain fruit juice. The conditions for labelling, along with information concerning how the juice content should be assessed, including the minimum Brix level considered acceptable for a range of single-strength juices, are given in the Code of Federal Regulations (CFR) 21 CFR.101.30 (Anon, 1990). This regulation also introduced the need for the producers to label the country of origin/s of the juice/juice blends used in the product, together with their percentages in the blend. There has been a lot of discussion in Europe over the last few years whether a similar approach should adopted in the EU. However, it is unlikely that this will be adopted.

The estimation of juice or fruit content is a difficult area, and a number of approaches have been suggested for its determination. These have involved assessing the ash content, levels of potassium ions, phosphorous content, total level of free nitrogen by measuring the formol value, levels of specific amino acids and a range of other components. The simplest of these procedures, the assessment of the total nitrogen content, phosphorous and potassium levels (N, P, K), was proposed by Hulme *et al.* (1965) as a method to assess the fruit content of comminuted orange squashes. This type of approach can also work for other juices where the levels of these components are well documented.

Useful references in this area are the RSK values (Anon, 1987b), *McCance and Widdowson's The Composition of Foods* (The Royal Society of Chemistry, 2015), Souci *et al.* (1981) and the AIJN code of Practice (Anon, 2011b), which give compositional information on a wide range of juices that can be used to estimate the juice or fruit content of a product. However, when this type of estimate is made, 'normal values' for each parameter should be used, and not the minimum value seen for each parameter. This latter approach can be used, but it would give an absolute maximum value for the juice content. It is also highly unlikely that all the parameters would be at their minimum values at one time. If there is additional information about the product (e.g. the origin is given), then more region-selective values can be used, such as those seen in the SGF database for that region. Only members of SGF can access this data.

If only a simple approach is used to establish the juice content of a drink, such as assessing the levels of nitrogen, phosphorus and potassium, it is easy for an unscrupulous supplier to circumvent. For instance, by the addition of potassium and ammonium phosphate salts to a product, the calculated juice content of the product would be enhanced artificially.

Alternative procedures have been suggested by other authors, such as the use of certain amino acids (Ooghe and Kastelijn, 1985) or organic acids (Wallrauch, 1995). However, because of the wide natural variation seen for all of these components, any calculation of a juice or fruit content should really only be considered as an estimate, rather than an absolute value. The situation is even more complicated for a soft drink which contains a mixture of juices. In this case, the parameters chosen for analysis should show fairly constant levels across the juices within the blend and, often, potassium or phosphate can be selected. If possible, components that are characteristic of a particular juice, such as quinic acid in cranberry juice, or isocitric acid in orange, grapefruit or lemon, should also be analysed. This type of specific marker can be used to assess the overall level of a particular juice in a blend. However, not all juices have a unique marker that can be used in this quantitative sense.

10.14 Conclusions

With the advent of computer-controlled production lines, there should be less risk of mistakes in product formulations. However, errors can still occur, so there should still be some checking of the product – including, in particular, its microbiological stability and regulatory compliance. Formulation checking is possible with HPLC systems that can be used near to the line now, as they require less highly skilled staff to operate. However, it is likely that, in the future, there may be automated systems that can use a spectroscopic technique to check rapidly if a product is within specification, such as supplied by FOSS for wine, which has been discussed above.

Fruit juice companies can do a lot of the testing on their own product to ensure that it meets 'normal' values. However, there can be problems if they are using concentrates from another supplier. They may not have the expertise necessary to assess the fine detail of the product, so the analyst may miss some subtle variations in the data that would be detected by an expert. Thus as it states in the AIJN COP, the assessment is best left to experts.

When setting up a scheme to control the authenticity and quality of raw materials, a range of tests should be used to assess samples, as these will provide the best chance of detecting any problems that exist. This does not mean that every conceivable test has to be carried out on every sample; a range of tests should be applied on a programmed basis, and this will give the best chance of detecting any deviations from normal.

One of the most critical features in the authenticity area is to know the supplier. If this is the case, problems are less likely. It is very hard and costly to develop a good reputation, but this can be lost overnight if a company is detected selling a sub-standard product.

References

Anon (a). Dionex application note number 25: The determination of inorganic ions and organic acids in non-alcoholic carbonated beverages. Available as a downloaded file from the Dionex website (http://www.Dionex.com).

Anon (b). Dionex application note number 67: The Determination of plant-derived neutral oligo- and polysaccharides. Available as a downloaded file from the Dionex website (http://www.Dionex.com).

Anon (c). Dionex application note number 54: The Determination of sulfite in food and beverages by ion exclusion chromatography with pulsed amperometric detection. Available as a downloaded file from the Dionex website (http://www.Dionex.com).

Anon (d). Dionex application note number 21: The Determination of organic acids in wines. Available as a downloaded file from the Dionex website (http://www.Dionex.com).

Anon (e). Dionex application note number 123: The Determination of Inorganic Anions and Organic Acids in Fermentation Broths. Available as a downloaded file from the Dionex website (http://www.Dionex.com).

Anon (f). Dionex application note number 159: Determination of Sucralose Using HPAE-PAD. Available as a downloaded file from the Dionex website (http://www.Dionex.com).

Anon (g). Dionex application note number 241: Determination of Steviol Glycosides by HPLC with UV and ELS Detections. Available as a downloaded file from the Dionex website (http://www.Dionex.com).

Anon (h). CLAS laboratory accreditation system by Campden BRI Research Association (http://www.campden.co.uk/services/clas.htm).

Anon (i). LABCRED laboratory accreditation system offered by Exova (http://www.lawlabs.com/Content.asp?PageID=71&Action=ViewCategory&CategoryID=3).

Anon (1964). UK Soft Drinks Regulations (as amended), Statutory Instrument No. 760, HMSO, London.

Anon (1985). Determination of glucose and fructose enzymatic No. 55. In: *International Federation of Fruit Juice Producers (IFU) Handbook of Analytical Methods*. The International Fruit Juice Union, Paris, France.

Anon (1985). Fibre/Total dietary fibre: Enzymatic-gravimetric method. *AOAC method No* 985.29.

Anon (1986). Quinic, malic, and citric acids in cranberry juice cocktail and apple juice; liquid chromatographic method. *AOAC method No. 986.13*.

Anon (1987a). *Methods of Biochemical Analysis and Food Analysis Using Test-Combinations*. Boehringer Mannheim, Mannheim Germany.

Anon (1987b). *RSK Values – The Complete Manual*, Flussiges Obst, Schonborn.

Anon (1990). Code of Federal Regulations. Percentage juice declaration for foods purporting to be beverages that contain fruit and vegetable juice, 21 CFR Ch.1.103.30, 4-1-96, 77.

Anon (1991a). Fibre/Insoluble dietary fibre: Enzymatic-gravimetric method phosphate buffer. *AOAC method No* 991.42.

Anon (1991b). Fibre Total/Insoluble dietary fibre and soluble fibre: Enzymatic-gravimetric method MES-TRIS buffer. *AOAC method No* 991.43.

Anon (1992). Analysis of Fruit Juice Adulteration with Medium Invert Sugars from Beets. *Dionex Application Note. No 82*. Available as a downloaded file from the Dionex website (http://www.Dionex.com).

Anon (1993a). Fibre soluble dietary fibre; Gravimetric method. *AOAC method no* 993.19.

Anon (1993b). Fibre/Total dietary fibre; Non-enzymatic-gravimetric method. *AOAC method no* 993.21.

Anon (1994a). Directive 94/35/EC on Sweeteners for Use in Foodstuffs. *EC Official Journal*, L178, 10.9.94, 3 and its amendments.

Anon (1994b). Directive 94/36/EC on Colours for use in foodstuffs. *EC Official Journal*, L237/13, 10.9.94, and its amendments.

Anon (1995a). UK Food Regulations (miscellaneous revocation and amendments). Statutory Instrument No. 3267, HMSO, London.

Anon (1995b). UK Sweeteners in Food Regulations 1995, Statutory Instrument No. 3123, HMSO, London (amended as follows: 1996, SI No. 1477, 1997, SI No. 814, 1999, SI No. 982, 2001, SI No. 2294, 2002, SI No. 379, 2003, SI No. 1182 & 2007, SI No. 1778).

Anon (1995c). UK Miscellaneous Food Additives Regulations 1995, Statutory Instrument No. 3187, HMSO, London (amended 1997, SI No. 1413, 1999, SI No. 1136, 2001, SI No. 60 & 3775, 2003, SI no.1008, 2005, SI No. 1099 & 2008, SI No. 42).

Anon (1995d). Directive 95/2/EC on Food Additives Other than Colours and Sweeteners. *EC Official Journal*, L0002, 15.8.95, 1 and its amendments.

Anon (1995e). Preservative (HPLC) test no. 63. In: *International Federation of Fruit Juice Producers (IFU) Handbook of Analytical Methods*. The International Fruit Juice Union, Paris, France.

Anon (1995f). AOAC method 990.31, Sulfites in foods and beverages, ion exclusion chromatographic method. In: Cunniff, J. (ed.) *AOAC Compendium of Methods*, 16th edition, vol. 2, pp.33–34. AOAC, Arlington, Virginia.

Anon (1995g). UK The Colours in Food Regulations 1995, Statutory Instrument No. 3124, HMSO, London (amended 2000 SI No. 481, 2001 SI No. 3442).

Anon (1995h). Codex General Standard for Food Additives, CODEX STAN 192-1995. Available as a download from the codex website: http://www.codexalimentarius.net/gsfaonline/CXS_192e.pdf.

Anon (1995i). Beet sugar in fruit juices site specific natural isotope fractionation – nu clear magnetic resonance (SNIF–NMR®) method. *AOAC Official Method* No 995.17.

Anon (1996a). The successful use of UV detection with the AccQ-Tag method. *Waters AccQ-Tag solutions* March, 2 pages. Available as a downloadable file from the Waters Website (http://www.waters.com).

Anon (1996b). Retention time table for amino acids using the AccQ-Tag method. *Waters AccQ-Tag solutions* July, 2 pages. Available as a downloadable file from the Waters Website (http://www.waters.com).

Anon (1996c). *British Herbal Pharmacopoeia*. Published by the British Herbal Medicines Association.

Anon (1997a). *Application Notes for the Separation of Permanent Gases in Headspaces*, No. 1245, 1247 and 1248. Chrompack Middleburg, The Netherlands.

Anon (1997b). EC Directive 97/4/EC Labelling, Presentation and Advertising of Foodstuffs. *EC Official Journal* **L43**, 14.2.97, 21.

Anon (1997c). Fructans in food products. *AOAC method No.* 997.08.

Anon (1998a). Determination of sucrose enzymatic. No. 56. In: *International Federation of Fruit Juice Producers (IFU) Handbook of Analytical Methods*. The International Fruit Juice Union, Paris, France.

Anon (1998b). Determination of anthocyanins by HPLC No. 71. In: *International Federation of Fruit Juice Producers (IFU) Handbook of Analytical Methods*. The International Fruit Juice Union, Paris, France.

Anon (1999a). *Foodstuffs – Determination of acesulfame-K, aspartame and saccharin – High performance liquid chromatographic method*. BS EN 12856:1999. Available from British standards or one of the other EU standards bodies.

Anon (1999b). *Foodstuffs – Determination of cyclamate – High performance liquid chromatographic method*. BS EN 12857:1999. Available from the British Standards Institute (BSI) or one of the other EU standards bodies.

Anon (1999c). *Fruit and vegetable juices. Determination of free amino acids content liquid chromatographic method*. BS EN 12742:1999. Available from the British Standards Institute (BSI) or one of the other EU standards bodies.

Anon (1999d). Measurement of total fructans in food. *AOAC Official method No.* 999.03.

Anon (2000). Determination of total sulfurous acid test no. 7a. In: *International Federation of Fruit Juice Producers (IFU) Handbook of Analytical Methods*. The International Fruit Juice Union, Paris, France.

Anon (2001a). Fruit Juice and related products Directive 2001/112/ EC. *Official Journal of the European Communities*, **L10/58**, 12.1.2002.

Anon (2001b). Enzymatic-Gravimetric Method, Liquid Chromatography Determination. *AOAC Official method* 2001.03.

Anon (2003a). Determination of Amino Acids in Cell Cultures and Fermentation Broths. *Dionex Application Note 150*. Available as a downloaded file from the Dionex website (http://www.Dionex.com).

Anon (2003b). *Fruit Juice and fruit nectars (England) regulations 2003*. 2003 SI 1564.

Anon (2003c). *Determination of vitamin C by HPLC*. BS EN 14130:2003. Available from the British Standards Institute (BSI) or one of the other EU standards bodies.

Anon (2004). Detection of syrup addition to apple juice by capillary gas chromatography recommendation No 4. In: *IFU Compendium of Analytical Methods for Fruit Juices and Nectars*. The International Fruit Juice Union, Paris, France.

Anon (2005a). Relative Density (Method using density meter) No. 1A. *IFU Compendium of Analytical Methods for Fruit Juices and Nectars*. The International Fruit Juice Union, Paris, France.

Anon (2005b). Total Monomeric Anthocyanin Pigment Content of Fruit Juices, Beverages, Natural Colorants, and Wines. *AOAC method* No. 2005.02.

Anon (2006a). *Guidelines for laboratories performing microbiological and chemical analyses of food and pharmaceuticals – An aid to interpretation of ISO/IEC 17025:2005*. AOAC International Gaithersburg, Maryland, USA.

Anon (2006b). EU regulation 2006, *1924 Nutrition and health claims regulations*.

Anon (2007). *Determination of benzoic acid – spectrophotometric method*. BS ISO 5518:2007. Available from the British Standards Institute (BSI) or one of the other EU standards bodies.

Anon (2008). Recommendation for colour measurement in cloudy juices. No 9 in *IFU Compendium of Analytical Methods for Fruit Juices and Nectars*. The International Fruit Juice Union, Paris, France.

Anon (2009). Total dietary fibre – Enzymatic-gravimetric-liquid chromatography. *AOAC method No* 2009.01.

Anon (2010a). *IFU Compendium of Analytical Methods for Fruit Juices and Nectars*. International Fruit Juice Union, France. (http://www.ifu-fruitjuice.com).

Anon (2010b). *Foodstuffs: Determination of acesulfame-K, aspartame and saccharin. High performance liquid chromatographic method*. BS EN 12856:1999. Available from the British Standards Institute (BSI) or one of the other EU standards bodies.

Anon (2010c). Measurement of the colour of clear and hazy juices (spectrophotometric method) No 80. In: *IFU Compendium of Analytical Methods for Fruit Juices and Nectars*. The International Fruit Juice Union, Paris, France.

Anon (2011a). Vitamin B_{12} in fortified bovine milk-based infant formula powder, fortified soya-based infant formula powder, vitamin premix, and dietary supplements: Surface Plasmon Resonance *AOAC method* No 2011.01.

Anon (2011b). *AIJN Code of Practice for fruit juices*. Available for a fee from AIJN, Rue de la Loi 221, boîte 5, B- 1040, Brussels or from their website (http://www.aijn.org).

Argoudelis, C.J. (1984). Isocratic liquid chromatography method for the simultaneous determination of aspartame and other additives in soft drinks. *Journal of Chromatography* **303**(1), 256–62.

Bergmeyer, H. U. and Brent, E. (1974a). Enzymatic analysis of sucrose. In: Bergmeyer, H.U. (ed). *Methods of Enzymatic Analysis*, 1176–79. Academic Press, London.

Bergmeyer, H.U. and Brent, E. (1974b). Enzymatic analysis of fructose. In: Bergmeyer, H.U. (ed). *Methods of Enzymatic Analysis*, 1304–1307. Academic Press, London.

Bergmeyer, H.U., Brent, E., Schmidt, F. and Stork, H. (1974c). Enzymatic analysis of glucose. In: Bergmeyer, H.U. (ed). *Methods of Enzymatic Analysis*, 1196–1201. Academic Press, London.

Blake C.J. (2007). Analytical procedures for water soluble vitamins in foods and dietary supplements: a review. *Analytical and Bioanalytical Chemisty* **389**(1), 63–76.

Cohan S. (1984). *A pre-column derivatisation method for amino acid analysis*. Waters Lab highlights No. 220. Available as a downloadable file from the Waters website (http://www.waters.com).

Coppola, E.D. and Starr, M.S. (1986). Liquid chromatographic determination of major organic acids in apple juice and cranberry juice cocktail: collaborative study. *Journal of the Association of Official Analytical Chemists* **61**, 1490–92.

Dias, M.G., Camões, M.F. and Oliveira, L. (2008). Uncertainty estimation and in-house method validation of HPLC analysis of carotenoids for food composition data production. *Food Chemistry* **109**, 815–824.

Egan, H., Kirk, R.S. and Sawyer, R. (1990a). Analysis of saccharin. In: *Pearson's Chemical Analysis of Foods*, 8th edition, 215. Longman, Harlow.

Egan, H., Kirk, R.S. and Sawyer, R. (1990b). Analysis of cyclamates. In: *Pearson's Chemical Analysis of Foods*, 8th edition, 216–17. Longman, Harlow.

Egan, H., Kirk, R.S. and Sawyer, R. (1990c). Analysis of benzoic acid. In: *Pearson's Chemical Analysis of Foods*, 8th edition, 216. Longman, Harlow.

Egan, H., Kirk, R.S. and Sawyer, R. (1990d). Analysis of caffeine. In: *Pearson's Chemical Analysis of Foods*, 8th edition, 219. Longman, Harlow.

Egan, H., Kirk, R.S. and Sawyer, R. (1990e). Analysis of Quinine. In: *Pearson's Chemical Analysis of Foods*, 8th edition, 219. Longman, Harlow.

Frazier, R.A., Inns, E.L., Dossi, N. Ames, J.M. and Nursten, H.E. (2000). Development of a capillary electrophoresis method for the simultaneous analysis of artificial sweeteners, preservatives and colours in soft drinks. *Journal of Chromatography A* **876**(1–2), 213–220.

Fry, J., Martin, G.G. and Lees, M. (1995). Authentication of orange juice. In: Ashurst, P.R. (ed). *Production and Packaging of Non-carbonated Fruit Juices and Fruit Beverages*, 2nd edition. Blackie Academic & Professional, Chapman & Hall, London.

German Pharmacopoeia (Deutsches Arznecbach (DAB), 10th edition and its amendments.

Gonzalez, J., Remaud, G., Jamin, E., Naulet, N. and Martin, G.G. (1999). Specific natural isotope profile studied by isotope ratio mass spectrometry (SNIP-IRMS): $^{13}C/^{12}C$ ratios of fructose, glucose, and sucrose for improved detection of sugar addition to pineapple juices and concentrates. *Journal of Agricultural and Food Chemistry* **47**(6), 2316–2321.

Grainger-Bisset, N and Wichtl, M. (1994). *Herbal drug & phytopharmaceuticals*. Published by Med Phram Stuttgart.

Grosspietsch, H. and Hachenburg, H. (1980). Analysis of acesulfame by HPLC. *Zeitschrift für Lebensmittel Untersuchung und Forschung* **171**(1), 41–43.

Hagenauer-Hener, U., Frank, C., Hener, U. and Mosandl, A. (1990). Determination of aspartame, acesulfame-K, saccharin, caffeine, sorbic acid and benzoic acid in foods by HPLC. *Deutsche Lebensmittel Rundschau* **86**(11), 348–51.

Hammond, D.A. (1990). Authentication of fruit juices using multi-component methods and a novel fingerprinting method pyrolysis-mass spectrometry. In: *Methods to detect adulteration of fruit juice beverages*, Volume 1, 112–119. AGScience Auburndale, Florida.

Hammond, D.A. (1996). Methods to detect the adulteration of fruit juice and purees. In: Ashurst, P.R. and Dennis, M.J. (eds). *Food Authenticity*. Blackie, Academic & Professional, Chapman & Hall, London.

Hammond, D.A. Correia, P., Day, M.P. and Evans, R. (1998). ^{13}C-IRIS – A refined method to detect the addition of cane/corn derived sugars to fruit juices and purees. *Fruit Processing* **3/98**, 86–90.

Harms, D, Nitzschi, F and Jansen, D. (2000). *Determination of carbon dioxide in carbonic acid containing drinks*. Poster presented at Pitcom 2000.

Hofsommer, H.J. (1994a). *Analysis of anthocyanins in fruit juices*. Paper presented as part of the SGF symposium: 'Progress in the authenticity-assurance for fruit juices', Parma, Italy, September 1994.

Hofsommer, H.J. (1994b). *Analysis of carotenoids in fruit juices*. Paper presented as part of the SGF symposium 'Progress in the authenticity-assurance for fruit juices', Parma, Italy, September 1994.

Holland, B., Welch, A.A., Unwin, I.D., Buss, D.H., Paul, A.A. and Southgate, D.A.T. (1991). *McCane and Widdowson's The Composition of Foods* (complied by the Royal Society of Chemistry and the Ministry of Agriculture Fishery Foods), Royal Society of Chemistry, Cambridge.

Hulme, B., Morris, P. and Stainsby, W.J. (1965). Analysis of citrus fruit 1962–1963. *Journal of the Association of Public Analysts* **3**, 113–17.

Jamin, E., Lees, M. Fuchs, G. and Martin, G.G. (2000). Detection of added L-malic acid in apple and cherry juice – site specific ^{13}C-IRMS method. *Fruit Processing* **11**, 434–436.

Jarmin, E., Guérin, R., Rétif, M., Lees, M. and Martin, G.J. (2003). Improved detection of added water in orange juice by simultaneous determination of the oxygen-18/oxygen-16 isotope ratios of water and ethanol derived from sugars. *Journal of agricultural & food chemistry* **51**(18), 5202–5206.

Junge, C. and Spandinger, C. (1982). Detection of the addition of l and d, l malic acids in apple and pear juice by quantitative determination of fumaric acid. *Flussiges Obst* **49**(2), 57–62.

Kirksey, S.T., Schwartz, J.O., Hutfliz, J.A., Gudat, M.A. and Wade, R.L. (1995). HPLC Analysis of polyphenolic compounds for juice authenticity. In: *Methods to detect adulteration of fruit juice beverages*, Vol. 1, 145–166.

Knights, A. (2000). Development and validation of a PCR-based heteroduplex assay for the quantitative detection of mandarin juice in processed orange juices. *Agro Food Industry Hi-Tech* **11**(2), 7–8.

Lawrence, P. (2002). *Development and validation of an HPLC method for the determination of folate in food*. FSA foodbase database: http://www.foodbase.org.uk/results.php?f_report_id=259.

Lawrence, J.F. and Charbonneau, C.F. (1988). Determination of seven artificial sweeteners in diet food preparations by RP-liquid chromatography with absorbance detection. *Journal of the Association of Official Analytical Chemists* **71**(5), 934–37.

Lea, A.G.H. (1988). HPLC of natural pigments in foodstuffs. In: Macrae, R. (ed). *HPLC Analysis of Foods*, 2nd edition, Academic Press, London.

Leenheer, A.P., Lambert, W.E. and Nelis, H.J. (1992). *Modern Chromatographic Analysis of the Vitamins*. Marcel Dekker Inc., New York.

Leon, L.J., Kelly J.D. and Downey G. (2005). Detection of Apple Juice Adulteration Using Near-Infrared Transflectance Spectroscopy. *Applied Spectroscopy* **59**(5), 593–599.

Low, N.H. and Hammond, D.A. (1996). Detection of high fructose syrup from inulin in apple juice by capillary gas chromatography with flame ionisation detection. *Fruit Processing* **4/96**, 135–141.

Lumley, I.D. (1993). Vitamin analysis in foods. In: Berry Ottaway, P. (ed). *The Technology of Vitamins in Food*. Blackie Academic & Professional, Glasgow, Scotland.

Macheix, J.-J., Fleuriet, A. and Billot, J. (1990). *Fruit Phenolics*. CRC Press, Boca Raton, Florida.

Martin, G. G., Jamin, E. Gonzalez, J., Remaud, G., Hanote, V., Stober, P. and Naulet, N. (1997). Improvement of the detection level of added sugar with combined isotopic and chemical analysis. *Fruit Processing* **9/97**, 344–349.

Marx, M., Schieber, A. and Carle, R. (2000). Quantitative determination of carotene stereoisomers in carrot juices and vitamin supplemented (ATBC). drinks. *Food Chemistry* **70**(3), 403–408.

McCann, D., Barrett, A., Cooper, A., Crumpler, D., Dalen, L., Grimshaw, K., Kitchin, E., Lok K., Porteous, L., Prince, E., Sonuga-Barke, E., Warner, J.O. and Stevenson, J. (2007). Food additives and hyperactive behaviour in 3-year-old and 8/9-year-old children in the community: a randomised, double-blinded, placebo-controlled trial. *The Lancet* **370**(9598), 1560–1567.

McKaren, K. (1980). Food Colorimetry. In: Walford, J. (ed). *Developments in Food Colours – 1*, 27–45. Elsevier Applied Science, London.

Monier-Williams, G.W. (1927). Determination of sulfur dioxide. *Analyst* **52**, 415–16.

Monsallier-Bitea, C., Jamin, E., Lees, M., Zhang, B-L., and Martin G.J. (2006). Study of the influence of alcoholic fermentation and distillation on the oxygen-18/oxygen-16 isotope ratio of ethanol. *Journal of Agricultural & Food Chemistry* **54**(2), 279–284.

Moreau, F. and Canivenc G. (2008). DNA as a tool for *citrus reticulata* adulteration detection and variety detection in commercial orange juices. *Fruit Processing* **2008**(3), 156–159.

Noppe, H., Abuin Martinez, S., Verheyden, K., Van Loco, J., Companyó Beltran, R., and De Brabander, H.F. (2009). Determination of bixin and norbixin in meat using liquid chromatography and photodiode array detection. *Food Additives and Contaminants* **26**, 17–24.

Ooghe, W. and Kastelijn, H. (1985). Amino acid analysis, a rapid, low cost, reliable screening test for evaluating commercially marketed fruit juices. *Voedingsmiddelentechnologie* **18**(23), 13–15.

Ough, C.S. (1988). Determination of sulfur dioxide in grapes and wines. In: Linskens, H.F. and Jackson, J.F. (eds). *Modern Methods of Plant Analysis: Wine Analysis*, Vol. 6, 342–43. Springer-Verlag, Berlin.

Rinke, P. (2008). Successful application of SGF-Profiling. *New Food* **1**, 18–23.

Schieber, A., Marx, M. and Carle, R. (2002). Simultaneous determination of carotenes and tocopherols in ATBC drinks by HPLC. *Food Chemistry* **76**(3), 357–362.

Scotter, M. (2010). *Review and evaluation of available methods of extraction and analysis for approved natural colour in foods and drinks*. UK Food Standards Agency report FD/10/02. Available from the FSA website: http://www.foodbase.org.uk//admintools/reportdocuments/529-1-925_A01074.pdf.

Smith, J.A. and Runtz, L.A. (1990). Chemometrics with attenuated total reflectance spectroscopy to detect common adulterants in orange juice concentrate. In: *Methods to detect adulteration of fruit juice beverages*, Vol. 1, 112–119. AGScience Auburndale, Florida.

Song W.O., Beecher, G.P. and Eitemiller R.R. (2000). *Modern analytical methodologies in fat and water soluble vitamins*. Wiley interscence, Hobken, NJ, USA.

Souci, S.W., Fachmann, W. and Kraub, H. (1981). In: Von Scharze, H. and Kloss, G. (eds). *Die Zusammersetzung de Lebensmittel-Nahruert Tabbellen 1981/82*. Wiss Verlagesges, Stuttgart.

Tanner, H. (1963). The estimation of total sulfur dioxide in drinks, concentrates and vinegars. *Mitteilungen Gebiete Lebensmitteluntersuchung und Hygiene* **54**, 158–74.

Thomas, F., Randet, C., Gilbert, A., Silvestre, V., Jamin, E., Akoka, S., Remaud, G., Segebarth, N. and Guillou, C. (2010). Improved characterization of the botanical origin of sugar by carbon-13 SNIF-NMR applied to ethanol. *Journal of Agricultural and Food Chemistry* **58**, 11580–11585.

Tsang, W.S., Clarke, M.A. and Parrish, F.W. (1985). Detection and determination of aspartame and its breakdown products in soft drinks by RP-HPLC with UV detection. *Journal of Agriculture and Food Chemistry* **33**(4), 734–38.

Twomey, M, Downey, G. and McNulty, P.B. (1995). The potential of NIR spectroscopy for the detection of the adulteration of orange juice. *Journal of the Science of Food and Agriculture* **67** (1), 77–84.

Valenti, L.P. (1985). Liquid-chromatographic determination of quinine, hydroquinine, saccharin and sodium benzoate in quinine beverages. *Journal of the Association of Official Analytical Chemists* **68**(4), 782–784.

Wadds, G. (1984). Analysis of synthetic food colours. In: Walford, J. (ed). *Developments in Food Colours – 2*, 23–74. Elsevier Applied Science, London.

Walker, B (2006). *Analysis of illegal dyes in paprika powder by LC-MSMS*. Available as a download from the LGC website: http://www.governmentchemist.org.uk/dm_documents/Analysis%20of%20illegal%20dyes%20in%20paprika%20powder%20by%20LC-MSMS_cwxn4.pdf.

Wallrauch, S. (1995). Arbeitsgruppe Fruchsäft und fruchtsafthaltige Getränke. *Lebensinittelchemie* **49**, 40–45.

Wedzicha, B.L. (1984). *The Chemistry of Sulphur Dioxide in Foods*. Elsevier Applied Science, London.

Williams, M.L. (1986). Rapid separation of soft drinks ingredients using HPLC. *Food Chemistry* **22**(3), 235–44.

Woidich, H, Gnauer, H. and Galinovsky, E. (1967). Thin layer chromatographic separation of some food preservatives. *Zeitschrift fuer Lebensmittel Untersuchung und Forschung* **133**, 317–22.

Woodward, B.B., Heffelfinger, G.P. and Ruggles, D.I. (1979). HPLC determination of sodium saccharin, sodium benzoate and caffeine in soda beverages: collaborative study. *Journal of the Association of Official Analytical Chemists* **62**(5), 1011–19.

Wrolstad, R.E, Hong, V., Boyles, M.J. and Durst, R.W. (1995). Use of anthocyanin pigment analysis for detecting adulteration in fruit juices. In: Nagy, S. and Wade, R.L. (eds). *Methods to Detect the Adulteration of Fruit Juice Beverages*. AGSCIENCE, Auburndale, Florida.

Zache, U. and Gruending, H. (1987). Determination of acesulfame-K, aspartame, cyclamate and saccharin in soft drinks containing fruit juice. *Zeitschrift fuer Lebensmittel Untersuchung und Forschung* **184**(6), 503–09.

CHAPTER 11
Microbiology of soft drinks and fruit juices

Peter Wareing
Principal Food Safety Advisor, Leatherhead Food Research, Leatherhead, UK

11.1 Introduction

The cold beverage market, represented by what are termed soft drinks and fruit juices, continues to expand in diversity and size, with an ever-expanding variety of products being released every year. This has increased the potential for spoilage problems. Formerly, the majority of soft drinks were nutrient-poor media which were spoiled by relatively few organisms – usually yeasts and a few acid-tolerant bacteria and fungi. Many were carbonated, which shifted the spoilage flora to those organisms that are tolerant of carbon dioxide. Many products are now still (not carbonated), and enhanced by the addition of low levels of fruit juice, which tends to allow similar spoilage flora to pure fruit juices. The use of ever more exotic raw ingredients may lead to the discovery of unusual spoilage organisms in the future. Yeasts in general, and *Zygosaccharomyces bailii* in particular, remain the key spoilage organisms, because of their overall physiology and resistance to organic acid preservatives (Stratford *et al.*, 2000).

Microbial problems within soft drinks and fruit juices can be divided into two groups:

1 Growth in, and deterioration of, the product by general organisms to produce spoilage;
2 Growth in, or contamination of, the product by pathogens to produce food poisoning.

There have been relatively few instances of food poisoning in fruit juices or soft drinks, but microbial spoilage is very common. Previous editions of this book gave excellent reviews of, and practical guides to, the identification of yeast spoilage problems in the soft drink industry – in particular, by the late Professor Davenport (1998), to which the reader is referred, if required.

Chemistry and Technology of Soft Drinks and Fruit Juices, Third Edition. Edited by Philip R. Ashurst.
© 2016 John Wiley & Sons, Ltd. Published 2016 by John Wiley & Sons, Ltd.

11.2 Composition of soft drinks and fruit juices in relation to spoilage

There is a huge range of soft drinks and fruit juices for sale, and many methods for their manufacture. Soft drinks can be non-carbonated, carbonated, with or without added fruit juice and, often, with the addition of organic acid preservatives. They can be filled on standard, clean fill lines or aseptic packaging lines. Fruit juices, fruit juice concentrates and fruit nectars may be fresh, unpasteurised and clean filled, or pasteurised, then hot-, aseptic-, or clean-filled (Stratford *et al.*, 2000; Stratford and James, 2003). Each of these products and processing types can be associated with a unique microbial flora.

Recent technology using high-pressure processing (HPP) has been used to produce 'cold pasteurised' fruit juices. These have the advantage of a fresh juice mouthfeel, but with destruction of pathogens and the majority of spoilage agents, enhancing the shelf life of an essentially fresh product (Mermelstein, 1999; Zook *et al.*, 1999). The synergistic effects of HPP and essential oils – for example, limonene and carvacrol – has been investigated to deliver an increased kill in fruit juices (Espina *et al*, 2013). Pulsed Electric Fields (PEF) has also similarly been used as a lighter processing method (Juvonen, 2011). Other non-thermal techniques for soft drinks include pulsed light and ultrasound (Ferrario *et al*, 2014), repetitive UV irradiation (Shamsudin *et al*, 2014) and ultrafiltration (Larko *et al*, 2013).

Recent trends for new and exotic natural flavours, sourced from increasingly remote supply bases, means that there is a greater likelihood of unusual spoilage organisms being imported with the flavours and juices. In addition, the 'clean label' movement also means that, where possible, soft drinks are preservative-free and, often, lightly processed, to minimise destruction of these new and often delicate flavours. Such products will often have to be chilled, due to light or no processing and being without preservatives. Exotic fruits may also be less acidic than those traditionally used, making them more susceptible to spoilage. Certainly, an increase in the number of microbial issues has been seen, in part due to an increase in the use of unusual ingredients.

The same processing and preservation issues arise with the rise in popularity of sports, health or added nutrition drinks, due to the nature of their ingredients. In addition, some products may be neutral pH in some cases, and often with low sugar or artificial sweeteners.

Previous chapters detail the types of products and outline the range of microbial issues to which they can be subjected. For example, there is a likelihood of unpreserved squashes and cordials with a short shelf life being subjected to microbial spoilage if they are not stored in the fridge after opening. They are not usually of sufficiently low water activity to prevent the growth of yeasts and moulds that have contaminated the product post-opening.

Simple soft drinks, such as orangeade and lemonade, are usually too acidic for the growth of most organisms, so that spoilage is generally by carbonation-resistant species such as *Dekkera anomala* (Stratford and James, 2003). Yeasts usually require a carbon source, such as a hexose sugar, a nitrogen source, such as amino acids or ammonium salts, simple salts (phosphate, sulphate, potassium and magnesium ions), trace minerals and vitamins. Some yeasts have particular sugar requirements – for example, *Z. bailii* and *Z. rouxii* cannot utilise sucrose (Pitt and Hocking, 2009; Stratford *et al.*, 2000).

Sugars have a protective effect on the heat resistance of yeasts and bacteria, and this is an important consideration at higher concentrations of sugar. Soft drinks are often nitrogen-poor and, thus, the addition of fruit juice greatly enhances the potential for spoilage. Some yeasts, for example, *Dekkera bruxellensis*, can use nitrate. Phosphate levels are often low while trace minerals are satisfactory, particularly in hard water areas. The low pH value of soft drinks and fruit juices, pH 2.5–3.8 (Table 11.1), inhibits most bacteria, but leaves yeasts unaffected. In soft drinks and fruit juices, oxygen levels are usually low, and CO_2 levels either low or very high (in carbonated soft drinks). Spoilage is, therefore, due to facultative anaerobes (organisms that can grow with or without oxygen). The addition of even small quantities of fruit juice to any product makes it more sensitive to microbial growth, as fruit sugars increase the risk of yeast growth.

Microbial growth can take some time to get started after processing and packing, if survivors occur or post-process contaminants get into the product. However, the shelf life of an ambient product is long enough to give time for growth to occur.

In carbonated drinks, mould and aerobic bacterial growth is very unlikely, as these organisms are very sensitive to CO_2. Generally, CO_2 is required at a

Table 11.1 Examples of fruit and vegetable juice pH and risk organisms.

	Approximate pH ranges	Risk organisms
Fruits		
Apples	2.9–3.91	Yeasts
Grapes	3.2–4.51	Yeasts
Oranges	3.20–4.3	Yeasts
Raspberries	3.12	Yeasts
Blackcurrants	2.48–3.60	Yeasts
Pineapples	3.3–3.7	Yeasts and bacteria
Mangoes	3.95–4.50	Yeasts and bacteria
Tomatoes	3.80–4.80	Yeasts, bacteria and moulds/bacteria
Vegetables		
Carrot	4.90–6.44	Bacteria
Celery	5.7–6.1	Bacteria
Cabbage	5.4–6.0	Bacteria
Pea	6.65–6.77	Bacteria

minimum of 1.7 volumes (3.3 g/l) and above to inhibit aerobic microorganisms. For example, most bacteria and moulds show increasing inhibition as CO_2 level increases to 3.0–3.5 volumes (6.5–6.9 g/l). At this upper level, all but yeasts are normally inhibited.

The recent trend towards the use of polyethylene terephthalate (PET) packaging presents its own problems. Although most PET containers can be hot filled at 85°C, some cannot, which means that a hot fill is not possible. The blowing process can generate a static charge on the bottle, which can allow airborne mould spores to attach to the bottle. If these are heat-resistant, they can survive the hot fill process, growing and causing spoilage over the shelf life of the product. The solution is to rinse the bottles with sterile air or water prior to filling. PET containers are also permeable to oxygen (Rodriguez *et al.*, 1992), which allows the growth of aerobic spoilage agents.

11.3 Background microbiology – spoilage

Many microorganisms are found in soft drinks as environmental or raw material contaminants, but relatively few can grow within the acidic and low-oxygen environment. Yeasts are the most significant group of microorganisms associated with spoilage of soft drinks and fruit juices. Spoilage will usually be noticed as the growth and production of metabolic by-products (e.g. CO_2), and the development of acidic and tainting compounds. These are likely to give rise to off-flavours or odours, or changes to the mouthfeel of the product. There may also be visible growth – for example, the development of a haze in an otherwise clear product. If fermentation is vigorous enough, there may be distention or rupture of containers. Finally, production of pectinases may lead to clearing or separation of fruit juices – although it should be noted that poor upstream juice processing can also give rise to similar effects. As noted above, most spoilage is, therefore, by yeasts and mould species, with yeasts generally the more important of the two groups and, finally, some spoilage by acid-tolerant bacteria (Hocking and Jensen, 2001; Jay and Anderson, 2001).

11.3.1 Sources
Fruit and fruit juices are commonly contaminated with yeasts and moulds from the environment, and also from insect damage. Fallen fruit should therefore be avoided where possible, for all the risks outlined below. Sugars and sugar concentrates are commonly contaminated with osmophilic yeasts (e.g. *Z. rouxii*). Growth is slow in concentrated solutions, but one cell per container of diluted stock is enough to cause spoilage (Davenport, 1996). Water and other chemicals are all potential sources of microbial contamination, as are flavourings and colouring agents. Flavourings are now mostly based on alcoholic or other non-aqueous solvents and are, consequently, rarely a source of contamination.

Process machinery and filling lines are particularly problematic, and strict hygiene is essential.

11.3.2 Yeasts

There are over 800 species of yeasts currently described (Barnett *et al.*, 1990), but only about ten are commonly associated with spoilage of foods prepared in factories that operate good standards of hygiene and use correctly applied chemical preservatives (Pitt and Hocking, 2009). Others are found if something goes wrong during manufacture – for example, incorrect preservative level, poor hygiene or poor-quality raw ingredients. Fermenting yeasts are very resistant to CO_2, whereas film forming yeasts are more CO_2-sensitive.

Typical effects of yeast spoilage include development of taints of acetic acid (vinegary), yeasty, ethanolic buttery, acetoin, and diacetyl characteristics. Gas production and the development haze or cloudiness are also often noted (Juvonen *et al.*, 2011).

Davenport (1996, 1997, 1998) described a simple classification scheme for yeasts causing spoilage in the soft drinks industry. He found that the yeasts isolated could be divided into four categories, designated Groups 1–4:

- *Group 1 yeasts* are described as spoilage organisms that are well adapted to growth in soft drinks and are able to cause spoilage from very low cell numbers (sometimes as few as one cell per container). The characteristics of Group 1 yeasts are osmotolerance, aggressive fermentation, resistance to preservatives (particularly weak organic acids) and a requirement for vitamins. *Z. bailii* is a typical example of this group, and this group corresponds closely with Pitt and Hocking's (2009) key spoilage yeasts.
- Davenport describes *Group 2 organisms* as spoilage/hygiene types, able to cause spoilage of soft drinks, but only if something goes wrong during manufacturing – for example, too low a level (or absence) of a preservative, ingress of oxygen, failure of pasteurisation or poor standards of hygiene. Group 2 organisms are common contaminants in factories, but can be severely restricted if good manufacturing practices (GMP) are under control.
- *Group 3 organisms* are indicators of poor hygiene standards. These yeasts will not grow in soft drinks, even if present in high numbers, and are typical of the yeasts found in many factories; the higher the count, the worse the hygienic state of the factory.
- *Group 4 yeasts*, called 'aliens' by Davenport, are those out of their normal environment. An example would be *Kluyveromyces lactis*, a dairy spoilage yeast, which ferments lactose, not glucose or fructose. Table 11.2 details the most common organisms in each group.

One of the most confusing aspects of mould and yeast taxonomy is the frequency with which names are changed. Many yeasts and moulds have sexual and asexual stages that may have different names. Sometimes, the association of a yeast with a particular genus is recognised as wrong, and a name change then occurs. The current trend is to name the organism by the sexual name, with the

Table 11.2 Examples of yeast species found in soft drink factory environments.

Group 1 – Fermentation- and preservative-resistant	Group 2 – Spoilage and hygiene	Group 3 – Poor hygiene	Group 4 – Aliens
Dekkera anomala	Candida davenportii	Aureobasidium pullulans	Kluyveromyces marxianus (dairy yeast)
Dekkera bruxellensis	Candida parapsilopsis	Candida sake	Kluyveromyces lactis (dairy yeast)
Dekkera naardenensis	Debaryomyces hansenii	Candida solani	
Saccharomyces (atypical strains)	Hanseniaspora uvarum	Candida tropicalis	
Saccharomyces exiguus	Issatchenkia orientalis	Clavispora lusitania	
Schizosaccharomyces pombe	Lodderomyces elongisporus	Cryptococcus albidus	
Zygosaccharomyces bailii	Pichia anomala	Cryptococcus laurentii	
Zygosaccharomyces lentus	Pichia membranifaciens	Debaryomyces etchellsii	
Zygosaccharomyces rouxii	Saccharomyces bayanus	Rhodotorula glutinis	
Zygosaccharomyces bisporus	Saccharomyces cerevisiae	Rhodotorula mucilaginosa	

Sources: Based on Davenport (1996); Stratford and James (2003).

asexual stage described as a synonym. Thus, *Candida famata* (asexual) becomes *Debaryomyces hansenii* (sexual). It is important to be aware of this variation when conducting literature searches. Also, names do not always reflect the spoilage potential. For example, *Saccharomyces cerevisiae* is an important yeast for the brewing and baking industries, but a disaster for the soft drinks and fruit juice industry. Table 11.3 details alternative names for important yeasts.

11.3.3 Bacteria

Bacteria that have been associated with spoilage in the soft drinks industry include *Acetobacter, Alicyclobacillus, Bacillus, Clostridium, Gluconobacter, Lactobacillus, Leuconostoc, Saccharobacter, Zymobacter* and *Zymomonas* (Vasavada, 2003). Most bacterial spoilage agents are found exclusively in still products, with the exception of lactic acid bacteria. Although *Clostridium* is an anaerobe, it cannot grow in carbonated products because the pH of these products is too low.

Gluconobacter, an acetic acid bacterium, is a common spoilage agent of fruit juices. It is a strict aerobe, requiring free oxygen (Stratford *et al.*, 2000). Spoilage by acetic acid bacteria can lead to a sour, vinegary off-note, haze, ropiness, production of gluconic acid and acetic acid and acetoin.

Lactic acid bacteria can produce lactic acid and spoilage off-notes such as diacetyl, which can give a cheesy or buttery flavour. Gas production, cloudiness, and ropiness may also be noted (Juvonen *et al*, 2011).

Table 11.3 Names of the most important yeasts within the soft drinks industry, with synonyms.

Current name	Synonym
D bruxellensis	Brettanomyces intermedius/Brettanomyces bruxellensis
D. naardenensis	Brettanomyces naardenensis
S. cerevisiae	Saccharomyces carlsbergensis
S. bayanus	Saccharomyces uvarum/Saccharomyces heterogenicus
S. exiguus	Candida/Torulopsis holmii
Z. bailii	Saccharomyces bailii
Z. bisporus	Saccharomyces bisporus
Z. lentus	Saccharomyces lentus
Z. rouxii	Saccharomyces rouxii
Torulaspora delbreuckii	Saccharomyces delbreuckii/Saccharomyces rosei/Candida colliculosa
Pichia anomala	Hansenula anomala
P. membranifaciens	Candida valida
Issatchenkia orientalis	Candida krusei
Kluyveromyces marxianus	C. kefyr
K. lactis	C. sphaerica
Debaryomyces hansenii	C. famata
Pichia guilliermondii	C. guilliermondii
Hanseniaspora uvarum	Kloeckera apiculata
Metchnikowia pulcherrima	Candida pulcherrima

Alicyclobacillus spoilage is an important issue in heat-treated fruit juices, and the bacterium is commonly found contaminating fruit in the field. Growth of the organism is associated with the production of antiseptic and 'smoky' taints within juice. The former is due to 2,6-dibromophenol (2,6-DBP), which gives a 'TCP' flavour (Jensen and Whitfield, 2003); the latter is due to guaiacol (2-methoxyphenol) (Jensen, 1999). Heat shock, the presence of taint precursors, incubation temperature and oxygen are all important factors in the production of taints (Jensen, 1999). Recent evidence suggests an association of high fructose corn syrup with *Alicyclobacilus*. This also applies with apple and orange juice (Yokata *et al*, 2007).

Alicyclobacillus acidoterrestris is a thermotolerant, acidophilic bacterium ('TAB'). It is also a spore-forming strict aerobe that can survive fruit juice heat treatments to grow in the juice post-processing. Spoilage can occur from inoculum levels as low as 1 spore per 10 ml (Pettipher and Osmundon, 2000; Walls and Chuyate, 2000a). Growth minima and maxima are at temperatures of 26–50°C, minimum pH 2.0, D^{95} of 2.4 min. Isolation methods have been compared (Pacheco, 2002; Pettipher and Osmundson, 2000; Silva and Gibbs, 2001; Walls and Chuyate 2000b). Effective control can be achieved by rapid chilling of juice to below 20°C after pasteurisation, the use of sorbate or benzoate, removal of oxygen or addition of ascorbic acid (Cerny *et al.*, 2000), or by using an appropriate pasteurisation regime (Silva and Gibbs, 2001). The organism will not grow in carbonated products. Surface disinfection of fruit using chlorine or 4% peroxide has also proved effective (Orr and Beuchat, 2000).

Propionibacterium cyclohexanicum is another heat-resistant spore-forming bacterium that has been associated with pasteurised orange juice, from which an off-flavour was noted (Kusano, 1997). It can survive thermal processes of 95°C for ten minutes (Walker and Phillips, 2007).

11.3.4 Moulds

Mould problems can be divided into two types: (i) growth of a variety of moulds due to poor hygiene within the factory or field environment; and (ii) growth of heat-resistant moulds within heat-processed juices. The former type can cause tainting, discolouration and other general problems associated with gross mould growth. The latter type can result in slow growth of the mould within the processed product. There is some overlap between the two groups. Xerophilic (highly sugar-tolerant) fungi are likely contaminants if hygiene is poor.

In the former case, any airborne mould could contaminate finished products, but heavily sporing genera would be expected to be more common (e.g. *Penicillium, Alternaria, Cladosporium, Aspergillus, Eurotium* and *Fusarium*).

Heat-resistant moulds (HRM) able to cause spoilage of fruit juices and soft drinks include *Aspergillus ochraceus, Aspergillus tamarii, Aspergillus flavus, Byssochlamys nivea, Byssochlamys fulva, Paecilomyces variotii, Neosartorya fischeri, Eupenicillium brefeldianum, Phialophora mustea, Talaromyces flavus, Talaromyces trachyspermus* and *Thermoascus aurantiacum*. Others include *Penicillium notatum, Penicillium roquefortii* and *Cladosporium* spp. (de Nijs *et al.*, 2000; Pitt and Hocking, 2009; Stratford *et al.*, 2000). They are able to survive typical thermal processes for many soft drinks and fruit juices – for example, 72–80°C for 10–20 minutes for tunnel pasteurisation, or a flash pasteurisation of 90–95°C for 30–90 seconds (Juvonen *et al*, 2011).

Mould spoilage within the factory is associated with poor hygiene. Various types of heat-resistant spores can be produced, including ascospores, chlamydospores and sclerotia. Warmer weather in the summer over the past few years in northern Europe has possibly led to an increase in problems associated with HRM in soft drinks fruit juices and fruit purees. This is also possibly associated with the increased requirements for low thermal processes for fruit juices and products with unusual ingredients. For example, incorporation of collagen and other functional ingredients may mean a less severe heat process is used. There is an apparent association of HRM with 'weak' or less acidic fruits, including raspberry, strawberry, peach, guava, nectarine, and passion fruit, from which more problems are seen.

Many of the above moulds are found on fruits pre- and post-harvest. Most moulds require oxygen to grow. It should be noted that many heat-resistant moulds, and *Penicillium roquefortii*, can grow using dissolved oxygen, and *Paecilomyces variotii* can grow under almost anaerobic conditions. However, even those that are tolerant of low oxygen/high carbon dioxide levels are not usually able to grow in carbonated products.

Growth is exhibited as surface mats, sometimes producing copious spores. Some moulds produce extracellular degradative enzymes, such as pectinases. Detection of heat-tolerant moulds is usually carried out by plating out a sample of heat-shocked juice.

Mould identification is carried out by reference to standard morphological texts, and it is difficult unless carried out by experienced personnel. An excellent scheme that combines basic physiology with simple morphological attributes was developed by Pitt and Hocking (1997, 2009). It uses growth on three media at three temperatures to differentiate moulds: a high sugar content (xerophilic) medium; a closely defined medium with nitrate as a nitrogen source and sucrose as hexose sugar; and a general medium at 5°, 25° and 37°C. This technique is particularly useful for the differentiation of *Penicillium* species.

11.3.4.1 Mycotoxins

Mycotoxins are toxic secondary metabolites produced by fungi growing within or on foods. They can be a serious threat to human and animal health (Nagler *et al.*, 2001). Table 11.4 details mycotoxins associated with soft drinks and fruit juice manufacture and raw materials. Patulin is the most common mycotoxin associated with fruit juice, and particularly apple juice (Pitt and Hocking, 2009). It commonly occurs if juice is produced from stored apples. Mould growth in infected apples increases with time, raising levels of patulin. The use of windfall apples for juice is also a factor. Avoidance of windfall apples, filtration of juice and pressing quickly after harvest are all methods to reduce the incidence of patulin in juice. Patulin can be destroyed by fermentation to cider, or by the addition of ascorbic acid (Marth, 1992). Within Europe, the European Union has set a limit of 50 µg/kg for patulin in both apple juice and cider. A recent survey of apple products in Chile found that 28% of samples of juice and concentrate exceeded this limit (Canas and Aranda, 1996).

Table 11.4 Examples of toxin-producing moulds associated fruit with fruit raw materials and soft drinks products.

Fruit(s)	Mould	Toxin(s)
Apple	*Penicillium expansum*	Patulin, citrinin, roquefortine C
Citrus	*P. digitatum*	Tryptoquivalins
Carbonated beverages	*P. glabrum*	Citromycetin
Bottled water	*P. roqueforti*	Roquefortine C
Diluted fruit/water Beverages	*P. roqueforti*	Isofumigaclavine A and B
Treated orange juice	*Fusarium oxysporum*	Oxysporone
Fruit juices	*Aspergillus versicolor*	Geosmin sterigmatocystin

11.4 Microbiological safety problems

E. coli and enterococci have been isolated from unpasteurised citrus juices, and unpasteurised apple juice has been associated with Cryptosporidiosis, caused by *Cryptosporidium parvum* (Deng and Cliver, 2000). Contamination by Hepatitis A and Norwalk-like viruses has been reported in fruit juices (Vasavada, 2003).

Medina *et al* (2007) reported that *E. coli*, *Salmonella enterica*, *L. monocytogenes* and *S. aureus* are not killed immediately if inoculated into carbonated colas. Sheth *et al* (1988) reported survival of up to 48 hour for *E. coli* and *Salmonella* in colas.

11.4.1 Escherichia coli

There have been a number of cases of food poisoning associated with contamination of fresh apple juice by pathogenic *E. coli* O157 (Battey and Schaffner, 2001; Czajka and Batt, 1996; Miller and Kasper, 1994; Parish, 1997, 2000; Splittsoesser *et al.*, 1996; Zhao *et al.*, 1993). This organism was found to be able to grow in apple juice with low pH, and it potentially has a low infective dose of 100 cells. This, coupled with the occasional severe side-effect of haemolytic uraemic syndrome, makes *E. coli* O157 a severe hazard within unpasteurised juices. Unpasteurised juices are generally not seen within the European marketplace for this reason.

11.4.2 Salmonella

Similarly, *Salmonella* is generally only a problem in fresh, unpasteurised fruit juices, due to its low thermal tolerance. There have been three major outbreaks of Salmonellosis, all from unpasteurised orange juice, in the United States, Canada and Australia, involving 62, 298 and 400 reported cases in 1995 and 1999 (Bell and Kyriakides, 2001). *Salmonella* Hartford, *Salmonella* Muenchen and *Salmonella* Typhimurium were involved. The main risk factors were associated with the fertilisation of agricultural produce. Crops were grown in orchards where sheep grazed, manuring the soil, and fallen fruit, allowing soil and faecal contamination, were often used. In addition, poor decontamination of fruit occurred in the factory; poor-quality wash water was used; pest control within the factory was poor; and the clean-up of fruit boxes and conveyors was poor. Data indicated that *Salmonella* could survive for up to 300–968 days in soil treated with animal slurry. Other studies showed that fresh juice had to be stored for 15–24 days at 4°C to achieve a 10^6 reduction from 10^7 to 10^1 (Bell and Kyriakides, 2001).

11.5 Preservation and control measures

Thermal and non-thermal preservation treatments are covered in other chapters within this book. The focus in this section is with the negative effects of microorganisms on those preservation systems, and the auxiliary systems that can be utilised to minimise the likelihood of microbial growth. All microorganisms show a pattern of growth from a lag phase, followed by exponential growth, a stationery phase, and finally a decline and death phase. The key is to extend the lag phase, or slow growth phase, for any microbes that survive process or recontaminate after processing, so that the intended shelf life can be achieved.

As noted above, HRM can survive normal acid pasteurisation conditions for soft drinks and fruit juices. For example, the D-values for *Byssochlamys fulva* range from 1–12 minutes at 90C, with z-values of 6–7 minutes (Pitt and Hocking, 2009). Similar values are noted for other HRM, with a typical pasteurisation probably capable of a 3-log reduction at best. A heavy contamination of raw materials could lead to survival.

Some preservative-resistant yeasts and moulds do not just grow in the presence of the organic acid preservatives sorbic acid and benzoic acid – they can also degrade the preservative, which can allow other spoilage organisms to grow as well.

Another group of preservative-resistant organisms metabolise sorbic acid to 1,3-pentadiene, producing a so-called 'kerosene taint'. A range of fungi, including *Trichoderma* spp., *P. variotii, Penicillium chrysogenum, Penicillium simplicissimum, Penicillium crustosum*, and *Aspergillus niger*, and the yeasts *Z. rouxii* and *D. hansenii*, have been associated with the problem (Pinches and Apps, 2007).

Benzoic acid is now not used as much in the soft drinks industry, due to the decarboxylation of benzoic acid that can occur in the presence of ascorbic acid, with a transition metal catalyst (iron or copper) under heat, to produce benzene. This has led to increased spoilage problems, since benzoic acid was a more effective preservative than sorbic acid. The reduction in use of benzoic acid is largely due to benzene production from reaction between benzoic acid, ascorbic acid and heat (Gardner and Lawrence, 1993) although the reaction is inhibited by sugars.

Control measures include maintaining the cleanliness of equipment, the control of storage temperature, hot water immersion, the use of chemical sanitisers, surfactants and surface waxes (particularly on oranges) and UV irradiation. The chemical preservative dimethydicarbonate (Velcorin) has been used to 'cold pasteurise' fresh juice products to reduce microbial loading, minimising the use of sulphur dioxide (Threlfall and Morris, 2002; Vasavada and Heperkan, 2002). HPP, PEF and ultraviolet light have also been investigated as control measures. Note that these are, at best, pasteurisation treatments, and they have no effects on bacterial spores. Modelling of yeast and bacterial growth has been used as a technique for determining effective preservative and other control regimes (Battey and Schaffner, 2001; Shearer *et al*, 2002).

Good hygienic practices and adherence to GMPs are the most effective control measures for microbial contamination in the soft drinks industry, particularly for yeasts. Sticky sugar and fruit residues are ideal food sources for yeasts and moulds. The Hazard Analysis and Critical Control Point (HACCP) approach has been adopted by food processors around the world. In the United States, HACCP is mandatory for fruit juice processors, with good agricultural practices (GAP) as the foundation of a successful HACCP system. In Europe, growers, distributors and packaging houses must meet the EUREGAP protocols if they wish to be certified to sell their products to certain markets or established buyers (Stier and Nagle, 2003). There is a need to set and keep to sensible specifications, to minimise the likelihood of microbial problems.

11.6 Sampling for microbial problems

In order to be able to understand and characterise the causes of any problems, effective sampling must be undertaken. Previous versions of this book gave excellent protocols for investigative analysis (Davenport, 1998; Wareing and Davenport, 2005). The main sites to be inspected for dominant microflora include: drains; raw material stores; filler heads and near the filler; soap lubrication from jets; electric motor fins (all areas); bottle tracks; ultraviolet insectocutors; soak baths; capping systems; swabs from brushes; squeegees; and floor cleaners.

11.7 Identification schemes and interpretation

11.7.1 Sample isolation

Once samples have been taken, it is important to take them to a laboratory for analysis as quickly as possible, to avoid any changes to the microbial flora. Chilling is often useful, to prevent any further growth, but it is also important to avoid a decreased microbial count as well.

Although molecular identification methods are now the norm, it is still generally necessary to isolate the tentative spoilage agents from the soft drink matrix, the ingredients, or the environment. Traditional methods are therefore still required, at least initially, and some skill is still required to obtain a 'clean' sample and purification plate.

For example, it can be very useful to make a microscope slide from a swab or scraping from a suspect area, or to make a wet mount slide from a bottle in which there is visual growth, or bottle distention, due to gas production. In this way, a tentative diagnosis can be made before any confirmation comes through or samples have been purified. It is recommended initially to isolate the culture on a general purpose medium before any more selective media are used, in case any of the microorganisms are damaged in any way. The exception to this would be if osmophiles or xerophiles are suspected, as they are not likely to grow well on low-sugar media, and they could be overgrown by other contaminating non-osmophiles. Some details of these methods is given below.

Spoilage source can be determined sometimes by the proportion of spoiled containers – if random, by chance, or survivors. For instance, if one in 5, 10, or 20 is spoiled, it can often due to filler head contamination.

11.7.2 Non-molecular methods

Each group of microorganisms requires media and growth conditions to optimise recovery and growth. For example, yeasts and moulds tend to be grown at 25°C, whereas HRM and most spoilage bacteria are grown at 30°C. While results can be obtained in a few days for most bacteria, yeasts and moulds usually require 5–7 days incubation, and extreme osmophiles/xerophiles may take longer before growth is observed.

If the samples are expected to be contaminated with a variety of microbial groups, but yeasts or moulds are the primary suspects, it can be useful to use semi-selective isolation media – for example, Dichloran Rose Bengal Chloramphenicol Agar (DRBC) and Dichloran 18% Glycerol Agar (DG18). These inhibit the growth of bacteria and fast-growing moulds, allowing an efficient extraction.

The forensic grouping approach was pioneered by the late Professor Davenport, and was explained in some detail in the previous two editions of this book (Davenport, 1998; Wareing and Davenport, 2005), to which the reader is referred if necessary. It is based around the four groups of spoilage yeasts, outlined in section 11.3.2. Forensic grouping involves consideration of simple behaviour patterns of microorganisms as evidence for decision-making. The data obtained indicate growth, significance and tentative recognition/identification. This approach to identification has been superseded, in part, by molecular identification techniques.

A similar general scheme was developed by Pitt and Hocking (2009) for yeasts associated with foods. They identified 13 species of spoilage yeasts, from ten genera, which have been found to cause the majority of all spoilage associated with foods, including soft drinks, jams, confectionery, sauces and pickles (see Table 11.5). They recommended using Czapek agar, malt extract agar (MEA), MEA at 37°C, malt yeast 50% glucose agar (MY50G) MEA acidified with 0.5% acetic acid, malt yeast 10% salt 12% glucose agar (MY10-12) to identify non-osmophilic, osmophilic and acid or salt tolerant yeasts.

11.7.3 Molecular identification

The advent of molecular identification techniques have simplified the diagnostic process. Yeasts are particularly difficult to identify by traditional methods, requiring growth, physiology, and some simple physical determinants to be measured. Now, once a pure culture has been obtained, an accurate identification is a simple process once the genetic material has been extracted and results compared with library data. However, it should be recognised that the skills of an experienced microbiologist are usually still required to interpret the data from the molecular analysis.

Table 11.5 Most common spoilage yeasts (after Pitt and Hocking, 2009).

Yeast	Characteristics
Debaryomyces hansenii	Growth at low water activities
Dekkera bruxellensis	Off odours in beer and soft drinks
Hanseniaspora uvarum	Spoilage of fresh and processed fruits
Issatchenkia orientalis	Preservative resistant, film formation
Pichia membranifaciens	Preservative resistant, film formation
Rhodotorula glutinis, R. Rubra	Common food contaminant, though not usually of fresh fruits
Saccharomyces cerevisiae	Common contaminant, sometimes fermentation of soft drinks
Saccharomyces exiguous	Moderate preservative resistance, spoilage of brined olives, sometimes found in juices and soft drinks
Schizosaccharomyces pombe	Preservative resistant, not commonly found
Zygosaccharomyces bailii	Extreme preservative resistance, fermentative spoilage of acid, liquid preserved products; soft drinks, juices, ciders and sauces
Z. bisporus	Preservative resistant, intermediate characteristics between *Z. bailii* and *Z. rouxii*
Z. rouxii	Growth at very low water activities, fermentative spoilage of juice concentrates, honey, jam, dried fruits, confectionery

11.8 Brief spoilage case studies

In this section, a brief outline will be given of some spoilage issues that we have noted from our laboratory over the past few years. The intention is to outline the diversity of problems faced within the soft drink industry.

Example 1:

A ready-to-drink product containing apple, strawberry and rhubarb juice spoiled from growth of *Byssochlamys nivea* after approximately one month's retail display at ambient. The three key ingredients were pasteurised and hot-filled into containers at different manufacturers. These were blended at a secondary processor to make a three-juice semi-concentrate, which was flash-pasteurised at 95°C for 30 seconds and clean-filled into foil-lined barrels. The finished product manufacturer then diluted this concentrate and blended with it with water and other ingredients to give the end product, which was tunnel-pasteurised at 75°C for 20 minutes.

A mycelial mat in the finished product was noted by customers approximately three to five weeks after manufacture. *B. nivea* therefore survived three successive pasteurisations. The likelihood was that vegetative cells were killed, but heat-resistant ascospores survived, received a heat shock, germinated and produced a sufficiently large crop of new ascospores before that product was, itself, heat-processed. Because each component was stored at cool ambient, apart from the finished product, spores that survived each successive heat treatment grew slowly enough to complete their life cycle without producing noticeable growth in the various ingredients. The source of the contamination was traced to the rhubarb juice, not the strawberry juice, as would normally be expected. Improved supplier assurance procedures reduced the likelihood of the occurrence.

Example 2:
A flexible pouch pack smoothie for children gave a greenish-black growth in the neck of the product, under the cap. This type of product is made from a foil pouch, with a resealable screw cap. The contamination was identified as *Cladosporium cladosporioides*; this was not a heat-resistant isolate. In this case, the packs were in-pack pasteurised, but the neck of the package was not properly heated, and the viscous product adhered to the relatively narrow neck. The mould, therefore, did not receive a correct heat process, which was relatively light to preserve flavour and nutrients for children. The solution was to properly immerse packs during the heat process.

Example 3:
Operational changes to the layout at a juice ingredient manufacturer meant that the line, post-holding tube but pre-pack, was shortened. This allowed survival of *Byssochlamys spectabilis* (sexual stage of P. *variotii*) ascospores. The thermal process was originally designed to destroy typical heat-resistant moulds found in the ingredients. *B. spectabilis* is slightly more heat-resistant than other moulds found in the ingredients. The residual thermal process that the product received as it was being cooled post-holding tube was formerly sufficient to kill ascospores of *B. spectabilis*. The new, more efficient cooling process reduced this residual thermal process. The solution was to recalculate the thermal process required in the holding tube, to correctly destroy ascospores of the mould.

Example 4:
Growth of *Penicillium* spp. occurred in the neck flange of a fruit juice-flavoured water. High-speed photography showed that contamination of the neck flange was due to splashing occurring during filling. *Penicillium* spp. grew on the trace levels of juice that were left on the outside of the bottle, caught in the flange. The solution was to improve the filling process, so that splashing of product during the filling process was minimised.

11.9 Conclusions

Many changes have occurred over the course of the last few editions of this book. There is an ever-increasing range of soft drinks, new and unusual ingredients, newer processing techniques, a requirement for lighter processing and fewer preservatives, with a concomitant increase in chilled products. At the same time, traditional relatively slow and complex identification techniques, that relied upon the expertise and judgement of an experienced microbiologist, have been largely superseded by molecular methods for microbial identification.

There is still a need, however, for the involvement of an experienced microbiologist, in order to be able to interpret the data from any troubleshooting exercise and any identifications made from isolations. Traditional methods still have a place for the isolation and purification phases, prior to any identification. It is vital that a pure culture is obtained for molecular identification or the results can be meaningless. One advantage can be that a rapid tentative diagnosis can be made, prior to confirmation.

The timescale for this approach ranges from a few minutes (sample data plus direct microscopy) up to a maximum of 7–21 days, if obtaining a pure culture and undertaking molecular ID. With conventional testing, a pure culture has to be used, to ensure that only the characteristics of the individual organism are determined. The conventional approach presumes that, once an organism is identified, other conclusions are readily obtained. It has, therefore, been seen as an essential part of the problem-solving process.

A troubleshooting approach can also be used to quickly determine the likelihood of any of the isolated components being able to cause spoilage in the finished product or juice ingredient. In this case, any of the isolated microbes are re-inoculated back into finished product, juice ingredient or diluted concentrates, and incubated at 25–30°C. Observation for gas, sediment, turbidity or clearing of the product indicates that the isolate is capable of causing spoilage, without necessarily undertaking a full identification. This approach has been given in some detail elsewhere (Davenport, 1996, 1997, 1998; Wareing and Davenport, 2005).

Control of contamination still relies upon the need for a hygienic working environment, which includes the fabric, equipment, ingredients, people and working practices. The increasing emphasis on clean labels means that preservative use is diminishing in the high value sector. This requires that ingredients are of good quality, with reduced levels of spoilage organisms, and are consistently sourced from quality crops that are handled well after harvest. The problem with potential benzene production from benzoic acid has exacerbated this trend.

The increased use of exotic ingredients and complex mixtures of fruits and vegetables for juices means that the microbiologist must be aware of the potential changes to the microflora that this could bring. Some of these fruits will be 'sensitive', with a greater likelihood of contamination with HRM or other problem organisms, and with a reduced natural preservative effect, due to their higher pH. These products also may undergo a lighter processing regime, to preserve the more delicate flavours that they contain. Climate change is leading to warmer summers, allowing a greater chance of HRM growth after processing if the moulds survive the thermal process. In addition, other microbes may become problematic, where previously they were unable to grow. Microorganisms are capable of adaptation, and lower levels of preservatives may, for example, allow the growth of organisms that are partially tolerant of organic acid preservatives.

Underpinning all of this is the food safety management tool, Hazard Analysis and Critical Control Point (HACCP), and its associated Prerequisite Programmes. This procedural-based approach to food safety relies upon prescribed processes being followed, in order to ensure food safety.

For example, supplier assurance and traceability are vital, in order to assure the quality and safety of ingredients. This can be difficult with an ever diverse and far-flung supply chain, and manufacturers may not have the resources to audit suppliers if they are located at some distance from their factory. This may mean that ingredients do not always meet specifications; ingredients are not always tested upon receipt, and it is the norm to rely upon the certificate of conformance supplied with the ingredient. Unexpected heat-resistant microorganisms may part of the flora from these new ingredients. The existing thermal process may not be sufficient to destroy them, leading to spoilage problems. It is very important to make sure that the HACCP plan is reviewed when any changes occur, such as new ingredients, or suppliers, or countries of origin, or changes to the process.

The ever-changing nature of the soft drinks market requires that the microbiologist or food safety specialist is involved with any new products or processes, from product inception right through to placing the product on the market.

References

Battey, A.S. and Schaffner, D.W. (2001). Modelling bacterial spoilage in cold-filled ready to drink beverages by *Acinetobacter calcoaceticus* and *Gluconobacter oxydans*. *Journal of Applied Microbiology* **91**(2), 237–247.

Bell, C. and Kyriakides, A. (2001). *Salmonella: A Practical Approach to the Organism and its Control in Foods*. Blackwell Science, Oxford, 336pp.

Canas, P. and Aranda, M. (1996). Decontamination and inhibition of patulin-induced cytotoxicity. *Environmental Toxicology and Water Quality* **11**, 249–253.

Cerny, G., Duong, H.-A., Hennlich, W. and Miller, S. (2000). *Alicyclobacillus acidoterrestris*: influence of oxygen content on growth in fruit juices. *Food Australia* **52**(7), 289–91.

Czajka, J. and Batt, C.A. (1996). A solid phase fluorescent capillary immunoassay for the detection of *Escherichia coli* O157:H7 in ground beef and apple cider. *Journal of Applied Bacteriology* **81**(6), 601–7.

Davenport, R.R. (1996). Forensic microbiology for soft drinks business. *Soft Drinks Management International* April, 34–5.

Davenport, R.R. (1997). Forensic microbiology II. Case book investigations. *Soft Drinks Management International* April, 26–30.

Davenport, R.R. (1998). Microbiology of soft drinks. In: Ashurst, P.R. (ed). *Chemistry and Technology of Soft Drinks and Fruit Juices*, 1st edition, pp. 197–216. Sheffield Academic Press, Sheffield, UK and CRC Press, Boca Raton, FL.

de Nijs, M., van der Vossen, J., van Osenbruggen, T. and Hartog, B. (2000). The significance of heat resistant spoilage moulds and yeasts in fruit juices – a review. *Fruit Processing* **10**(7), 255–9.

Deng, M.Q. and Cliver, D.O. (2000). Comparative detection of *Cryptosporidium parvum* oocysts from apple juice. *International Journal of Food Microbiology* **54**(3), 155–162.

Espina, L., Garcia-Gonzalo, D., Lagaoui, A., Mackey, B.M. and Pagán, R. (2013). Synergistic combinations of high hydrostatic pressure and essential oil or their constituents and their use in preservation of fruit juices. International *Journal of Food Microbiology* **161**, 23–30.

Ferrario, M., Alzamora, S.M. and Guerro, S. (2014). Study of the inactivation of spoilage microorganisms in apple juice by pulsed light and ultrasound. *Food Microbiology* **xxx**, 1–8.

Gardner, L.K. and Lawrence, G.D. (1993). Benzene production from decarboxylation of benzoic acid in the presence of ascorbic acid and a transition-metal catalyst. *Journal of Agricultural and Food Chemistry* **41**, 693–695.

Hocking, A.D. and Jensen, N. (2001). Spoilage of various food classes: soft drinks, cordials, juices, bottled water and related products. In: Moir, C.J. and Waterloo, D.C. (eds). *Spoilage of Processed Foods: Causes and Diagnosis*, pp. 91–100. Australian Institute of Food Science and Technology Incorporated Food, Microbiology Group, AIFST Inc.

Jay, S. and Anderson, J. (2001). Spoilage of various food classes: fruit juice and related products. In: Moir, C.J. and Waterloo, D.C. (eds). *Spoilage of Processed Foods: Causes and Diagnosis*, pp. 187–98. Australian Institute of Food Science and Technology Incorporated Food Microbiology Group AIFST Inc.

Jensen, N. (1999). *Alicyclobacillus* – a new challenge for the food industry. *Food Australia* **51**(1–2), 33–6.

Jensen, N. and Whitfield F.B. (2003). Role of *Alicyclobacillus acidoterrestris* in the development of a disinfectant taint in shelf-stable fruit juice. *Letters in Applied Microbiology* **36**(1), 9–14.

Juvonen, R., Virkajarvi, O. Priha, A. and Laitila, A. (2011). *Microbiological spoilage and safety risks in non-beer beverages*. VTT Research Notes, 2599, pp1–119. VTT Research Centre of Finland.

Kusano, K., H. Yamada, M. Niwa, and K. Yamasoto. (1997). *Propionibacterium cyclohexanicum* sp. nov., a new acidtolerant ω-cyclohexyl fatty acid containing *Propionibacterium* isolated from spoiled orange juice. *International Journal of Systematic Bacteriology* **47**, 825–831.

Larko, A., Tongchitpakdee, S. and Youravong, W. (2013). Storage quality of pineapple juice non-thermally pasteurised and clarified by microfiltration. *Journal of Food Engineering* **116**, 554–561.

Marth, E.H. (1992). Mycotoxins: production and control. *Food Laboratory News* **8**, 34–51.

Medina, E., Romero, C., Brenes, M. and De Castro, A. (2007). Antimicrobial activity of olive oil, vinegar, and various beverages against foodborne pathogens. *Journal of Food Protection* **70**, 1194–1199.

Mermelstein, N.H. (1999). High-pressure pasteurization of juice. *Food Technology* **53**(4), 86–90.

Miller, L.G. and Kasper, C.W. (1994). *Escherichia coli* O157:H7 acid tolerance and survival in apple cider. *Journal of Food Protection* **57**(7), 645.

Nagler, M.J, Coker, R.D., Pineiro, M., Wareing, P.W. Myhara, R.M. and Nicolaides, L. (2001). *Manual on the Application of the HACCP System in Mycotoxin Prevention and Control*. FAO Food and Nutrition Paper 73, FAO/IAEA Training and Reference Centre for Food and Pesticide Control, FAO, Rome, 113pp.

Orr, R.V. and Beuchat, L.R (2000). Efficacy of disinfectants in killing spores of *Alicyclobacillus acidoterrestris* and performance of media for supporting colony development by survivors. *Journal of Food Protection* **63**(8), 1117–22.

Pacheco, C.P. (2002). Sensibility and specificity of methods for *Alicyclobacillus* detection and quantification: a collaborative study. *Fruit Processing* **12**(11), 478–82.

Parish, M.E. (1997). Public health and non-pasteurised fruit juices. *Critical Reviews in Microbiology*, **23**(2), 109–19.

Parish, M. (2000). Relevancy of *Salmonella* and pathogenic *E. coli* to fruit juices. *Fruit Processing* **10**(7), 246–50.

Pettipher, G.L. and Osmundson, M.E. (2000). Methods for the detection, enumeration and identification of *Alicyclobacillus acidoterrestris*. *Food Australia* **52**(7), 293–5.

Pinches, S.E. and Apps, P. (2007). Production in food of 1,3-pentadiene and styrene by *Trichoderma* species. *International Journal of Food Microbiology* **116**, 182–187.

Pitt, J.I. and Hocking, A.D. (1997). *Fungi and Food Spoilage*, 2nd edition. Blackie Academic and Professional, London.

Pitt, J.I. and Hocking, A.D. (2009). *Fungi and Food Spoilage*, 3rd edition. Springer, Dordrecht.

Rodriguez, J.H., Cousin, M.A. and Nelson, P.E. (1992). Oxygen requirements of *Bacillus licheniformis* and *Bacillus subtilis* in tomato juice; ability to grow in aseptic packages. *Journal of Food Science* **57**, 973–6.

Shamsudin, R., Adzahan, N.M., Yee, Y.P. and Mansur, A. (2014). Effect of repetitive UV irradiation on the physic-chemical properties and microbial stability of pineapple juice. *Innovative Food Science and Emerging Technologies* **23**, 114–120.

Shearer, A.E.H., Mazzotta, A.S., Chuyate, R. and Gombas, D.E. (2002). Heat resistance of juice spoilage microorganisms. *Journal of Food Protection* **65**(8), 1271–75.

Sheth, N.K., Wisniewski, T.R. and Franson, T.R. 1988. Survival of enteric pathogens in common beverages: An *in vitro* study. *American Journal of Gastroenterology* **83**, pp. 658–660. ISSN 00029270.

Silva, F. and Gibbs, P. (2001). *Alicyclobacillus acidoterrestris* spores in fruit products and design of pasteurization processes. *Trends in Food Science and Technology* **12**(2), 68–74.

Splittsoesser, D.F., McLellan, M.R. and Churney, J.J. (1996). Heat resistance of *Escherichia coli* O157:H7 in apple juice. *Journal of Food Protection* **59**(3), 226–9.

Stier R.F. and Nagle N.E. (2003). Ensuring safety in juices and juice products: good agricultural practices. In: Foster, T. and Vasavada, P.C. (eds). *Beverage Quality and Safety*, pp. 1–7. CRC Press, Boca Raton, FL.

Stratford M. and James S.A. (2003). Non-alcoholic beverages and yeasts. In: Boekhout, T. and Robert, V. (eds). *Yeasts in Food: Beneficial and Detrimental Aspects*, pp. 309–45. Woodhead Publishing Ltd, Cambridge.

Stratford, M., Hofman, P.D. and Cole, M.B. (2000). Fruit juices, fruit drinks, and soft drinks. In: Lund, B.M., Baird-Parker, T.C. and Gould, G.W. (eds). *The Microbiological Safety and Quality of Food*, Vol. 1, pp. 836–69. Aspen Publishers, Gaithersburg.

Threlfall, R.T. and Morris, J.R. (2002). Using dimethyldicarbonate to minimize sulfur dioxide for prevention of fermentation from excessive yeast contamination in juice and semi-sweet wine. *Journal of Food Science* **67**(7), 2758–62.

Vasavada, P. and Heperkan, D. (2002). Non-thermal alternative processing technologies for the control of spoilage bacteria in fruit juices and fruit-based drinks. *Food Safety Magazine* **8**(1), 8–13, 46–8.

Vasavada, P.C. (2003). Beverage quality and safety. In: Foster, T. and Vasavada, P.C. (eds). *Microbiology of Fruit Juice and Beverages*, pp. 95–123. CRC Press, Boca Raton, FL.

Walker, M. and Phillips, C.A. (2007). The growth of *Propionibacterium cyclohexanicum* in fruit juices and its survival following elevated temperature treatments. *Food Microbiology* **6**, Vol. 24, pp. 313–318.

Walls, I. and Chuyate, R. (2000a). Spoilage of fruit juices by *Alicyclobacillus acidoterrestris*. *Food Australia* **52**(7), 286–88.

Walls, I. and Chuyate, R (2000b). Isolation of *Alicyclobacillus acidoterrestris* from fruit juices. *Journal of AOAC International* **83**(5), 1115–20.

Wareing, P.W. and Davenport, R.R. (2005). Microbiology of soft drinks and fruit juices. In: Ashurst, P.R. (ed). *Chemistry and Technology of Soft Drinks and Fruit Juices*, pp 279–299. Blackwell Publishing.

Yokata, A., Fuji, T. and Goto, K. (eds, 2007). *Alicyclobacillus: Thermophilic Acidophilic Bacteria*. Springer Verlag, Japan, 282 pp.

Zhao, T., Doyle, M.P. and Besser, R.E. (1993). Fate of enterohemorrhagic *Escherichia coli* O157:H7 in apple cider with and without preservatives. *Applied and Environmental Microbiology* **59**, 2526–30.

Zook, C.D., Parish, M.E., Braddock, R.J. and Balaban, M.O. (1999). High pressure inactivation kinetics of *Saccharomyces cerevisiae* ascospores in orange and apple juices. *Journal of Food Science* **64**(3), 533–5.

Further reading

Ashurst, P.R. (ed, 1995). *Production and Packaging of Non Carbonated Fruit Juices and Fruit Beverages*, 2nd edition. Blackie Academic & Professional, Chapman & Hall, London.

Barnett, J.A., Payne, R.W. and Yarrow, D. (eds, 1990). *Yeasts: Characteristics and Identification*, 2nd edition. Cambridge University Press, Cambridge.

Chesworth, N. (ed, 1997). *Food Hygiene and Auditing*. Blackie Academic & Professional, Chapman & Hall, London.

Desrosier, N.W. (1959). *The Technology of Food Presentation*, AVI, Westpoint, CT.

Gravesen, S., Frisvad, J.C. and Samson, R.A. (1994). *Microfungi* (Munksgaard).

Kreger-Van Rij, N.J.W. (1984). *The Yeasts, A Taxonomic Study*, 3rd edition. Elsevier Science, Amsterdam.

Onions, A.H.S., Allsopp, D. and Eggins, H.O.W. (eds, 1981). *Smith's Introduction to Industrial Mycology*, 7th edition. Edward Arnold, London.

Pitt, J.I. and Hocking, A.D. (1997). *Fungi and Food Spoilage*, 2nd edition. Blackie Academic and Professional, London.

Pitt, J.I. and Hocking, A.D. (1985). *Fungi and Food Spoilage*. Academic Press, New York.

Priest, F.G. and Campbell, I. (eds, 1987). *Brewing Microbiology*. Elsevier Applied Science, London.

Spencer, J.F.T. and Spencer, D.M. (eds, 1997). *Yeasts in Natural and Artificial Habitats*. Springer Verlag, Berlin.

Tressler, D.H., Joslyn, M.A. and Marsh, G.L. (1939). *Fruit and Vegetable Juices*. AVI, Westpoint, CT.

Wolf, C. (ed, 1996). *Non-conventional Yeasts in Biotechnology: A Handbook*. Springer-Verlag, Berlin.

CHAPTER 12

Functional drinks containing herbal extracts

Ellen F. Shaw[1] and Stuart Charters[2]

[1] *Flavex International Limited, Kingstone, UK*
[2] *Bwlch Garneddog, Gwynedd, UK*

12.1 History

The Shanidar Cave is a valued archaeological site, as it contained some of the earliest physical evidence to support the theory that herbs have been used medicinally for as long as 45 000–50 000 years (Schwartz, 1993). The cave overlooks the Great Zab River of Northern Iraq, a fertile environment in which plants can thrive. Excavations of the cave by Ralph Solecki in the early 1950s uncovered nine Neanderthal skeletons, which were named Shanidar I–IX. They were all different in age, level of articulation and state of preservation (Whitaker, 2010).

Of all the skeletons found in the cave, it is the remains of Shanidar IV, a 30–45 year old male, which provides the best evidence that Neanderthals were aware of the medicinal properties of herbs. Termed 'the flower burial', Shanidar IV had been ritually laid to rest along with a variety of wildflowers, including bachelor's buttons and grape hyacinths (Schwartz, 1993). In addition, yarrow, cornflower, St Barnaby's thistle, ragwort, woody horsetail and hollyhock pollens were identified, all of which have been used in traditional medicines as stimulants, astringents, diuretics and anti-inflammatory agents (Wilde, 2009). It has been speculated that the medicinal properties of these flowers had also been known to the Neanderthals of Shanidar (Schwartz, 1993).

The earliest known herbal drinks are said to have originated in China, with written formulae dating as far back as the 3rd century BC. The Chinese believe we are sustained by the life force *qi* (Victor, 2010). Wolfberries, ginseng and Ling Tzi are three ancient cherished *qi* tonics.

Wolfberries (or Goji berries) are said to sustain health in various ways, including nourishing the *yin*, supporting the kidneys, liver and blood, as well as strengthening muscles, bones and eyes – all of which is believed to boost *qi*. This phenomenon has been noted in China for centuries, yet it is only recently that the importance of the wolfberry has been apparent to the western world. A daily

---

Chemistry and Technology of Soft Drinks and Fruit Juices, Third Edition. Edited by Philip R. Ashurst.
© 2016 John Wiley & Sons, Ltd. Published 2016 by John Wiley & Sons, Ltd.

intake of wolfberries is said to account for the astonishing health and vitality of the elderly residents of Ningxia Province, many of whom are over 100 years old and still enjoy an active and fulfilling life (Ning Xia Red History, 2009).

Ginseng is documented as one of the oldest *qi* tonic herbs, and was thought of as an overall curative agent (Bown, 2003). It was believed that its properties settled the spirits, established the soul, alleviated fear, banished evil, brightened the eyes, revealed the emotions of the heart, aided knowledge and, when taken regularly, energised the body and lengthened life. The Penobscot people used ginseng to increase fertility by immersing the root in water and drinking the liquid. The Creeks drank an infusion of the root to cure symptoms of shortness of breath, coughs, croup and fevers. The Halmas people boiled the root, producing a drink to prevent vomiting or, by adding a little whisky to the infusion, they concocted a treatment for rheumatism (Pokladnik, 2008).

The renowned herb Ling tzi *(Ganoderma lucidum)* has been used by Chinese emperors in traditional Chinese medicines for thousands of years. Ling tzi is mentioned in the writings of Shen Nong's *Ben Cao Jing*, written during the Han dynasty, and was credited to be the most significant herb in existence. Mushroom-like in appearance, and similar to ginseng, it matures gradually, absorbing the essence and *qi* of nature. Even now, this rare herb is highly valued in China and throughout Asia for prolonging life (Lu, 2001).

In China, tea is the primary beverage; it is not only considered refreshing, but is also used as a therapeutic. The majority of other tea drinking cultures consume the herbal beverage for hydration, to satisfy their taste-buds and for the stimulating effects of its caffeine content. Chinese history suggests that tea drinking was first introduced by Shen Nong in 2737 BC. Tea brewing led to the highly popular method of decoction, a technique where herbs are boiled in water, making it particularly effective for treating acute disorders, as the herbs are quickly absorbed into the bloodstream (Joiner, 2001).

There is a wide variety of medical research that documents the health benefits of tea. The leaves contain three types of polyphenol flavonoid antioxidants; catechins, theaflavin and quercetin. The potency of the polyphenols varies with the different tea manufacturing techniques. Green tea, in particular, is a rich source of xanthine antioxidants, and it has been suggested that these molecules can help to lower cholesterol, and also act as an anti-inflammatory, anti-irritation and cellulite-reducing agents. The greatest concentration of naturally occurring catechins are found in Chinese white and Japanese green teas, as these varieties are the least processed, which effectively preserves the natural actives. Recent studies have shown green tea to increase the metabolic resting rate, making it a potential ingredient for weight-loss products.

White tea leaves are less mature, still holding their buds, and undergo the much more delicate method of steaming. Research has shown that white tea has potent anti-carcinogenic properties and, with a rapid surge in ready-to-drink tea varieties, it will not be long before exotic white tea is a 'must-have' product.

Oolong and black tea have a lower leaf count, and are fully oxidised (fermented) to alter the colour and obtain a richer flavour, which results in the production of more complex theaflavins. Black tea is mainly used as a beverage, although studies have concluded that the flavonoids present can protect cells and tissues from oxidative damage by free radicals. Tea extracts developed with higher polyphenol content contain the greatest concentration of active ingredients, and thus provide the greatest health benefits.

The practice of macerating herbs in alcohol is one of the oldest and simplest methods of making medicinal herbal drinks. The alcohol within the medicinal wine known as Yao Jiu is itself regarded as nourishing, with blood-invigorating properties, and it enhances the therapeutic effects of herbs. Second only to water, alcohol is the solvent of choice for most herbal preparations (Joiner, 2001).

Alcoholic elixirs created by alchemists during the Middle Ages consisted of every conceivable plant species. Invented by Benedictine monks in France during the 19th century Benedictine, DOM falls into the category of a herbal flavoured liqueur. French Carthusian monks also concocted their own variety, containing 130 herbs, named Chartreuse, while other nations have their own variations, including Italy's Strega and Gallino and Spain's Cuaenta Y Tres. All were once noted for their medicinal properties whereas, today, the main focus is on the appealing tastes and intoxicating effect of these alcoholic beverages.

Surprisingly, coffee did not make its way into European drinking habits until the 17th century. Originally, the beverage was seen as controversial, due to its stimulating side-effects, but now it is one of the most popular natural stimulants across the globe. Caffeine acts upon the nervous system, causing nerve fibres to release a variety of neurotransmitters around the body. As a result, the consumer experiences feelings of increased alertness and higher energy levels (Weil and Rosen, 2004).

Caffeine is believed to work by blocking adenosine receptors in the brain and other organs. The ability of adenosine to bind to these receptors is consequently reduced, which causes an increase in cellular activity. The stimulated nerve cells release the hormone epinephrine (a form of adrenaline), which increases the heart rate, blood pressure and blood flow to muscles, decreases blood flow to the skin and organs, and causes the liver to release glucose. Caffeine also increases the brain's levels of the neurotransmitter dopamine. The effects are short-lived, as caffeine is quickly metabolised (Helmenstine, 2012).

The world's most popular soft drink started life promoted as a non-carbonated brain and nerve tonic (Pendergast, 1994). It consisted of lime, cinnamon, coca leaves and cola seeds. Cola nuts are caffeine-containing seeds produced by the tropical cola tree. The nuts have a bitter, aromatic flavour and can be chewed in order to release their stimulating effects. This bottled soft drink is related to the cola plant, although the modern version contains only small traces of cola nut and does not incorporate the nutty taste. These products do contain caffeine, but it is usually synthetic caffeine, or caffeine extracted from coffee beans, that is used today.

Coca leaves contain 14 active ingredients, one of which is cocaine. When cocaine enters the area of the brain where the neurotransmitter dopamine is located, it blocks the reuptake pumps that remove the dopamine from the synapse of the nerve cell. This causes dopamine to gather at the synapse, and feelings of intense pleasure result. This feeling continues until the cocaine is removed naturally from the system (Hoegler, 2008). The original cola recipe did not omit coca from its formulation until 1905, and it would be unrecognisable to the consumer of today (Weil and Rosen, 2004).

During the mid-1980s, a new era of herbal drinks emerged, accompanied by a message of sophisticated healthiness, just as the drink-drive laws began to take effect in many countries and lifestyle became a significant issue in affluent societies.

12.2 The extraction process

Botanical extraction is the process of isolating medicinally bioactive components naturally found in plant cells. Selective solvents and varying environmental conditions allow the separation of soluble plant metabolites from the insoluble cellular remains. It involves a series of diffusion or mass transfer of molecules through the cellular plant matrix into the solvent medium. Such procedures are designed to obtain the therapeutically desirable, relatively complex mixtures of metabolite, and to facilitate elimination of the inert material.

There are five key steps that occur within the plant cells during the extraction process:

1 Solvent diffuses into the plant's matrix.
2 Dissolution of various plant compounds into the solvent.
3 Internal diffusion, where solutes are transported through the plant matrix via membranes to the surface of the cell. This mechanism is driven by concentration gradients.
4 External diffusion, where solutes are transported from the cell surface into the bulk solvent surrounding the cell. This is driven by concentration gradients.
5 Solvent displacement – the relative movement of solvent compared to the solids.

The rate of extraction can be improved by increasing the temperature and providing a larger contact area for diffusion, a shorter diffusion path, low solvent viscosity and a higher concentration gradient.

With the increasing demand for herbal functional drinks, extract manufacturers are using efficient, high-quality extraction technologies to produce extracts of defined quality to promote minimal variation between each batch.

12.2.1 Extraction heritage

Since the 19th century, there have been significant advances in the extraction procedures of botanical preparations. Classical extraction procedures can still be classified into three main groups: distillation, solvent extraction and compression. Such extraction techniques and processes were highly valued in the phytochemical field, and led to the standardisation of extracts for therapeutic purposes.

Traditionally, most extracts were made from dried herbs, as these were easier to keep in good condition from one harvest to the next. Dried herbs are still the preferred raw material today. When fresh herbs are used, the increased water content, which can be as much as 90%, needs to be taken into consideration, to avoid dilution of the final product.

Galenical is the term used when referring to the products derived from such natural extraction processes (although please note, this term is also used for natural animal tissue extraction products). These classes of preparations include infusions, decoctions, liquid extracts, tinctures, semi-solid extracts and soft extracts (Collins English Dictionary, 2011). It is important to note that familiarity with the descriptions of the various galenicals may help to avoid misunderstandings when contacting extract manufacturers.

The following subsections briefly describe the preparation, properties and characteristics of the most important industrial extraction procedures. It also looks at the applications appropriate to the various types of extract available.

12.2.1.1 Infusion

Infusion is a general term for preparation of aqueous extracts. These are traditionally prepared by pouring boiling water over the plant raw material (Wren, 1988). Modern-day infusions for beverage applications are often carried out under ambient temperature conditions, with the addition of ethanol. There is no set value for the plant : extract ratio, as the maximum strength is dependent on the individual bulk density, which is specific to the raw material in question.

A general method for producing 500 g/ml of a strong infusion is detailed as follows:
- Weigh 200 g of raw material.
- Place herb into appropriately sized plastic containers.
- Cover with diluted ethanol (20%).
- Mix thoroughly.
- Cover with a lid and leave to soak for 24 hours

- Measure the volume of the liquid by filtering through fluted filter paper in a stainless steel funnel.
- Collect the filtered extract into a graduated cylinder to check the volume and weight.
- A further 72–96 hours later, the extract needs to be filtered again using the same technique.
- The final product is poured into a suitable container and the herb is discarded.

Strong infusions yield approximately 2.5 times the amount of product, compared to the quantity of starting raw material used.

A few medicinal herbs, such as Calumba and Quassia, which contain water-soluble constituents can be infused using water alone.

Infusions made by a local extract manufacturer are normally the extract of first choice for ready-to-drink beverages. Infusions are the simplest types of herbal extracts, and the first stage of most other extracts. They are the least processed and, therefore, the most economical to use for beverage applications. Unless there are strong economic reasons to use a more concentrated extract, it does not make sense to take an infusion, concentrate it at extra cost, possibly damage it in the process, and then return it to a dilute form when the product is manufactured.

Concentrated infusions are sometimes economically viable if they are off-the-shelf products. For instance, because of high demand, liquid guarana extracts are available, containing standardised amounts of caffeine. The demand for these is enough to warrant large-scale production in the country of origin. At this point, concentration becomes desirable, because the extract requires shipment over long distances, and transporting solvent is a costly business.

12.2.1.2 Decoction

Decoction is the method employed for the extraction of harder materials, such as roots and barks (Wren, 1988). This is a variation on the general infusion process, in that the finely divided herb is steeped in boiling water, to which heat is then supplied to keep it constantly at boiling point. The concoction is left simmering gently, rather than boiling vigorously. Water is added to compensate for evaporative losses. When cool, the decoction is filtered and the herb is discarded.

Essentially, the principles involved in making a decoction are the same as brewing a cup of tea. Heat and moisture cause the plant cells to swell, the cell walls expand, and hydro-diffusion of plant constituents occur through the swollen membranes.

12.2.1.3 Percolation

Percolation is a solvent extraction procedure used as an alternative to the infusion process. Varying proportions of alcohol and water solvents are trickled on to the top of the raw material, which is contained in either a glass or metal column known as a percolator. The gradual addition of the solvent to the top of the column causes a slow down-flow through the crushed raw material. Liquid containing the soluble plant components is collected as it emerges from the bottom of the percolation vessel. This may then be recycled again, to achieve higher yields and faster extraction rates than a static soaking process. Percolation involves the mechanistic processes of osmosis and diffusion in and out of the cell. This gentle method can be used in the production of natural flavours.

12.2.1.4 Liquid extract

Liquid extracts provide more permanent and convenient forms of preserving the constituents of botanicals in higher concentrations (Wren, 1988). The extract traditionally produces a 1 : 1 yield (i.e. one kilo of extract from 1 kilo of botanical). It is made by soaking the raw material for two to three soaks in water and/or ethanol, followed by filtration. Extraction proceeds with a succession of fresh batches of extracting liquid. These batches are combined and concentrated back to the original weight of the herb, by evaporation under reduced pressure. Liquid extracts are able to replace dried herbs on a weight-for-weight basis in medicinal formulations prescribed by herbalists.

This form of extract typically contains 20% alcohol, which acts as an effective preservative. Some extracts are specified at higher alcohol strengths, to produce the optimum extraction of the actives contained in the raw material. As a result of heating involved in the concentration stage, a liquid extract is often dark brown in colour, and may have a caramelised odour.

12.2.1.5 Soft extract

The first step in soft extract preparation involves soaking the raw material in solvent. Two to three soaks may be required, which are filtered and evaporated down, one after the other. The initial preparation is the same as a liquid extract, but the concentration stage is then continued until the resultant extract has a moisture content of 30% or less. The final product is usually very viscous, similar to a syrup or soft thick paste, and is usually dark brown in colour, with a caramelised flavour.

The main application for a soft extract in a soft drink is when the product label claims a relatively high level of plant active in its formulation. The extract may need to contain a specified concentration of assayed active within it to guarantee the advertised claim is met. The technical challenge is often to produce a formulation that can mask the colour, aroma and taste of the extract. A soft extract may be used to blend with ingredients of a flavour system for spray- or freeze-drying when the system is required to contain botanical components. An infusion would be unsuitable for this process, due to the high fluid content. Soft extracts usually contain a much larger quantity of the active component from the botanical from which they are prepared, and in higher concentrations than the extracts left in liquid form.

12.2.1.6 Dry extract

A dry extract is made by replacing the moisture content in a soft extract with an equal amount of a carrier substrate, such as calcium phosphate, starch or maltodextrin. Soft extracts tend to contain about 70% solid matter, and can be mixed into a slurry with the carrier. The moisture is normally removed using a vacuum oven, to avoid excessive thermal degradation of the extract. After drying, extracts may be milled to the required consistency. The level of active substances in a dry extract may be varied and, if required, standardised by further addition of carrier. Another function of the substrate is to prevent the dried extract from reabsorbing moisture, which would reduce it back into a hard or sticky mass.

Dry extracts can be further blended with other ingredients in a powdered drink formulation.

12.2.1.7 Tincture

Traditionally, a tincture is an ambient temperature extract. It is made with a high alcohol level, the extraction liquid typically being 60–70% or more. The solvent acts selectively without the addition of heat.

Tinctures, like liquid extracts, are permanent preparations. They are particularly suitable for extracting drugs containing resinous and volatile principles, because the alcohol precipitates unwanted gums and albuminous matter. This enables tinctures to be filtered to yield clear, elegant preparations which are well preserved from deterioration. Tinctures have broad applications, and can be used for a wide variety of raw materials (e.g. dried plant, leaves, roots, bark and flowers). The raw material is usually soaked a minimum of twice in a mixture of water and ethanol (Wren, 1988). The soaks are gathered and filtered. Different strengths are available: 1–10 equates to 10 kgs of extract obtained from 1 kg of raw material, and a concentration of 1–5 means 5 kg are recovered from 1 kg of raw material. The number of soaks can vary from one extract to another.

12.2.1.8 Countercurrent extraction (CCE)

Countercurrent extraction (CCE) is recognised as an efficient technique which can be used as a batch or continuous process. The operation involves the movement between solvent and raw material. Fresh raw material is introduced to the solvent at its highest solute concentration, while the exhausted solids can be introduced to the fresh solvent step-by-step or continuously. In comparison with co-current flow, CCE supplies a larger overall driving force for

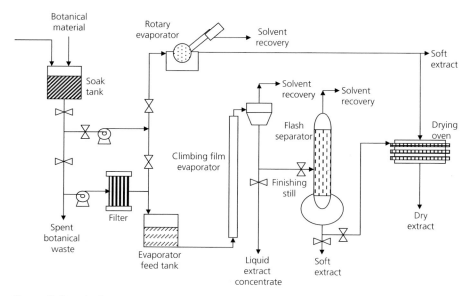

Figure 12.1 Typical extraction process. (*Source*: Courtesy of J. Ballard).

mass transfer and provides a higher recovery of soluble solids (>90%) producing a highly concentrated extract (Ming, 2007). In CCE, wet raw material is often pulverised to a fine slurry using 'toothed disc disintegrators'. The slurry is moved in one direction within a cylindrical extractor, which contains the solvent. The further the slurry moves within the extractor, the more concentrated the resultant extract becomes. When the flow rate of solvent and slurry are optimised, it is possible to achieve complete extraction, making this a highly efficient method.

CCE can provide significant advantages when compared to previously mentioned methods:
- A smaller volume of solvent is required for extraction.
- Ambient, room temperature conditions protect thermolabile constituents from exposure to heat.
- As pulverisation occurs under wet conditions, the heat energy generated is absorbed and neutralised by the solvent, again minimising thermo-degradation to the active components.

12.2.1.9 Supercritical fluid (SCF) extraction

Over the last two decades, a major technology has emerged which has provided an alternative to the conventional solvent extraction of natural products. Supercritical fluid (SCF) extraction is safe, clean and environmentally friendly. It uses non-polluting solvents, and the energy costs are lower compared with traditional methods of extraction. Consequently, SCF extraction is becoming more widely used, is receiving increased attention, and is gaining in importance among natural extract manufacturers.

Compression of a gas to a high pressure results in change of state into a liquid. However, if the gas is heated above a particular temperature, no level of compression will transform the hot gas into a liquid. This required temperature is known as the critical temperature, and the corresponding pressure is termed the critical pressure. These values define a critical point that is unique to the individual substance. When both the temperature and pressure exceed the critical point, the substance is called a supercritical fluid (SCF). It now demonstrates properties belonging to both liquids and gases. At this point, the maximum solvent capacity and major variations in solvent properties can be attained with small adjustments in temperature and pressure.

SCF extraction can show very desirable properties, due to the favourable viscosity, surface tension and diffusion, rate plus other physical characteristics of the solvent. Its low viscosity and surface tension enables its molecules to penetrate the plant cells, and its high diffusivity assists rapid mass transfer. This allows the active to be extracted at a faster rate, compared with conventional solvent methods. SCF is used selectively to extract target molecules from a complex.

Carbon dioxide (CO_2) is the most desirable of all SCF solvents for application in beverage markets. This relatively inert gas is readily available, is relatively cheap, odourless, tasteless, generally regarded as safe (GRAS), and no solvent residue is left behind in the extracted product. The near-ambient critical temperature of 31°C means that CO_2 is ideal when extracting thermo-sensitive active substances, and its low latent heat of vaporisation produces the most natural-tasting and -smelling extracts.

Other solvents suitable for SCF include propane, benzene, toluene, ammonia, nitrous oxide, ethane, ethylene, trichlorofluoromethane and chlorotriflouromethane (Mukhopadhyay, 2000).

12.2.1.10 Hexane and acetone

The flavour and modern phytopharmaceutical industries have made big changes to traditional extraction processes. Whereas ethanol was really the only significant solvent apart from water used by the early pharmaceutical extractors, solvents such as hexane and acetone are now being used by flavour companies to make soft-extract oleoresins for natural-flavour components. Special care needs to be taken when using hexane, as it has been graded as highly toxic and is categorised as a hazardous air pollutant (HAP) by the US Environmental Protection Agency (Saxena *et al.*, 2011). Hexane acts as a non-polar solvent and so is effective, for example, when used to extract lupeol from *Crataeva nurvala*. Modern concentration methods and processes, such as reverse osmosis, spray drying and freeze-drying, can yield extracts with less colour and caramelisation problems, but this benefit does come with a price penalty. Botanicals processed by hexane extraction are not permitted for use in organic products by the Soil Association or Ecocert (Bowyer, 2011).

12.2.1.11 Ultrasonic extraction

The ultrasonic extraction technique incorporates the use of ultrasound at frequencies ranging between 20–2000 kHz. Ultrasonic extraction is used to increase permeability of plant cell walls and the formation of cavitation. This process involves the rapid implosion of cavities in a liquid – in other words, the making of and bursting of bubbles. This often occurs when a liquid experiences rapid pressure changes, resulting in the formation of cavities where pressure is relatively low.

Cavitation provides the following improvements:
- Mass transfer intensification. A major mechanical effect of ultrasound is greater mass transfer. This is due to the collapse of cavitation bubbles close to, or on, walls and interfaces. This results in microparticle synthesis and emulsification, which makes a larger surface area available for the solvent, plus microdispersion of one solvent into another.
- Disruption. Plant cell walls can be broken by ultrasound, causing the release of their contents and creating a greater bulk of plant material. As a result, more plant cells come into contact with the extracting solvent, leading to more efficient extraction of the active components. The use of ultrasound aids dispersion and particle size reduction, and this can improve extraction rates.
- Penetration. Extraction yield can benefit when ultrasound is used to promote penetration of the solvent into the plant cells. As cavitation bubbles collapse close to the cell wall, an ultrasonic jet is initiated, which forces the solvent at the cell wall. The effect acts as a micro-pump, with the solvent pushing its way into the cell, dissolving its contents and carrying the components out of the cell.
- Capillary effect. The capillary effect is described as the rise of liquid in a capillary tube under the influence of ultrasound. This effect can be used to extract components found in the capillary systems of plants, in particular polar and ionic molecules.

One disadvantage of ultrasonic extraction is the possibility of free radical formation, resulting in undesirable changes in the active molecules. Free radicals can occur due to the deleterious effect of ultrasound, when the energy used is above 20 kHz (Handa *et al.*, 2008).

12.3 An extraction operation

This section briefly presents a typical process for manufacturing an infusion, and will focus on the factors that make extracts different from single-chemical components.

Extracts by their nature are complex mixtures of (often) diverse active compounds contained within a plant, which are brought into solution by the extraction process. The aim of the extractor is to produce, over a period of time, batches of an extract meeting a customer's individual specification, with as little variation as possible. There are parameters over which the extractor has some control, and these can be used to help achieve product consistency, and also to fine-tune an extract to particular customer's needs.

12.3.1 Raw materials

The process starts with the herb, which itself starts as a growing plant. To achieve maximum potency, raw materials are collected at different times throughout the year.

There are many variable factors that require consideration, including climate, terrain and husbandry. Wild herbs tend to be more potent than cultivated varieties. However, herbs from cultivated sources are in greater demand, due to the feasibility of control over the growing environment by way of irrigation, fertilisation and, possibly, shelter (Rayburn, 2007). The downside to cultivated herbs is they are more likely to be treated with pesticides and herbicides. Wild herbs gathered directly from their natural growing habitat are less likely to have been treated with chemicals, but are subject to the full force of the weather. Wild varieties are also more likely to be subject to post-harvest adulteration or substitution. Differences in soil composition also affect the growing herb, producing raw materials of varying potencies and appearances, as seen with grapes grown for making wine.

Herbs of both types are best collected during dry weather, as they contain a greater concentration of oils and resins. Rainy conditions can reduce the therapeutic effects of the active components contained within the leaves. Also, wet herbs deteriorate more quickly than if they were picked in dry conditions. Early mornings, once the dew has evaporated, or early evening, before the dew has formed, are optimum times of the day for harvest.

It is also important to consider the transit and storage conditions from field to factory, as this can have a critical effect on the raw material quality. Suitable drying is very important, in order to preserves the colour and properties of a herb. Freshly harvested herbs usually have moisture contents between 50–80% and, unless used on the same day as the harvest, they require drying prior to storage. Most dried herbs maintain their full potency for approximately 12 months.

A proportion of the herbs used in functional drinks are grown or collected from the wild in less developed nations. The crop is spread out on the ground to dry in the sun, which is a good and cheap method of gentle drying. The problem arises when the drying area is shared with domestic and farmyard animals, and this may explain why some dried herb deliveries have a very high microbiological content.

In cooler growing regions, where sun-drying is not possible, air-drying in ovens is used. The danger with this method is that, in the interests of efficient usage of expensive drying machinery, a higher temperature than ideal may be used to obtain a faster rate. This can result in the excessive loss of volatiles, or damage to thermally liable actives (Handa *et al.*, 2008).

The best drying methods are:
- Outdoor drying. The herbs are spread across a drying screen and left in a shaded, well-ventilated area until the leaves crumble easily.
- Indoor drying. The herbs are arranged in bunches and bound at the stem. Each bunch is hung upside down away from direct sunlight until the leaves crumble easily.
- Oven drying. This method requires constant monitoring, as it is a very rapid method. Low temperatures, no greater than 60°C, are ideal.

The extract manufacturer tries to exercise as much control as possible over the raw material, but this is not easily done. Best practice would be to source material from single wholesalers who obtain their product from a single merchant from a particular growing region. These merchants should, in turn, purchase from a single locality. This not cost-effective unless the project is large, with sufficient funding for the administrative efforts and travel costs necessary to set up a controlled and traceable supply line. Even if this were possible, the variation in weather from year to year will affect concentrations of actives, making quality difficult to predict.

The maturity of the plant at harvest can also implicate active content of the raw material. If it is possible to control the supply line, then it may be possible to control harvesting of the crop at a specified maturity. Otherwise, it is likely that the crop will be harvested by the local people, to optimise the yield of the bulk crop for greatest financial return.

Some control can be built in by specifying a minimum active concentration, but this standard is likely to slip, due to commercial expediency in times when the harvest produces active concentrations lower than the specification. During such times, it is likely that raw materials meeting the specifications will carry a price premium, and it may also mean acceptance of sub-quality material into the market, owing to scarcity. This has been exemplified by the shortage of ipecacuanha in the last decade.

If the raw material selected for a given development project is one for which the market demand is insufficient to warrant a commercial growing operation, it will probably be supplied from the wild, and classified as a spot market purchase, subject to variability in price and availability. Single wholesaler sourcing would be the best option here. Consideration has to be made for 'Good Agricultural Practice' on the part of any reputable purveyor, though this may prove difficult in practice in some developing country markets.

When the raw material requirement is of sufficient size, it is possible to set up a contract for the growing of a suitable amount of the desired raw material. This contract can be directly with the grower, if conditions and resources allow, or through a herbal wholesaler with good contacts in a region where the requested herb grows well. This will remove some of the supply variables, and should enable a certain degree of control over the husbandry and harvesting of the crop. The difficulty comes with the initiation of a new project, when it is usually very difficult to predict the volume of sales of a new product and, hence, the quantity of raw materials required.

Raw material prices affect the price of a finished extract. The normal rules, where larger volumes attract a bulk discount, do not always apply here. Supply and demand are the governing factors. When a raw material is selected for a new development project, it is advisable to ascertain whether the herb is readily available in the quantities required for the forecast sales volumes. If the anticipated annual demand for the selected herb represents a significant proportion of the prevailing harvest quantity then, as a result, there will initially be increase by the supply chain, until subsequent harvests are increased to meet the greater demand.

In summary, the inherent variability in the raw material supply chain, prior to the extraction process, introduces great inconsistencies into the subsequent processes and, as a consequence, influences product costs.

12.3.2 Extraction

The complex details of equipment and procedures are of interest primarily to those manufacturing the extracts, rather than to those using them. Therefore, the stages of the extraction process will only be discussed briefly below:
- Receiving the raw material and subjecting it to acceptance to chemical examination and where practical biological (through cell structure microscopy) by quality control (QC).
- Preparation of the raw material, through stages such as milling, in order to produce a suitable particle size for extraction.
- Steeping the raw material in the chosen solvent, preferably in a closed container of stainless steel, glass or plastic, capable of resisting attack by the solvent and plant components. The volume of measured solvent needs to account for the extra quantity absorbed by the dry plant material, to ensure that the desired yield is met.
- The product needs to be left to stand for a set period of time specified by the extraction method, and at a specified, controlled temperature for the raw material in question, before the liquid can be removed from the spent herb. Subsequently, this may be pressed to recover the greatest quantity of extract possible.
- Using the resulting basic crude infusion directly or subjecting it to further concentration under reduced pressure to make a soft extract which, in turn, can be dried to a powdered extract. For the majority of functional drinks, the infusion is the preferred method of extraction, as it is the simplest to dispense and mix into the end product.

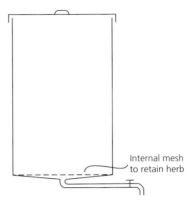

Figure 12.2 A typical infusion vessel.

- To finish the crude infusion, filtering is required. Different mesh sizes are often necessary, in sequence, to remove material of different sizes. At the final stage, a fine membrane filter (0.2–0.4 μm) is used to remove impurities such as microbiological organisms.
- When packing the finished extract, it is important to select a suitable container. The container must not be subject to degradation by the active components or solvent, and must be generously sized, to allow for any possible expansion once sealed.

Usual storage conditions required are a regulated cool temperature, preferably in darkness and away from direct sunlight.

Within the extract process itself, it is widely known that efficiency is affected by critical processing parameters that can impair quality and consistency of the final product. These include raw material particle size, extraction time, extraction temperature and choice of solvent. The difference between these variables, and those related to the raw materials previously mentioned is, to a certain extent, controlled by the manufacturer. Control over these variables leads to a consistently fine-tuned extract to meet the customer's specific requirements.

There are many different factors, both in nature and within the extraction process itself, that produce variability. It is perhaps a tribute to the extract manufacturers' combination of technology and art, built up over the years, that the extracts produced and supplied to customers are as consistent as they are. The effect the controllable parameters have on the final extract is discussed in the subsequent paragraphs.

12.3.2.1 Particle size

The particle size of the raw material used in an extraction process is a compromise between rapid extraction times and ease of filtration once removed from the spent herb. Solubility is dependent upon particle size. When particles are smaller than about 0.01 mm in diameter, solubility increases greatly as the particles become smaller. This is due to the increasing role played by surface effects.

Powdering the plant material causes its organs, tissues and cell structures to become ruptured. This releases the active components and exposes them to the extracting solvent. Additionally, reduction in particle size maximises the surface area and accelerates mass transfer of the actives from plant cells into the surrounding solvent. The finer the raw material particles, the greater the chance of reaching extraction completion at a faster rate. This, in turn, yields consistent batch strengths from identical batches of raw materials (Handa et al., 2008).

Conversely, coarse particles require a longer soak time for the soluble components to diffuse out into the extract, and causes an increased likelihood of batch-to-batch variation. Fine particle matter blocks filter pads or membranes more quickly than coarse matter, leading to potential wastage of extract and extended processing times. Some extract manufacturers choose to purchase their raw materials in the whole, or uncut, form, and mill them on site. This is dependent on the herbs passing the initial quality control (QC) acceptance tests, which typically assess confirmation of identity, moisture content, soluble extractive content and microscopic examination. Testing these variables helps to control batch-to-batch particle size consistency. By purchasing the uncut herb extract, manufacturers may be able to avoid the use of adulterated and wrongly labelled herbs, as visual detection and identification is made easier. If the herb is bought ready-cut or as a powder, it is essential to specify the particle size required to minimise batch-to-batch variability.

12.3.2.2 Time

The process of extraction involves penetration of the solvent into the plant cells of the dried herb, causing the soluble components to dissolve and diffuse out into the free solvent surrounding the plant particles. This process eventually reaches a state of equilibrium (theoretically, complete equilibrium is only achieved at infinite time, and the approach to completion is an asymptotic curve).

In practice, around 90% of available solids come into solution within 24 hours for a typical leaf herb at ambient temperature. However, harder material, such as dried woody roots and barks (e.g. quassia root), will have a much longer extraction time to achieve a satisfactory degree of extraction.

The time allotted by a manufacturer for the extraction of a herb is usually a commercial compromise between obtaining near complete extraction, which helps achieve consistency, and maximising effective use of costly extraction equipment and operators. To achieve batch-to-batch consistency, extraction time is usually well defined within the manufacturer's standard operating procedure. For example, an extraction should not be left soaking over the

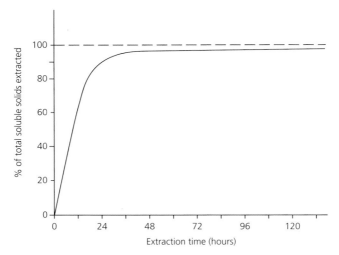

Figure 12.3 Typical extraction curve.

weekend if a 24-hour extraction time is specified. The reason for this is that different individual soluble components will diffuse out of the plant matrix at different rates. Therefore, an extract that is left to stand for longer than specified may have a higher soluble solids content and, quite possibly, a different ratio of individual components, from those required in the standard specified extract.

It can prove advantageous to shorten extraction times specified for complete extraction, in cases where the active being sought from the herb is readily soluble and reaches equilibrium before undesirable components such as brown colours from degraded chlorophyll or bitter-tasting tannins are fully dissolved.

The power of the active components increases with increasing extraction time and decreasing particle size. This increase in time and the surface area available for molecular transport contributes to greater mass transfer of solutes between phases (Handa *et al.*, 2008).

12.3.2.3 Temperature

Increasing the temperature of an extraction increases the rate of reaction. This shortens the extraction time, which is a desirable advantage for the manufacturer. Conversely, the colder the process, the slower the diffusion rate, which results in a longer extraction time. This may give rise to a weaker extract, due to incomplete extraction within the specified time. Consequently, consistency between batches can be maintained more easily by controlling the extraction temperature.

Another effect of raising extraction temperature is to increase the solubility of the less soluble plant components, which may lead to haze formation and sedimentation of the extract. The components will slowly come out of solution when the extract is left to stand during storage. This side-effect may not be an issue if the extract is to be used in a cloudy drink.

12.3.2.4 Solvent

Aqueous ethanol is the most commonly used solvent in the extraction of herbs for soft drinks. Usually, a concentration of 20% alcohol by volume (ABV) is required. The 1980s saw the introduction of the newly emerging 'adult soft drinks' sector, where manufacturers sought to avoid using any synthetic ingredients, thus enabling 'all natural ingredients' and 'preservative-free' claims to be advertised on product labels. Extract preservation is essential, and a concentration of 20% ethanol not only makes a very efficient extracting solvent, but can also act to help preserve the extract, as long as microbiological counts have been reduced by filtration or other non-chemical means.

Aqueous extracts must be preserved, unless the extract has been sterilised and aseptically packed. This is because unpreserved extracts provide an ideal growth medium for microbiological organisms.

In general, reactions in aqueous solution proceed rapidly, as they involve interactions between ions. Solvents are classified as ionising solvents if they produce solutions where solutes are ionised. Non-ionising solvents are those that produce solutions in which the solutes are not ionised. Water is known as a common ionisation solvent, and is characterised by its ability to produce hydrogen (hydroxonium) ions:

$$2H_2O = H_3O^+ + OH^-$$

When choosing a herb for the specific active components it contains, it is important to select an appropriate solvent that is effective in dissolving the desired actives. The properties of common solvents are listed in Table 12.1.

Water is an extremely polar solvent, and acts by dissolving polar materials such as salts and sugars. Ethanol, propylene glycol and glycerine are less polar, although any water present will tend to dominate the solvent polarity unless in relatively small proportions.

Even less polar solvents, such as acetone or hexane, dissolve water-insoluble non-polar components including oils, fats and resins. Water acts as the main carrier medium within plant systems, which means that most plant actives found within the plant's cell structure are, to some extent, soluble in water. Similarly, the human body has a water-based system, and so requires the plant actives to be water-soluble in order to prevent digestion wastage if the actives cannot be absorbed.

Using solvents at the non-polar end of the spectrum offers a means of selectively extracting specific actives. An example from the nutraceutical sector is the selective extraction of ginkgo biloba, using a high proportion of acetone in an acetone/water solvent system. This produces a 50 : 1 extract (50 kg of dried herb produces 1 kg of dried extract), standardised on 24% flavone glycosides and 6%

Table 12.1 Solvent comparison.

	Polarity (*E*)	Boiling point (°C)	Latent heat evap. (cal/g)	Viscosity (cP)	
				0°C	20°C
Carbon dioxide	0.00	−56.6	42.4	0.10	0.07
Acetone	0.47	56.2	125.3	0.40	0.33
Benzene	0.32	80.1	94.3	0.91	0.65
Ethanol	0.68	78.3	204.3	1.77	1.20
Ethyl acetate	0.38	77.1	94.0	0.55	0.46
Hexane	0.00	68.7	82.0	0.40	0.33
Methanol	0.73	64.8	262.8	0.82	0.6
Dichloromethane	0.32	40.8	78.7	—	0.43
Pentane	0.00	36.2	84.0	0.29	0.24
Propan-2-ol (IPA)	0.63	82.3	167.0	—	2.43
Toluene	0.29	110.6	86.0	0.77	0.59
Water	0.73	100.0	540.0	1.80	1.00
Propylene glycol	0.73	187.4	170.0	—	56.00
Glycerol	0.73	290.0	239.0	12,110.00	1,490.00

Source: Moyler (1988).

terpenes. Sub- and supercritical forms of carbon dioxide produce very good quality extracts and here, again, there is the possibility of using differences in polarity between the sub- and supercritical states to assist in selective extraction (Ming, 2007).

To overcome the fact that a single solvent rarely enables complete extraction of all the potentially available components from a herb, multi-solvent extracts have been developed. This process involves initial extraction of the herb using a non-polar solvent, followed by extraction with a more polar solvent. The resultant extracted components are combined, using a suitable emulsifier system to hold the two types of incompatible components together without separating. This will render the extract miscible, if not almost completely water-soluble, without the possibility of settling out in the final product.

The maximum permitted level of ethanol in a soft drink in the UK is 0.5% ABV, and this limits use of an extract containing ethanol at 20% ABV in a soft drink to a maximum of 2.5% by volume. Local regulations should be consulted when ethanolic extracts are required.

12.3.2.5 pH

In some instances, where the desired actives to be extracted are of a basic nature (as opposed to acidic), extraction can be selectively enhanced by adjusting the pH of the extraction solvent with a suitable organic acid, such as citric acid. Ethanol extractions are not usually affected by pH.

12.3.3 Organic extracts

Consumer interest in the quality of what they eat and drink is increasing, which has led to a large growth in the demand for the supply of organic products. Materials produced through organic farming certify that no inorganic fertilisers or pesticides have been used in the growing of foodstuffs. Organic certification fetches high demand and prices internationally. Organic farming methods are seen to have roots established in all things natural, by making use of only organic materials. Its main objectives include mulching, crop rotation, animal waste manuring, composting, bio-gas slurry, bio-fertilisers and organic recycling (Joy et al., 1998).

It is possible to produce a range of extracts that can be certified as organic. Firstly, the herb supplied must be certified as organic. This does not normally pose much of a problem, as most herbs suitable for functional drink manufacture are readily available from herb suppliers that specialise in offering organically certified raw materials. The main problem arises with the extraction process. Other than water, any solvents utilised also need to be certified as organic. Currently, organic ethanol is available, and is deemed the best option for beverage extracts, unless it is specified that alcohol is unsuitable in the product (for example in the Islamic markets). Kosher markets can be more complex, as organic ethanol is required to be produced from organic grain, such as grape alcohol, and as a result may cause issues with Kosher certification. An alternative option is to use aqueous extracts containing preservatives which have been permitted by a local organic certifying body. Throughout the United Kingdom, the Soil Association permits the use of sodium benzoate and potassium sorbate preservatives for this purpose.

It is possible to produce organic glycerine by steam-splitting organic vegetable oils. Glycerine is a good solvent for organic and inorganic substances, due to its strong hydrogen-bonding properties, giving it a high solvent activity. Its use can eliminate part, or all, of the alcohol normally required for such preparations. In the beverage market, glycerine is used in the production of soft drinks, tea and coffee extracts, and is found as a natural ingredient in wine and beer (The Glycerine Producers' Association, 2007).

Steam-splitting hydrolysis involves the hydrolysis of the vegetable oil into the corresponding fatty acid and glycerine.

It is also possible to produce a low-polarity organic solvent from natural organic raw material stock by physical means which, in turn, should enable the production of organic concentrated soft extracts reasonably similar to those produced using acetone and hexane.

12.3.4 Extract costs

There are fixed and variable costs involved in the extraction process. In general terms, the raw material handling costs, plant overheads, and technical service

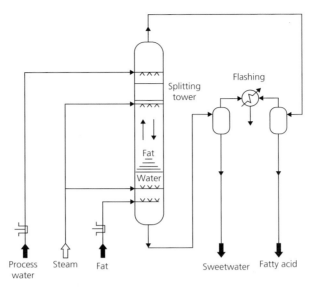

Figure 12.4 Typical fat-splitting process (*Source*: Bernardini, 2012. Courtesy of SPEC Engineers & Consultants Pvt. Ltd).

costs such as QC, are relatively independent of batch size. The main costs that vary with batch size are the raw materials, energy consumption, labour, packaging and delivery. The basic infusion process, using readily available herbs, with a large enough volume demand to enable larger batch sizes (of around 250 kg), will result in a more economically priced extract.

Smaller one-off batches carry a premium. The higher the raw material price, the bigger the percentage of the overall extraction cost it represents and, consequently, the lower the influence on cost savings made by larger batch sizes.

The yield of the extract is an important factor in the price of concentrated extracts. Low-yield extracts, which have a high concentration of actives in them, will usually be priced much higher. Firstly, the amount of raw material required to produce a specified weight of extract is much higher, compared with a simple infusion. Secondly, the process itself may be more complex, involving greater costs for solvents and concentration energy. The equipment used may also be complex and may attract significantly higher production overheads.

Certified organic extracts often carry a significant raw material price premium over their non-organic counterparts. Sub- and supercritical carbon dioxide extracts are usually the most expensive, due to the capital cost of high-pressure equipment required.

12.4 Extract characteristics and their problems

12.4.1 Specifications

It is important to understand the factors that make extracts different from single-component ingredients, as this can help to avoid frustration during use. Extracts are often complex mixtures of components, and their corresponding specifications demonstrate this.

They do not, for example, have a straightforward melting point, or meaningful boiling points. They do not have simple spectra or other readily quantifiable properties. However, they do have subjective parameters, such as aroma, taste and colour. Some extracts may carry an assay of one or more active components (e.g. ginsenosides in a ginseng extract), but this does not necessarily give an accurate assessment for the quality of the whole extract. In the case of ginseng, the ginsenosides tend to be preferentially located in the hair of the roots and rootlets, rather than in the bulk of the main root. This means that extracts with higher levels of ginsenosides can be made from these plant sections, rather than the more expensive whole root that contain other components of the herb.

12.4.2 Stability

Liquid extracts, especially infusions, will often produce a fine sediment over time. Plant extracts naturally produce sedimentation, the degree of which is unique to the raw material in question. For example, cocillana and liquorice liquid extracts are prone to excessive sedimentation, as they are both prepared from the hard root of the plant. More than likely, there should be reference to this in the specification and on the container label. If the final product into which the extract is due to be incorporated is a cloudy drink or opaque-like fruit juice, then the container and its contents can simply be shaken each time to re-disperse the sediment prior to dispensing. If it is essential for the extract to be clear, then it must be carefully decanted when being weighed for use as a production batch.

Extracts in storage will often change colour over time. This is why specifications include colour parameters within the shelf life of the extract. Otherwise, situations may arise where a customer purchases an extract from the manufacturer and it passes QC colour tests on delivery, but it is later rejected by QC after the extract has been left to stand in storage, particularly if the drum has been previously opened and part-used.

Stability is also concerned with the stability of the active components within the extract, as it is common for degradation to occur over a period of time. This is a particular problem seen in tinctures of Squill. Here, the main active, proscillaridin A, is easily hydrolysed into two separate components that are not easily distinguishable chemically.

Stability indications are an essential part of the specification and testing is usually carried out under recommended guidelines, such as those in Good Manufacturing Practice (GMP). The initial step involves sampling of the extract under question into suitable containers – that is, containers made out of the same material the extract will be contained in during its storage period (for example, high-density polyethylene plastic containers). These samples are then stored at two different temperatures, which are usually a maximum of 25°C to represent standard storage conditions, and 40°C to represent extracts stored at high temperatures. The higher temperature represents conditions in tropical countries, and also enables the gathering of accelerated results for faster shelf life predictions. As a general guide, stability testing should be carried out four times during the first year of storage, and twice in subsequent years. This is continued until the extract sample falls out of specification, thus providing an expiry date.

12.4.3 Hazing

Another consideration is whether the extract demonstrates physical stability within the finished product of the customer. If the product containing the extract is a cloudy beverage, then there may not be an issue. If, however, the end product is clear, and packaged in a clear, see-through container, hazing can detract from its appearance.

Hazing may be due to a difference in pH between the product and the extract itself. Typically, an infusion will have an normal pH of around 5.5–6.5, whereas soft drinks are more acidic, typically pH 3.0–4.0. If hazing arises due to pH variations and causes problems in the final product, the extract can be pre-conditioned to match that of the final product. This means that any proteins or other components precipitated by the acidic conditions can be brought out of solution and allowed to settle, prior to the extract being filtered for the final time. This allows the unstable components to be removed by the extract manufacturer, rather than ending up in the finished product and creating a haze.

Another cause of hazing can be due to the presence of significantly differing polarities between the extracting solvent and the product matrix. This is known as polarity mismatch and is likely to occur when an extract with a high alcoholic content is used in an aqueous soft drink. Resins, and other non-polar components such as fixed oils, which all require high-strength ethanol to enable solubility for extraction, are known to come out of solution when mixed with water-based systems, again producing a haze.

Seeds and barks are the main materials containing non-polar components, and so also require larger volumes and greater strengths of ethanol, compared with softer parts of the plant. The kola nut is a good example of this as it was traditionally extracted using 60% ethanol, which managed to dissolve a significant quantity of the resinous material. This extract needed to be stabilised with glycerine to keep the resins in suspension, once the ethanol was removed, after concentration procedures and subsequent dilutions into drinks and other preparations.

12.4.4 Availability

The large number of herbs that are available for customers, combined with the relatively low volumes required, may mean it is not economically viable for an extract manufacturer to carry stocks of large numbers of extracts, in case demand diminishes and shelf life expires. The usual practice is for extract manufacturers to carry a fairly wide range of dried raw materials in sample quantities so that, when a customer requests a sample, it can be produced within a reasonably short time. Once a new herbal drink has been developed using samples and, perhaps, a pilot batch, a production-size batch of extract will be made for the product launch. Following that, if the product sells and there is a demonstrable demand pattern, it may be agreed by the extract manufacturer for production of a batch for stock to be called off by the drinks manufacturer.

12.5 Incorporation of extracts in beverages

In most applications, the extract is incorporated at the pre-mix or syrup stage, along with other critical ingredients such as flavourings, colours, high-intensity sweeteners and preservatives. Some of the factors related to the incorporation of extracts into beverages are as follows.

12.5.1 Fruit juice-based and fruit-flavoured drinks

Some herbal infusions have two features that may not facilitate their use in soft drinks. Most have their own flavour, which is not usually pleasant or supportive of the desired product flavour, and they are often unfavourably pigmented, with a fairly strong brownish colour.

Fortunately, the levels at which infusions are normally added to fruit juice drinks are low and, thus, their flavour contribution can usually be masked easily, negating the potential problem. Likewise, the colour contribution does not interfere with the majority of juice drinks, as the light straw-brown colour generated by a dilute extract is usually disguised by the oranges, yellows and reds of most products. Most herbal raw materials are appropriate to this category of drink, and a theme that has emerged is the use of flower extracts, which convey a general image of naturalness and wholesomeness without having an overtly functional message. Elderflower, ginger and, more recently, hibiscus flower, are good examples of these types of beverages.

Not-from-concentrate (NFC) fruit juices have become increasingly popular. The idea of fruits with added benefits, such as 'superfruits,' have been welcomed by consumers. Juices containing acai and goji are among the most purchased superfruit varieties and, since 2007, they having been seen on supermarket shelves among market leading fruit juices. Juices produced from superfruits have advantages when it comes to producing palatable and visually stimulating functional beverages. These fruits naturally contain appetising flavours, aromas and colours, while also providing noticeable health benefits.

Fruit smoothies are also in demand, and help to drive trends in consumption of tropical and berried fruit blends by contributing to the consumer's 'five-a-day' target. The smoothie market is likely to remain a key area for growth.

12.5.2 Mineral-water based and flavoured water drinks

Mineral water with added functional ingredients is a fast-growing category within the functional drinks sector. Generally, these products have a more focused functional proposition, aimed at commonly recognised lifestyle issues, such as coping with stress, boosting the immune system, relaxation, calming the mind, promoting sleep and reviving tired minds and bodies. Such products tend to contain a cocktail not only of herbs, but also vitamins and minerals appropriate to the advertised benefits of the product. As examples, ginseng is well recognised as helping the body deal with stress, echinacea is widely seen as an immune-system boosting herb, and chamomile, passion flower and valerian are all well-known herbs that calm the mind and prepare the body naturally for sleep.

Flavoured waters have steadily provided consumers with an alternative to plain water, delivering variety to help with hydration, with the addition of functional health benefits.

12.5.3 Carbonated and dilutable drinks

Carbonated drinks are the most popular category within the soft drink market, and with the addition of functional ingredients, with diet and low calorie varieties. Carbonates will more than likely be soon to hold a prestigious place within the functional drink sector.

Similarly, dilutable drinks also account for a large proportion of soft drink consumption and, coupled with health and wellbeing trends towards functional beverages, dilutables are keeping pace. Here, superfruits, with the addition of vitamins and minerals, can be incorporated to produce cheaper and more convenient in-home functional drink varieties.

12.5.4 Energy and sports drinks

Nearly all energy drinks have one thing in common other than their calorific value – a high caffeine content. Not all products in this category contain herbal extracts, with brand leaders both in Europe and Japan sticking to energy benefits alone.

Herbal extracts that are incorporated into energy drinks are generally from stimulant herbs of one kind or another. Most commonly used is guarana, which is a natural source of caffeine and is native to Brazil. Other herbs that contain similar levels of caffeine are kola nut, coffee, tea and maté. The cocoa bean, which is the seed of *Theobroma cacao*, is in the same family of plants as the kola tree and contains the stimulants theophylline and theobromine, which are similar in structure to caffeine.

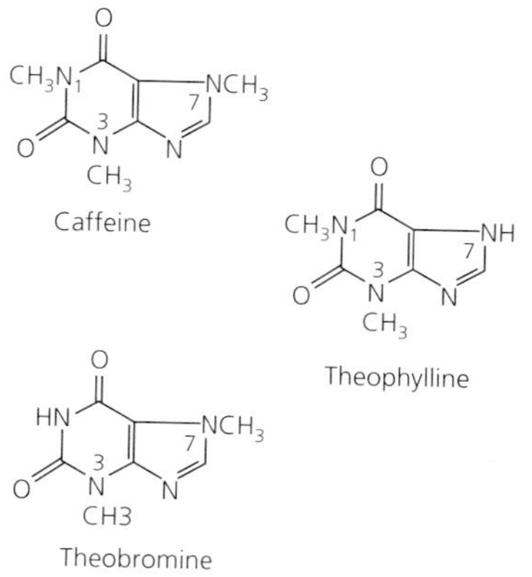

Figure 12.5 Caffeine structural comparison to theophylline and theobromine.

Extracts of coffee bean and cocoa bean have been produced experimentally that contained about 3% of caffeine and 3% theobromine, respectively, without the disadvantage of flavour concentration typically associated with those raw materials. These are potentially alternative natural sources of effective stimulants to guarana and kola. Other supplementary herbs that have a place in energy drinks are those that support the concept of exercise or vitality, such as ginseng, or alleged aphrodisiacs, such as damiana and muira puama. These are often incorporated in drinks designed for social occasions and are primarily for sale in clubs and bars.

Sports drinks, once the domain of athletes and gym-goers, have seen manufacturers modify their target market to also introduce everyday replenishment. Newly emerging energy and sports drinks are aimed at focusing the mind and providing both physical and mental alertness, whilst using natural stimulants, to help consumers deal with their daily energetic demands.

12.5.5 Regulatory issues

Regulations differ from country to country, and vary over time as new legislation is enacted. Therefore, it is very difficult to give any universal guidance, other than to suggest that product development technologists use the local food legislation experts, whose role it is to keep up to date with the latest developments in their national regulations.

In Europe and the United Kingdom, there are two main issues to address. First, herbal beverages should avoid incorporating levels of herbal extracts that are high enough to deem them liable for consideration as herbal remedies or medicines. This means using volumes and concentrations of extracts that avoid delivering cumulative therapeutic levels of active components when consumed on a typical daily basis. Individual specifications are available for the botanical herbs in question that will state appropriate dosages for application.

The second issue is that of product claims. This subject is difficult to write upon definitively, as new or amended regulations are constantly being updated. However, the trend seems to be that claims made by the manufacturer will require a degree of substantiation based upon scientific reports. Folklore and anecdotal information about herbs are steadily being withdrawn more and more from product labels, advertisements and point-of-sale literature. More information on such matters can be found by visiting the European Commission Health and Consumers (EUROPA) website, which dictates regulatory issues surrounding food, and pays particular attention to proper labelling and advertisement claims.

Another useful reference for European and the United Kingdom is *Natural Sources of Flavourings*, published by the Council of Europe. It is commonly known as the 'Blue Book', and is gradually being superseded by a series of updated versions, containing greater detail on safety of use. To date only the first report, consisting of two volumes, has been published. The report provides information on the toxicological evaluation of chemically-defined flavouring substances and the safety in use of natural flavouring source materials for food and drink application.

The second report, which contains information on natural sources and source materials of natural flavourings for food and drink, was due to be published in 2012. So far, the most commonly used plant materials have been mentioned in these reports, with many of the plants listed being traditional herbs that were used prior to the introduction of modern medicines for their curative or tonic benefits. The plants listed as positive and non-hazardous to ingest (there are a selection of plants listed as hazardous for use in food and drink) are divided into six categories, which are defined as follows:

- Category 1 – plants, animals and other organisms or parts thereof consumed as food in Europe: no restriction is made on the parts used, under the usual conditions of consumption
- Category 2 – plants, animals and other organisms, and parts thereof, including herbs, spices and seasonings, not commonly used as foodstuffs in Europe and considered not to constitute a risk to health in the quantities used.
- Category 3 – plants, animals and other organisms, and parts of these products thereof, normally consumed as food items, herbs or spices in Europe, that contain defined 'active principles' or 'other chemical components' requiring limits on usage levels.
- Category 4 – plants, animals and other organisms, and parts of these or products thereof, and preparations derived therefrom, not normally consumed as food items, herbs and spices in Europe that contain defined 'active principles' or 'other chemical components' requiring limits on usage level.
- Category 5 – plants, animals and other organisms, and parts of these or products thereof, and preparations derived therefrom, for which additional toxicological and/or chemical information is required.
- Category 6 – plants, animals and other organisms, and parts of these or products thereof, and preparations derived therefrom, that are considered to be unfit for human consumption in any amount.

12.6 Some commonly used herbs

This section gives information about some of the more commonly known herbs that have been used in drinks. This list generally excludes herbs and spices that are associated primarily with culinary or flavour use, although many of these will also have some health benefits. Some of the herbs listed below are of European or American origin and are listed in the 'Blue Book'. The names of listed herbs are followed by their category. Several herbs that have been used in the past, such as St John's Wort, Ma huang (Ephedra) and Kava-kava, have been omitted as they are currently withdrawn from the European market due to health concerns by the regulators.

Artichoke

Botanical name: *Cynara scolymus*.
Family: Asteraceae.
Synonyms: Globe Artichoke, Bur Artichoke, *Cynara cardunculus* Moris.
Region of origin/habitat: Indigenous Mediterranean area, but also grown as a garden vegetable.
Description: A 2 m high perennial, with a flower characteristic of thistle, overlapping fleshy bracts tapering to a greenish or purplish tip. Bluish-white coloured petals. Pinnate and prickly leaves, a hairless upper surface, and grey coloured under surface.
Part of herb used: The fresh or dried basal leaves. The flower is used as a vegetable.
Main actives: Caffeic acid derivatives ($\approx 1\%$) including chlorogenic acid, flavonoids ($\approx 0.5\%$) and sesquiterpene lactones (up to 4%) – the major component is cyanaropicrin.

Medicinal use/benefits: The herb is claimed to stimulate the gall bladder and to detoxify and regenerate the liver tissues. It has been used to treat dyspeptic problems. It has also been shown to reduce blood lipids, serum cholesterol and blood sugar levels. The high insulin content makes it a valuable vegetable for diabetics.

Folklore: Artichokes were grown as vegetables by both the Greeks and the Romans. It is only relatively recently that the artichoke has become medicinally interesting, with the discovery of its beneficial action on the liver. (Bown, 2003; Gruenwald *et al.*, 2002; Tierra, 1998; Williamson, 2003).

Burdock

Botanical name: *Articum lappa*.

Family: Asteraceae.

Synonyms: Lappa, Bardane, Great or Thorny Burt, Beggar's Buttons, *Arctium majus*.

Region of origin/habitat: Grows in hedges and ditches in Europe, parts of Asia and North America. Flowers June and July.

Description: Large with broad, blunt, cordate leaves. Purple flower-heads with hooked bracts forming burrs. The fruits are brownish-grey and wrinkled.

Parts of the herb used: Dried aerial parts and roots.

Main actives: The aerial parts contain flavonoids and the root contains bitter components polyacetylenes and inulin.

Medicinal use/benefits: In traditional terms, burdock is used as a blood purifier; it is claimed that it helps the kidneys to remove toxins from the blood. Orexigenic properties may be useful functionally.

Folklore: Burdock was traditionally used to treat skin complaints such as eczema and psoriasis, as well as inflammations.

The roots can be eaten raw in salads or cooked like carrots and the young leaf stalks can be scraped and cooked like celery. (Bown, 2003; British Herbal Medicine Association, 1983; Hutchens, 1973; Shealy, 1998; Tierra, 1998; Williamson, 2003).

Clover (red)

Botanical name: *Trifolium pratense*.

Family: Fabaceae.

Region of origin/habitat: The herb is indigenous to Europe, Central Asia, India and North Africa.

Description: Egg-shaped flowers, sometimes paired, reddish-purple.

Parts of the herbs used: Flowering tops.

Main actives: Volatile oil, including benzyl alcohol, isoflavonoids, coumarin derivatives and cyanogenic glycosides. It has also been shown to contain genistein, a mildly oestrogen-like compound.

Medicinal use/benefits: Red clover is traditionally believed to have relaxant (sedative), expectorant and wound healing properties.

Folklore: Clover was used to treat coughs and chest problems, especially whooping cough. It was also used to help with skin complaints. Clover was a very important fodder crop, to the extent that, even in the medieval period, varieties had been cultivated to improve the persistence and flowering time of the herb. The solid extract of clover has been used as a flavouring agent in a range of food products. (Bown, 2003; British Herbal Medicine Association, 1983; Gruenwald *et al.*, 2002; Hutchens, 1973; Tierra, 1998; Williamson, 2003).

Damiana

Botanical name: *Turnera diffusa*.

Family: Turneraceae.

Region of origin/habitat: Mexico, Central America, and North America's more southerly states.

Description: Wedge-shaped leaves, short stalked with a few serrated teeth. Fig-like flavour.

Parts of the herbs used: Dried leaves.

Main actives: Volatile oil, tannins, resins and glycosides.

Medicinal use/benefits: A bitter, aromatic herb with a fig/date-like flavour that is claimed to be a nerve stimulant and was used to treat nervous exhaustion and anxiety of a sexual nature. It is claimed to be a mild irritant of the genito-urinary tract.

Folklore: The Mayans knew this plant as an aphrodisiac. (Bown, 2003; British Herbal Medicine Association, 1983; Hutchens, 1973; Tierra, 1998; Williamson, 2003).

Dandelion

Botanical name: *Taraxacum officinale*.

Family: Asteraceae.

Region of origin/habitat: Europe extending north to the Arctic region, east to the Orient and south to North Africa. Widely distributed through most of the world as a weed.

Parts of the herb used: The fresh and dried root and leaves.

Main actives: Sesquiterpene lactones, which are bitter-flavoured, triterpenes, steroids, flavonoids, mucilages and an inulin content that varies from 2–40% in the autumn.

Medicinal use/benefits: The bitter components were used to promote the flow of digestive juices in the upper intestinal tract.

Folklore: The French name for this herb is 'Pissenlit', which is self-explanatory. The herb is regarded as a good diuretic to help purify the system by removing toxins. For some time now the roots have been roasted and then extracted to make a caffeine-free dandelion coffee. Dandelion has also been used in root beers and soft drinks such as Dandelion and Burdock. (Bown, 2003; British Herbal Medicine Association, 1983; Gruenwald *et al.*, 2002; Hutchens, 1973; Shealy, 1998; Tierra, 1998; Williamson, 2003).

Echinacea

Botanical name: *Echinacea purpurea*.

Family: Asteraceae.

Synonyms: Coneflower, Purple Coneflower, Black Sampson.

Region of origin/habitat: Indigenous to North America, cultivated widely in Europe.

Description: Dried rhizome is greyish-brown. Transverse section shows a thin bark, a yellowish porous wood flecked with black.

Parts of the herbs used: Fresh or dried rhizomes and roots, aerial parts and juice.

Main actives: Water-soluble polysaccharides and glycoproteins, volatile oil (up to 2%), caffeic and ferulic acid derivatives including cichoric acid (0.6–2.1%), alkamides (0.01–0.04%), polyynes and pyrrolizidine alkaloids.

Medicinal use/benefits: Echinacea is held to be one of the most effective detoxifying herbs in Western medicine for a range of ailments, and is now used in ayurvedic medicine. The polyynes and cichoric acid components are reported to have antibacterial and virostatic effects. Echinacea also demonstrates an anti-inflammatory effect, due to the alkamides component. As an immune stimulant, it significantly raises immunoglobulin M levels. Antiviral activities against both the herpes simplex virus Type I and the influenza-A virus have been observed.

Folklore: *Echinacea purpurea* was used by native North Americans to treat wounds. Its use was greatly promoted by the Eclectic movement from the 1850s, until the movement declined in the 1930s. (Bown, 2003; Gruenwald *et al.*, 2002; Hutchens, 1973; Shealy, 1998; Tierra, 1998; Williamson, 2003).

Elderflower

Botanical name: *Sambuccus nigra*.

Family: Caprifoliaceae

[fruit: Category 3 (with limits on hydrocyanic acid); flowers and tips: Category 1; leaves and extracts: Category 5 (with limits on hydrocyanic acid)].

Synonyms: Black Elder, European Elder.

Region of origin/habitat: Europe, commonly growing in hedges and on waste ground.

Parts of herb used: The dried or fresh flowers, berries, leaves and bark.

Main actives: Flavonoids such as rutin, isoquercetin and quercetin, chlorogenic acids and volatile and fixed oils.

Medicinal use/benefits: Elderflowers have traditionally been used for colds and fevers, as their main action is claimed to be to induce sweating and reduce temperature.

Folklore: Many old superstitions surround the elder. It was considered most unwise to cut down an elder tree without first seeking permission of the 'elder mother' spirit in the tree. (Bown, 2003; British Herbal Medicine Association, 1983; Gruenwald *et al.*, 2002; Hutchens, 1973; Shealy, 1998; Tierra, 1998; Williamson, 2003).

German chamomile

Botanical name: *Matricaria recutita*.

Family: Asteraceae

Synonyms: Single chamomile, Hungarian chamomile.

Region of origin/habitat: Indigenous to Europe and north-western Asia.

Description: Flower heads much smaller than those of the Roman variety, and only have one row of florets which are usually bent backwards when dried.

Parts of herb used: Flowers and flowering tops.

Main actives: Volatile oil containing bisabolol compounds, flavonoids, coumarin compounds and mucilages.

Medicinal use/benefits: A bitter, aromatic herb, traditionally used for its gentle sedative, calming properties. It is also used to calm the digestive system. It is a mild herb that has been used for children's complaints.

Folklore: German Chamomile is used in toiletry and cosmetic preparations as a hair conditioner and lightener. (Bown, 2003; British Herbal Medicine Association, 1983; Gruenwald *et al.*, 2002; Hutchens, 1973; Shealy, 1998; Tierra, 1998; Williamson, 2003).

Ginkgo

Botanical name: *Ginkgo biloba*.

Family: Ginkgoaceae

Synonyms: Maidenhair-tree.

Region of origin/habitat: China.

Parts of herb used: Dried leaves. The fruits are eaten in China.

Main actives: Proanthocyanidins (8–12%), flavonoids (0.5–1.8%), biflavonoides (0.4–1.9%), diterpenes and sesquiterpenes.

Medicinal use/benefits: Ginkgo is highly regarded as a tonic for the circulatory system. Studies have shown that it improves blood flow to the extremities of the body, by reducing blood viscosity and dilating the blood vessels. Ginkgo has been shown to help retard the degenerative effects of Alzheimer's disease on cognitive functions, due presumably to improved blood flow in the cerebral capillaries. Other studies have shown that ginkgo improves memory and learning capabilities. One report has also claimed that ginkgo helps to counter loss of libido caused by antidepressants. It has been used by mountaineers to help counter altitude effects, and by mountaineers to keep their extremities warm.

Folklore: The ginkgo tree comes from a very ancient order of plants, and the modern single species is almost identical to fossilised plants that were growing 65 million years ago – before mammals evolved. Chinese medicine used the seeds of the ginkgo tree, but it was only in the latter half of the 20th Century that Western medicine started to research the properties of the leaves. (Bown, 2003; Gruenwald *et al.*, 2002; Shealy, 1998; Tierra, 1998; Williamson, 2003).

Ginseng

Botanical name: *Panax ginseng*.

Family: Araliaceae

Synonyms: Chinese Ginseng, Korean Ginseng, American Ginseng, Himalayan Ginseng, Siberian Ginseng.

Region of origin/habitat: Indigenous from Nepal to Manchuria.

Description: The root is spindle-shaped, pale brownish-yellow.

Parts of the herbs used: The main root, side roots and rootlets.

Main actives: Ginseng contains a complex mixture of triterpene saponins (0.8–6.0%) and also many ginsenosides, of which the predominant ones are Rb1 (0.15–1.2%), Rb2 (0.06–0.8%), Rc (0.1–1.2%), Rd (0.04–0.7%),

Re (0.15–1.5%) and Rg1 (0.2–0.6%). Ginseng also contains water-soluble polysaccharides and some polyynes.

Medicinal use/benefits: Ginseng's main action is to help the body fight off physical, chemical and biological attacks, by raising the body's own defence mechanisms. In human studies, there was a visible benefit in terms of physical and mental performance. It was also shown to reduce blood sugar levels in Type II diabetics.

Folklore: Ginseng is one of the oldest known tonic herbs, having been in use for about 5000 years in China. It was introduced into Europe several times from the end of the 9th Century onwards, but did not become established until the middle of the 20th Century, following studies by Russian scientists that established its adaptogenic properties. Most ginseng is now cultivated, since very few plants are found in the wild. South Korea grows a large amount of ginseng under government control to ensure that quality standards are observed. (Bown, 2003; British Herbal Medicine Association, 1983; Gruenwald et al., 2002; Hutchens, 1973; Shealy, 1998; Williamson, 2003; Yeung, 1983).

Guarana

Botanical name: *Paullinia cupana*.

Family: Sapindaceae.

Synonyms: Brazilian Cocoa *Paullinia sorbilis*.

Region of origin/habitat: Amazon Basin, namely Brazil and Venezuela.

Description: The paste is formed from the pulverised and roasted seeds, formed into rolls or bars and dried.

Parts of the herb used: The seeds, which are purple-brown to black with a characteristic white 'eye', contained within a red-orange fruit about the size of a hazelnut.

Main actives: Guarana has the highest known caffeine content of any herb, at 3.6–5.8%. It also contains small amounts of theophylline and theobromine, the other stimulant purine alkaloids similar to caffeine. Besides these, guarana contains about 12% tannins and some saponins.

Medicinal use/benefits: The caffeine content makes guarana a strong central nervous system stimulant. It is traditionally used as a tonic for fatigue, and to allay hunger and thirst. It also has short-term diuretic effects. The tannin content gives guarana an astringent effect, and it has been used to treat diarrhoea.

Folklore: Guarana is traditionally prepared by roasting the seeds to enable the shells to be removed, after which the seeds are crushed and ground into a paste that is fashioned into stick form and dried over a smoking aromatic charcoal fire. The Guaramis and other Amazon Indian groups grated a little of this dried guarana paste into water to make a stimulating drink that enabled them to overcome fatigue and hunger on long hunting trips. (Bown, 2003; Gruenwald et al., 2002; Tierra, 1998; Williamson, 2003)

Hops

Botanical name: *Humulus lupulus*.

Family: Cannabinaceae.

Region of origin/habitat: Europe, but now grown in temperate regions around the world.

Description: The female flower is yellowish-green, cone-like, formed from two membranous scales, one of which contains the small seed-like fruit at the base.

Parts of herb used: The dried strobile (female inflorescence).

Main actives: About 10% of α and β bitter acids including humulone and lupulone, volatile oil (0.3–1.0%).

Medicinal use/benefits: Hops are a traditional sedative and soporific (sleep promoter). The bitter acids have been shown to be antibacterial and antifungal, and also to stimulate the secretion of gastric juices. Hormonal anaphrodisiac effects have also been reported.

Folklore: Hops are primarily associated with the brewing of beer, but many other herbs were used in brewing long before hops came to prominence. Hops were traditionally used to stuff pillows as a way of promoting sleep. (Bown, 2003; British Herbal Medicine Association, 1983; Gruenwald *et al.*, 2002; Hutchens, 1973; Williamson, 2003).

Horehound (White)

Marrubium vulgare.

Family: Lamiaceae

Synonyms: White Horehound, Hoarhound.

Region of origin/habitat: Indigenous from the Mediterranean region to Central Asia. Grown wild throughout Europe.

Description: Leaves cordate-ovate, bluntly serrate, wrinkled and stalked. Small white flowers, in dense whorls.

Parts of herb used: The fresh or dried aerial parts of the plant (herb).

Main actives: Diterpene bitter components including marrubiin ($\approx 1\%$), caffeic acid derivatives including chlorogenic acid, flavonoids and a trace of volatile oil.

Medicinal use/benefits: The herb is an aromatic bitter that has been used to stimulate digestive juices. It is also a traditional expectorant.

Folklore: This herb was used as far back as ancient Egyptian times as a cough remedy. More recently, it has been made into candy cough sweets. At one time, horehound ale was brewed, particularly in the East Anglia region of the United Kingdom. The leaves have also been used in liqueurs. (Bown, 2003; British

Herbal Medicine Association, 1983; Gruenwald *et al.*, 2002; Hutchens, 1973; Williamson, 2003).

Kola

Botanical name: *Cola nitida* and *C. acuminata*.

[kola and kola nut extract: Category 4 (with limits on caffeine)].

Region of origin/habitat: West Africa.

Parts of herb used: The seed after removal of the testa.

Main actives: Purine alkaloids – mainly caffeine (0.6–3.7%), with some theophylline and theobromine, tannins, proanthocyanidins, and 45% starch.

Medicinal uses/benefits: Besides the nervous stimulant effect due to the caffeine, kola is used in its indigenous region to stimulate the digestive system, since it is claimed that it helps to break down fat. It has also been shown to have a mild diuretic effect, which is consistent with its caffeine content.

Folklore: Kola is traditionally used in tonics for exhaustion and poor appetite. The tannins have an astringent effect in cases of diarrhoea. In the countries of origin, the seed is ground as a condiment for food and chewed before meals to promote good digestion. (Bown, 2003; British Herbal Medicine Association, 1983; Gruenwald *et al.*, 2002; Tierra, 1998).

Lemon balm

Botanical name: *Melissa officinalis*.

Synonyms: Balm, Sweet Balm.

Regions of origin/habitat: Eastern Mediterranean and Western Asia. Flowers in spring and early summer.

Description: Low-growing plant with small white flowers that appear in summer and bloom until autumn. A member of the mint family, it is a hardy bush with crinkly, serrated, heart-shaped leaves.

Parts of herb used: Aerial parts of carefully dried herb.

Main actives: Complex mixture of volatile oils, glycosides of the alcoholic and phenolic volatile components, caffeic acid derivatives, flavonoids and triterpene acids.

Medicinal use/benefits: *In vitro*, the herb has been shown to have antibacterial and antiviral effects. It has also been used as a calming sedative for nervous indigestion, and is one of the herbs that are given to children for stomach upsets.

Folklore: The oil has insect repellent properties. The herb is used in cooking to impart a lemon flavour to the food. It is an ingredient of a melissa cordial made by Carmelite nuns, as well as being included in other liqueurs, such as benedictine and chartreuse. Traditionally, the herb was seen as an antidepressive. (Bown, 2003; British Herbal Medicine Association, 1983; Gruenwald *et al.*, 2002; Tierra, 1998).

Limeflower

Botanical name: *Tilia cordata* and *T. platyphyllos*.

Family: Tiliaceae.

Synonyms: Lindenflowers.

Region of origin/habitat: The tree is common throughout northern temperate zones.

Parts of herbs used: The dried flowers, extracts of which have a honey-like flavour with astringency.

Main actives: Mucilages (about 10%), flavonoids, tannins, chlorogenic acid and volatile oils.

Medicinal use/benefits: An aromatic mucilaginous herb that has traditionally been used for its diuretic and expectorant properties. It is claimed to calm the nerves, lower blood pressure and improve digestion.

Folklore: Limeflowers were thought to cure epilepsy if the sufferer sat under the tree. The wood of the lime tree is valued for its pale colour, and its suitability for turning and carving. It is used in the manufacture of musical instruments. (Bown, 2003; British Herbal Medicine Association, 1983; Gruenwald *et al.*, 2002; Shealy, 1998; Tierra, 1998, Williamson, 2003).

Maté

Botanical name: *Ilex paraguariensis*.

Family: Aquifoliaceae

Synonyms: Yerba Mate, Jesuit's Brazil or Paraguay Tea.

Region of origin/habitat: South America, between 20° and 30° latitude.

Parts of herb used: The dried leaf and leaf stems.

Main actives: Caffeine (0.4–2.4%) and theobromine (0.3–0.5%), caffeic acid derivatives including chlorogenic acid and neochlorogenic acid, flavonoids including rutin and isoquercitrin, saponins and volatile oil.

Medicinal use/benefits: The herb has stimulant effects, due to its caffeine and chlorogenic acids. It is also diuretic and reportedly has lipolytic (fat-burning) effects.

Folklore: In South America, a tea brewed from the herb (also called 'maté') is served on social occasions as a communal recreational beverage that is very stimulating. The maté is prepared in a bowl that is passed around the assembled people; it is drunk from the bowl by means of a silver straw, with a strainer on the lower end to prevent the leaves being ingested. (Bown, 2003; British Herbal Medicine Association, 1983; Gruenwald *et al.*, 2002; Williamson, 2003).

Meadowsweet

Botanical name: *Filipendula ulmaria*.

Family: Rosacea

Synonyms: Queen-of-the-Meadow, Bridewort.

Region of origin/habitat: Europe, parts if Asia and cultivated in North America.

Parts of herb used: The flowering tops and leaves (Herb).

Main actives: Salicin. It was from this plant that in 1838 salicylic acid was first isolated.

Medicinal use/benefits: Meadowsweet has traditionally been used for its astringent and antacid properties. It has long been held that it soothes and relieves pain, especially in joints and the digestive tract.

Folklore: Along with vervain and watermint, meadowsweet was one of the most important herbs for the Druids. It was also a popular strewing herb in medieval times. (Bown, 2003; British Herbal Medicine Association, 1983; Shealy, 1998; Williamson, 2003).

Nettle

Botanical name: *Urtica dioica*.

Family: Urtucaceae.

Synonyms: Stinging Nettle.

Region of origin/habitat: Common throughout the temperate zones of the world.

Parts of herb used: Fresh or dried aerial parts of the plant. Root can also be used though more for medical purposes.

Main actives: The fresh leaves and stems are rich in vitamins A and C and iron. They also contain histamine, serotonin, acetylcholine and formic acid in the stinging hairs. The dried herb contains flavonoids (0.7–1.8%), including rutin and isoquercitrin, silicic acid (1–5%), a trace of volatile oil and potassium and nitrate ions.

Medicinal use/benefits: Significant anti-rheumatic and anti-arthritic actions have been demonstrated in several studies, some with large groups of participants. Diuretic properties have been reported in connection with prostate problems and in cases of lower urinary tract infections.

Folklore: The name urtica is believed to be derived from the Latin verb 'urere', to burn, most probably referring to the stinging action of the plant. Nettle is a fibrous plant and was used in cloth manufacture from the Bronze Age until the early 20th Century. The fresh young plant tops have been cooked as a spinach-like vegetable dish, and used to be brewed into a nettle beer in certain parts of the United Kingdom. The herb was known as a blood purifier which, in current terms, is a detox herb. It has a high chlorophyll content and has been used as a source for extraction of this natural colour. (Bown, 2003; British Herbal Medicine Association, 1983; Gruenwald *et al.*, 2002; Hutchens, 1973; Shealy, 1998; Tierra, 1998; Williamson, 2003).

Passionflower

Botanical name: *Passiflora incarnata*.

Family: Passifloraceae.

Synonyms: Maypop.

Region of origin/habitat: Wild in the south-eastern United States to Argentina and Brazil. Cultivated in Europe.

Description: A climber, reaching up to 9 m in length. Ovate or cordate leaves with white, cross-shaped flowers.

Parts of herb used: The aerial tops of the stems, comprising leaves flowers and fruit.

Main actives: Flavonoids.

Medicinal uses/benefits: The flavonoids have led to this herb's longstanding use as an effective, non-addictive sedative that does not cause drowsiness. Passionflower is an ingredient in many herbal sedative remedies.

Folklore: Spanish missionaries in South America regarded the flower of this herb as a symbol of Christ's passion, the three stigmas representing the nails, the five anthers the wounds and the ten sepals the apostles present. The herb was used in native North American medicine, especially by the Houma tribe, who put it into drinking water as a tonic. It became popular as a treatment for insomnia in the 19th Century and was included in the US National Formulary from 1916 to 1936. (Bown, 2003; British Herbal Medicine Association, 1983; Gruenwald *et al.*, 2002; Hutchens, 1973; Tierra, 1998; Williamson, 2003).

Rooibos

Botanical name: *Aspalathus linearis*.

Region of origin/habitat: Western Cape region of South Africa.

Parts of herb used: Leaves and stems of new growth shoots.

Main actives: A range of polyphenols.

Medicinal use/benefits: Japanese studies in the 1980s showed that the polyphenol components had antioxidant properties similar to superoxide dismutase

(SOD), an enzyme that is a free radical scavenger and is thought to slow down the ageing process. The herb has less tannin than oriental tea, so it tastes less bitter, and it is caffeine-free.

Folklore: Locally, the herb is used in cases of allergy, such as eczema, hay fever and asthma. It is also used in Schnapps and liqueurs. (Bown, 2003).

Rosehip

Botanical name: *Rosa canina*.
Region of origin/habitat: Europe.
Parts of herb used: Fruit.
Main actives: Vitamin C (0.2–2.4%), fruit acids (3%), pectins (15%), sugars (12–15%), carotenoids and flavonoids.
Medicinal use/benefits: A source of natural vitamin C that has been used in cold and influenza preparations.

Folklore: Rosehips are also used to make syrups for babies and young children. Traditionally, rosehips were made into preserves to retain their health benefits into the winter months. (Bown, 2003; British Herbal Medicine Association, 1983).

Sarsaparilla

Botanical name: *Smilax regelii, S. aristolochiaefolia, S. febrifuga*.
Family: Smilacaceae.
Region of origin/habitat: Tropical and subtropical Central America and the West Indies.
Description: Narrow roots, very long and cylindrical. The external surface varies from greyish- to yellowish- or reddish-brown.
Parts of herb used: The dried rhizomes and roots.

Main actives: Steroidal saponins (0.5–3%).

Medicinal use/benefits: Sarsaparilla has long been used for skin complaints such as psoriasis. It is also believed to be a good diuretic and diaphoretic, so it has been used as a blood purifier and for kidney complaints.

Folklore: Although there are steroidal compounds present in sarsaparilla, the rumoured presence of testosterone, which made it of interest to body-builders, has not been substantiated. The root has been used in soft drinks and root beers. Sarsaparilla was introduced into Europe following the Spanish colonisation of South America. It was regarded as a cure-all and was established in pharmacopoeias until the early 20th Century. (Bown, 2003; British Herbal Medicine Association, 1983; Gruenwald *et al.*, 2002; Tierra, 1998; Williamson, 2003).

Schisandra

Botanical name: *Schisandra chinensis*.

Family: Schisandracae.

Region of origin/habitat: China.

Description: Berries are small, bright red and occur in clusters. When dried, they are wrinkled, dark reddish-brown, with a sticky pulp and a yellow or brown kidney-shaped seed.

Parts of herb used: Fruit.

Main actives: Essential oil, fruit acids, sugars and resin; the seeds contain schizandrins, sitosterol, vitamins C and E, resin, tannins and sugars.

Medicinal uses/benefits: A sweet and sour herb that is claimed by Chinese herbalists to control the secretion of body fluids, thus moistening dry and irritated tissues, and to act as a tonic for the nervous system and the circulatory system.

Folklore: The herb was mentioned in Chinese texts during the Han dynasty (ad 25–220). Its Chinese name is Wu Wei Zi, which means five-flavours fruit, because it has both sweet and sour flavours in the fruit skin and flesh, and acrid, plus bitter and salty flavours in the seeds. It was used by both men and women as a sexual tonic, and by women to improve the complexion. (Bown, 2003; Shealy, 1998; Williamson, 2003; Yeung, 1983).

Siberian ginseng

Botanical name: *Eleutherococcus senticosus*.

Region of origin/habitat: Siberia and northern parts of China, Korea and Japan.

Parts of herb used: The dried root and root bark.

Main actives: The main constituents are triterpene saponins, steroid glycosides, hydroxycoumarins, caffeic acid derivatives, lignans, steroids and polysaccharides.

Medicinal use/benefits: The actions of Siberian ginseng are similar to those of *Panax ginseng*, but stronger. Indigenously, it is used as a tonic for strength and revitalisation. The polysaccharides in the herb have been shown to have a good immunostimulatory action.

Folklore: Siberian ginseng was brought to prominence when Russian researchers were investigating *Panax ginseng*, and looked at other plants in the same family to see whether they had similar properties. Several *Eleutherococcus* species have been used in Chinese medicine for 2000 years. (Bown, 2003; Gruenwald *et al.*, 2002; Shealy, 1998; Tierra, 1998; Yeung, 1983).

Superfruits

Acai berry – Small, dark purple berries that are seen to increase mental alertness, increase energy, restore memory and help to prevent cancer. Grown in the Amazon rainforest of Brazil.

Chia – *Salvia hispanica* is a member of the mint family from Mexico and South America. The seeds are the parts used, and they contain more omega-3 fatty acids than salmon, along with a wealth of antioxidants, minerals, proteins and fibre. Chia is claimed to reduce inflammation, improve heart health and stabilise blood sugar levels.

Goji berry – see Wolfberry.

Guava – Native to Mexico, the Caribbean, Central America and Northern South America. It is full of antioxidants, including cancer-fighting lycopene, and is high in vitamins and calcium.

Mangosteen – The 'queen of fruits' is native to South east Asia. It contains xanthones, which are said to have potent antioxidative properties, including anti-bacterial, anti-inflammatory and anti-histamine actives. Mangosteen is used as a natural skin remedy, and for dysentery and diarrhoea.

Noni – From the South Pacific region, this is known for its ability to strengthen the immune system, reduce fevers and ease digestion problems. Noni is being studied for its possible cancer-fighting and anti-inflammatory properties.

Papaya – Native to the tropics of the Americas, it offers rich sources of antioxidants.

Passion fruit – Native to Brazil, it is full of Vitamins A, C and potassium.

Pomegranate – A Mediterranean fruit full of antioxidants, vitamins, minerals and phytochemicals.

Star fruit – Native to Southeast Asia, it is now found in warm climates such as Florida and Hawaii. An excellent source of vitamin C. (Jones, 2011; Kite, 2011; Everitt, 2012).

Valerian

Botanical name: *Valeriana officinalis*.

Family: Valerianaceae.

[roots: Category 5].

Region of origin/habitat: Europe and temperate regions of Asia.

Parts of the herbs used: Fresh or carefully dried rhizomes and roots.

Main actives: Valepotriates (0.5–2.0%), volatile oil (0.2–1.0%) and valeric acid (0.1–0.9%).

Medicinal use/benefits: Valerian has been used as a daytime sedative to reduce anxiety and stress, and it has been demonstrated to reduce the time it takes to fall asleep. Valerian root extracts and volatile oils are used as components in the flavour industry, especially in alcoholic beverages such as beers and liqueurs, and in soft drinks such as root beers. They have also been used in tobacco flavours.

Folklore: The aroma of valerian is very attractive to cats and rodents, and it has been used as bait in traps. It is thought that valerian was the basis for the story of the Pied Piper of Hamelin ridding the city of rats. (Bown, 2003; British Herbal Medicine Association, 1983; Gruenwald *et al.*, 2002; Hutchens, 1973; Shealy, 1998; Tierra, 1998; Williamson, 2003).

Vervain

Botanical name: *Verbena officinalis*.

Family: Verbena.

Synonyms: Verbena

Region of origin/habitat: Mediterranean region.

Description: Lobed leaves, rough and toothed. Lilac flowers with five petals in long slender leafless spikes.

Parts of herbs used: Aerial parts of the herb.

Main actives: Iridoids, flavonoids and caffeic acid derivatives.

Medicinal use/benefits: The herb has been used for its astringent, cough suppressant and lactation-promoting properties. Traditionally, it has been used as a diuretic, and to calm the nerves and improve the liver and gall bladder functions.

Folklore: In Western medicine, vervain has been used mainly for nervous complaints. Vervain is one of the herbs most commonly used to make herbal teas, and is also an ingredient in liqueurs. (Bown, 2003; British Herbal Medicine Association, 1983; Gruenwald *et al.*, 2002; Shealy, 1998; Tierra, 1998; Williamson, 2003).

Wolfberry

Botanical name: *Lycium barbarum*.

Synonyms: Goji berry.

Region of Origin/Habitat: China.

Parts of herb used: Fruits, which are bright red when first dried but darken with age.

Main actives: Carotenes, Vitamins B1, B2, B3 and C, β-sitosterol and linoleic acid.

Medicinal use/benefits: In traditional Chinese medicine, wolfberry is used as a nourishing herb for convalescence, and it is also used in cases of impotence. It has been reported to lower blood pressure and blood cholesterol levels. It is also reported to be a liver and kidney tonic, inhibiting the deposition of fat in liver cells and promoting the regeneration of liver cells.

Folklore: Although of Chinese origin (referred to as Gou Qi Zi as early as 200 BC) the plant has been established in Europe for some centuries and, in Britain, was known as the Duke of Argyll's tea-tree. It was also known as matrimony vine because, if planted near the home, it was said to create discord between husband and wife. In traditional Chinese medicine, it is combined with Schisandra and, if taken for one hundred days, is believed to develop sexual stamina. (Bown, 2003; Shealy, 1998; Tierra, 1998).

Wormwood

Botanical name: *Artemisia absinthium*.

Family: Asteraceae.

[Herb: category 4 (with limits on eucalyptol, methyleugenol and thujone); essential oil: category 4 (with limits on eucalyptol, methyleugenol and thujone)].

Region of origin/habitat: Mediterranean zones. Parts of herb used: The upper shoots and leaves.

Main actives: Volatile oil (0.2–1.5%) containing thujone and bitter sesquiterpene compounds, including absinthin (0.20–0.28%) and artabsin (0.04–0.16%).

Medicinal use/benefits: The herb is described as an aromatic bitter and, as such, is used to stimulate the appetite. It has been shown that the bitterness on the palate automatically stimulates an increase in secretion of digestive juices in the stomach. Also, it has been shown to increase liver function in patients with liver disorders.

Folklore: Wormwood has traditionally been used for digestive problems, including expelling intestinal worms. The essential oil of wormwood was used in the distillation and production of the aperitif absinthe, starting around the end of the 18th Century. The thujone in absinthe created some problems in Europe and the United States, as it emerged that it was addictive and could cause hallucinations in cases of overindulgence. This led to the drink being outlawed in certain countries. (Bown, 2003; British Herbal Medicine Association, 1983; Gruenwald *et al.*, 2002; Hutchens, 1973; Tierra, 1998; Williamson, 2003).

References

Ashurst, P.R. (ed, 1998). *Food Flavourings*, 3rd edition. Aspen Publishers Inc, Gaithersburg MD.

Bernardini, C.M. (2012). *Process Technology for OLEO Chemical Industry*. SPEC Engineers & Consultants Pvt. Ltd. http://www.specengineers.com/cm_bernardini_oleo_chemical_process.htm

Bowyer, N. (2011). *Why Soil Association and Ecocert do not permit most commercially available Meadowfoam Seed Oils*. Alfa-chemicals, Alfa News, http://www.alfa-chemicals.co.uk

Bown, D. (2003). *RHS Encyclopedia of Herbs and Their Uses*. Dorling Kindersley, London.

British Herbal Medicine Association (1983). *The British Herbal Pharmacopoeia* 1983. BHMA Publications, Dorset, United Kingdom.

Collins English Dictionary (2011). ISBN-10: 0007437862; ISBN-13: 978-0007437863.

Everitt L. (2012). *The Chia Craze*. BBC News Magazine.

Gruenwald, J. (2009). *Novel Botanical Ingredients for Beverages*. Analyze and Realize AG, Clinics in Dermatology, Elsevier Inc, pg 210–216.

Gruenwald, J., Brendler, T., Jaenicke, C. and Smith, E. (2002). *Plant Based Ingredients for Functional Foods*. Leatherhead Publishing, Surrey, UK.

Handa, S.S., Khanuja, S.P.S., Longo, G. and Rakesh, D.D. (2008). *Extraction Technologies for Medicinal and Aromatic Plants*. United Nations Industrial Development Organization and the International Centre for Science and High Technology.

Helmenstine, A.M. (2012). *What is Coffee and How Does it Work?* Caffeine Chemistry, About.com Guide, http://chemistry.about.com/od/moleculescompounds/a/caffeine.htm.

Hoegler, M. (2008). *Cocaine in the Brain*, Biology 202, Serendip Update, http://serendip.brynmawr.edu/exchange/node/1846.

Hutchens A.R. (1973). *Indian Herbalogy of North America*, Shambala, Boston, MA.

Joiner T.R. (2001). *Chinese Herbal Medicine Made Easy: Natural Effective Remedies for Common Illnesses*, 1st edition. Hunter House Inc. Publishers, ISBN 0-89793-276-5 – ISBN 0-89793-275-7.

Jones H.K. (2011). *Super Fruits (for Super Health)*. Real Jock, http://www.realjock.com/article/892.

Joy, P.P, Thomas, J., Mathew, S. and Skaria, B.P. (1998). *Medicinal Plants, Aromatic and Medicinal Plant Research Station*. Kerala Agriculture University.

Kite. K. (2011). *Types of Superfruits, Health Solutions for you*. Kelly Kite Health and Medicine, Health Solutions for You, http://www.kellykite.com/1245/types-of-superfruits.html.

Lu, D. (2001). *Wilde Ling Zhi – Immune System Protector*. American Healing Technologies (AHT) Inc.

Ming, O.S. (2007). *Comparative Study on Optimization of Continuous Countercurrent Extraction for Licorice Roots*, chapter 1–4, pp. 3–35. Department of Pharmacy, National University of Singapore.

Moyler, D.A. (1988). In: Ashurst P.R. (ed). *Food Flavourings*, 3rd edition. Aspen Publishers Inc, Gaithersburg, MD.

Mukhopadhyay, M. (2000). *Natural Extracts using Supercritical Carbon Dioxide*, pg 2–5. CRC Press, LLC, ISBN 0-8493-0819-4.

Ning Xia Red History (2009). *Why are Ningxia Wolfberries Considered a Chinese National Treasure*. Health Benefits, http://ningxiaredjuice.com/health-benefits/ning-xia-red-history.

Pendergast, M. (1994). *For God, Country, and Coca-Cola: The Unauthorized History of the Great American Soft Drink and the Company That Makes It*. Phoenix, London, UK.

Pokladrik, R.J. (2008). *Roots and Remedies of Ginseng poaching in Central Appalachia*, pp. 53–71. Antioch New England Graduate School, Department of Environmental Studies, UMI 3325848.

Rayburn, D. (2007). *Let's Get Natural with Herbs*. Ozark Mountain Publishing.

Saxena, D.K., Sharma, S.K. and Sambi, S.S. (2011). Comparative Extraction of Cottonseed Oil by n-Hexane and Ethanol. *ARPN Journal of Engineering and Applied Sciences* **6**(1). University School of Chemical Technology Asian Research Publishing Network, ISSN 1819-6608.

Schwartz, J.H. (1993). New York: Henry Holt What Bones Tell Us. *American Journal of Physical Anthropology* **94**(3), 11–13.

Searby, L. (2012). Milking the Benefits, IFI International Food Ingredients. *The Official Magazine of Hi & Fi Europe*, pp 19–24.

Shealy, C.N. (1998). *The Illustrated Encylopedia of Healing Remedies*. Element Books, Boston, MA.

Steyn, N. (2011*). Development and Characterization of a Functional Beverage from Red-fleshed Japanese plums* (Prunus salicina L.). Master of Science in Food Science, Stellenbosch University, pp. 5–7.

Tierra, M. (1998). *The Way if Herbs*. Pocket Books, New York, NY.

The Glycerine Producers' Association (2007). *Uses of Glycerine*. ACI Science, http://www.aciscience.org/docs/uses_of_Glycerine.pdf

Victor, D. (2010). *Use Chinese Herbal Medicine to Create Harmony and Health*. Natural News, http://www.naturalnews.com/027936_Chinese_herbs_health.html#ixzzlsHvZS400

Weil, A. and Rosen, W. (2004). *From Chocolate to Morphine: Understanding Mind-Active Drugs, Kinds of Stimulants, Coffee and other caffeine containing plants*, pp. 42–49, Library of Congress Cataloguing-in-Publication Data, ISBN 0-618-48379-9.

West, T. (2012). Buzz Lines, Energise your Soft Drinks Sales, Product File Energy Drinks. *Wholesale News Magazine*, 33–38.

Whitaker, A. (2010). *Ancient Wisdom, Shanidar Cave*. http://www.ancient-wisdom.co.uk/iraqshanidar.htm

Wilde, M. (2009). *80,000 evidence of herbal medicine at Shanidar Iraq, Wilde in the Woods*. http://monicawilde.wordpress.com/2009/12/26/8000_year_burial_flowers_iraq/

Williamson, E.M. (2003). *Potter's Herbal Cyclopaedia of Botanical Drug & Preparations, The Authoritative Reference work on Plants with a Known Medicinal Use*. Saffron Walden C. W. Daniel Company Ltd.

Wren, R.C., Williamson, E.M. and Evans, F.J. (1988). *Forms of Medicinal Preparations, Potter's New Cyclopedia of Botanical Dugs and Preparations*, pp. 296-298. Saffron Walden, the C.W. Daniel Company Ltd, Potter's (Herbal Supplies) Ltd, ISBN 0852071973.

Yeung, H. (1983). *Handbook of Chinese Herbs*, 2nd edition. Institute of Chinese Medicine, Rosemead, CA.

The 2003 Zenith Report on International Functional Soft Drinks, Zenith International Publishing Ltd, Bath, United Kingdom.

CHAPTER 13

Miscellaneous topics

Philip R. Ashurst[1] and Quentin Palmer[2]

[1] *Dr. P R Ashurst and Associates, Ludlow, UK*
[2] *Schweppes Europe, Watford, UK*

13.1 Introduction

The previous chapters in this book cover a wide range of most of the key elements that relate to the markets, background science and technology and manufacture of both soft drinks and fruit juices. There are, nevertheless, many areas that, while not justifying a chapter in their own right, are likely to be of interest to readers of this work. This chapter aims to include such topics.

Some nutritional aspects of products are included. There are concerns about the consumption of sugar and its impact on dental caries, as well as obesity and illnesses such as diabetes. Manufacturers in many countries are under pressure to reduce carbohydrate content, particularly sugar in beverages. One area of growing importance is that of drinks that are specifically designed for use in sports nutrition.

Bulk-dispensed soft drinks and fruit juices are an important sector of the hospitality industry, and a section on that topic is included. Consumer issues also feature, and the classification and handling of their complaints and other related matters are dealt with.

13.2 Nutrition

Soft drinks and fruit juices are usually consumed to provide hydration and thus quench thirst, or as a sociable activity, rather than for their nutritional content. However, they always provide dietary water and, depending on their formulation, can also contribute significant quantities of carbohydrate, vitamins, minerals and, sometimes, protein. To improve consumer information, most countries now require the declaration of nutritional information on

Chemistry and Technology of Soft Drinks and Fruit Juices, Third Edition. Edited by Philip R. Ashurst.
© 2016 John Wiley & Sons, Ltd. Published 2016 by John Wiley & Sons, Ltd.

labels. In EU member states, where the information is now obligatory, such it may be provided by:
(a) from analysis of the beverage;
(b) by calculation of the actual average values of the ingredients used; or
(c) from generally established and accepted data.

13.2.1 Nutritional components
13.2.1.1 Water
Sedentary adults in a temperate climate need to consume about two litres of water per day, a figure that can substantially increase when high temperature and/or physical activity causes significant sweating. Soft drinks and fruit juices provide an important water source, as shown in Table 13.1.

13.2.1.2 Carbohydrates
With the exception of soda waters and most low-calorie drinks, all soft drinks rely at least partially on sugars to provide sweetness and body.

Although different sugars have different sweetness factors, relative to sucrose, they are all identical in energy content, on a dry basis. Thus, to achieve a high energy content without cloying sweetness, sugars or carbohydrates of low sweetness must be used. This is illustrated in Table 13.2.

Many soft drinks contain less carbohydrate than fruit juices, although this may not be the perception of the consumer (Table 13.3).

Table 13.1 Water content of soft drinks and juices.

Drink	Typical water content (% v/v)
Soda water	99.8
Lemonade	94–97
Cola	93.5
Orange juice	88

Table 13.2 Sugars, sweetness and energy.

Carbohydrate	Approximate sweetness, relative to sucrose	kcal required to provide the same sweetness as 1 g sucrose
Fructose	1.3 ×	3
Sucrose	1.0 ×	4
Dextrose	0.8 ×	5
Glucose syrup (50 DE)	0.6 ×	6.7
Maltodextrin	0.05 ×	80

Table 13.3 Carbohydrate content of soft drinks and fruit juices.

Beverage	Typical carbohydrate content (% w/v)
Low-calorie lemonade	0
Lemonade	5–10
Grapefruit juice	8.5
Orange juice	9.1
Cola	10.6
Pineapple juice	10.8

Table 13.4 Fibre and protein contents of fruits and fruit juices.

	Fibre (% w/w)		Protein (% w/w)	
	Fruit	Juice	Fruit	Juice
Apple	1.8	trace	0.4	0.1
Grape	0.7	trace	0.4	0.1
Grapefruit	1.3	trace	0.8	0.4
Orange	1.7	0.1	1.1	0.5
Pineapple	1.2	trace	0.4	0.3

Source: Food Labelling Data for Manufacturers (1992), published by the Royal Society of Chemistry and the Ministry of Agriculture, Fisheries and Food.

13.2.1.3 Protein

Unless intentionally added, soft drinks contain negligible amounts of protein, and fruit juices typically contain between 0.2–0.6% (Table 13.4). However, the protein content is normally considered to be of no significance.

Protein-fortified soft drinks have been developed and marketed, but there are few, if any, mainstream brands in the world that are currently successful. Generally, the protein source used is dairy based, often whey or skimmed milk. Great care must be taken to prevent precipitation, as the protein is acidified through its isoelectric point to a microbiologically safe pH, and it is necessary to use stabilisers for long-term cloud stability.

The high cost of ingredients and processing often mitigates against commercial success, and against such drinks providing a practical nutritional supplement in Third World and famine situations.

13.2.1.4 Fat

The fat content of soft drinks is negligible, and any present will usually derive from essential oils arising during processing, or added as constituents of flavourings, citrus comminutes or clouding emulsions.

Processed fruit juices are usually manufactured to a maximum oil content of 0.03% v/v at natural strength.

Drinks fortified with fat are usually dairy-based, which puts them outside the scope of this book.

13.2.1.5 Fibre

This term is used to refer to food constituents not metabolised by humans, and analysis has shown the majority of such constituents to be polysaccharides. In the UK, fibre is defined as 'non-starch polysaccharides' (NSP) and, for consistency, the Englyst analytical method has been adopted (Englyst and Cummings, 1988).

The only sources of NSP in soft drinks are fruit materials, gums, and stabilisers such as sodium carboxymethylcellulose (CMC) and pectins. Of these items, only fruit juices are used in significant quantities. The NSP content of most fruits fall within the range 0.9–3.6% w/w but, as Table 13.4 shows, very little remains in processed juices. Citrus comminutes are conventionally considered to have the same NSP content as the corresponding fruit but, if they are known to contain substantial amounts of peel extracts, this should be accounted for in calculations.

The recommended daily intake of NSP is 18 g/day, so conventional soft drinks provide an insignificant contribution. High levels of fortification can be achieved, but high-fibre drinks launched in the UK do not have a good record of successful sales, despite good taste, texture and skilful marketing.

13.2.1.6 Vitamins

The only vitamins likely to be found in unfortified soft drinks are Vitamin C (either added as an antioxidant or deriving from fruit materials) and Vitamin A precursor (beta-carotene, added as a colour). However, soft drinks provide a good medium for vitamin fortification, the limitations being solubility (for fat-soluble vitamins), flavour impairment (for example, the 'meaty' notes of thiamine) and stability.

It is very important to minimise the air content of vitamin-containing drinks to reduce oxidative damage, and to ensure that sufficient 'overage' has been added to enable any label claims to be substantiated throughout the life of the product.

13.2.1.7 Minerals

Soft drinks will contain various minerals from the water employed. Sodium (from water and added sodium salts, such as benzoate, saccharin, citrate, CMC), may be of particular concern to specific dietary groups. Calcium and magnesium are also often present (from water and fruit materials), as are potassium and phosphate (from fruit materials). The presence of excess calcium can give rise to gelling problems if pectin in fruit materials has been degraded.

Adding minerals is easily achieved, but excesses can give rise to salty or astringent notes, metallic taints and laxative effects.

13.2.1.8 Energy
Palatable high-energy drinks are readily formulated by selecting the appropriate carbohydrate blend (Table 13.2).

13.2.1.9 Low-calorie drinks
Paradoxically, the nutritional component which is now often of greatest commercial significance is the absence of components with calorific value. This is achieved by substituting carbohydrate sweeteners with non-nutritive sweeteners, such as saccharin and acesulfame-K, or the amino acid sweetener aspartame.

13.2.2 Calculation and declaration of nutrition information
Within the European Union, the provision of nutrition information is now obligatory. Until recently, this was the situation only if nutrition claims were being made or implied for the product. The information provided must be derived in the prescribed manner (see EU regulation 1169/2011).

The descriptor 'low calorie' is specifically exempted from necessitating nutrition information, but reference to the following terms would act as a trigger:
- diet;
- slim (used alone or as part of a brand name);
- high energy;
- no added sugar;
- protein-rich.

The minimum requirement is to declare, per 100 ml (or 100 g), the amount of the following in the product:
- energy (both as kJ and kcal);
- protein (g);
- carbohydrate (g);
- fat (g).

If claims relating to specific nutrients are made (e.g. 'no added sugar'), then the following additional data must be given:
- carbohydrate – to be subdivided into sugars, polyols and starch, as appropriate;
- fat – to be subdivided according to saturation, if appropriate;
- fibre (g);
- sodium (g);
- vitamins;
- minerals.

If vitamins and/or minerals are claimed, they must be declared, both quantitatively and as a percentage of the recommended daily allowance specified in the regulations.

Table 13.5 Energy factors per gram.

	kJ	kcal
Carbohydrate	17	4
Polyols	10	2.4
Protein	17	4
Fat	37	9
Ethanol	29	7
Organic acids	13	3

Values must be quoted in the units specified and are to be average values. They may be based on an analysis of the drink or may be calculated from generally established and accepted data, such as McCance and Widdowson's The Composition of Foods, 7th Edition (McCance and Widdowson, 2015).

For energy calculations, the factors shown in Table 13.5 must be used.

Example

Calculate the full nutrition information for an orange drink with the following composition:

Ingredient	Quantity per 1000 litres	Nutrition contribution (l)
Sugar	60 kg	carbohydrate
Sodium benzoate	0.16 kg	sodium
Orange juice	50 l	carbohydrate, protein, fibre
Acesulfame-K	0.1 kg	—
Aspartame	0.1 kg	protein
Citric acid anhydrous	2.2 kg	organic acid
Sodium citrate dihydrate	0.8 kg	organic acid, sodium
Flavouring (50% v/v ethanol)	1.0 l	ethanol

Total acidity: 0.27% w/v, as citric acid anhydrous

Orange juice composition:

	% w/w
Carbohydrate	8.7
Protein	0.5
Fat	0.02
Fibre	0.15
Apparent density	1.042 g/ml (so 50 l = 52.1 kg)

Calculations

		Per 1000 l	g/100 ml
(i)	Carbohydrate content		
	added sugar	60.0 kg	
	ex juice, 52.1 × 8.7/100	3.53 kg	
		63.53 kg	6.35
(ii)	Protein content		
	aspartame (100% protein)	0.1 kg	
	ex juice, 52.1 × 0.5/100	0.26 kg	
		0.36 kg	0.036
(iii)	Fat content		
	ex juice, 52.1 × 0.02/100	0.01 kg	
	ex flavour	Negligible	
		0.01 kg	Trace
(iv)	Fibre content		
	ex juice, 52.1 × 0.1/100	0.05 kg	0.005
(v)	Organic acids		
	total acid content, 1000 × 0.27/100		2.7 kg
	sodium citrate, 0.8 × 192.13/294.1		0.52 kg
		3.22 kg	0.322
(vi)	Ethanol		
	1 l flavour @ 50% v/v = 0.5 l		
	ethanol, apparent density = 0.789 g/ml		
	mass of ethanol	0.39 kg	0.039
(vii)	Sodium		
	sodium benzoate, 0.16 × 23/144		0.026 kg
	sodium citrate, 0.8 × (23 × 3)/294.1		0.188 kg
		0.214 kg	0.021
(viii)	Energy content per 100 ml (Table 13.5)		
		kJ	kcal
	due to 6.35 g carbohydrate	108.0	25.40
	due to 0.036 g protein	0.6	0.14
	due to 0.322 g organic acids	4.2	1.00
	due to 0.039 g ethanol	1.1	0.27
	total	113.9	26.81

The full declaration per 100 ml will thus be as follows, rounded off appropriately as:

Energy	114 kJ (26.8 kcal)
Protein	0.04 g
Carbohydrate	6.4 g
(of which sugars	6.4 g)
Fat	0.0 g
Fibre	Trace
Sodium	0.02 g

13.3 Sports drinks

In this section, sports drinks are defined, the bodily need for them is discussed and aspects of their formulation are considered.

13.3.1 Definition and purpose

Sports drinks serve to provide water, energy and electrolytes in a palatable and readily assimilable form, suitable for consumption before, during and after sporting and other strenuous physical activities. Some brands also contain vitamins and minerals. Although these products are designed and marketed for those engaging in strenuous activity, sports drinks are often consumed by those merely aspiring to such activities, or who just enjoy watching them.

13.3.2 Physiological needs

The performance of physical work involves expenditure of energy and leads to loss of water and electrolytes in the form of sweat. The progressive depletion of energy reserves and the loss of water both accelerate the inevitable onset of fatigue.

13.3.2.1 Energy

To provide energy, the body predominantly metabolises fat and carbohydrate. As the rate of work increases, carbohydrate metabolism plays an increasingly important role and, by the time the body is operating at 85% of its maximum volumetric oxygen uptake (85% $VO_{2\,max}$), virtually all energy used derives from carbohydrates – either muscle or liver glycogen, or directly from blood sugars. Thus, for sustained sporting activities, energy must be provided in the form of carbohydrates. Many papers have been published on the relative merits of different carbohydrates but, apart from considerations of sweetness and molecular weight (hence osmotic contribution), there appears to be little to choose between any of the commercially available sugars. The exception is fructose which, being metabolised in the liver, is not rapidly available for utilisation.

13.3.2.2 Water

In a sedentary state, body temperature can be regulated by varying the amount of clothing worn, or by adjusting the surrounding temperature. However, these mechanisms are inadequate when strenuous work is undertaken, and temperature regulation must then be achieved by the evaporation of sweat from the surface of the skin.

Water loss by sweating is influenced by:
1. the rate of work – an adult generates 70 W when sedentary, but 1.1 KW when running a marathon in 2 hours, 30 minutes;
2. the duration of the work;
3. the external temperature and humidity.

Sweat loss of up to 2 litres per hour is common, rising to 3 l/h in hot and humid conditions. Loss of significant amounts of water reduces plasma volume. This impairs the delivery of blood to muscles and skin, eventually leading to loss of temperature control and heat exhaustion. When dehydration reaches 2% of body weight, performance is impaired; at 5%, work capacity falls by about 30% (Saltin and Costill, 1988). Water is also lost in respiration, but this is of less significance than sweating during strenuous activity.

13.3.2.3 Electrolytes

Sweat is not just water, but a dilute solution of electrolytes, mainly sodium and chloride ions. The actual composition varies from individual to individual and within individuals, according to circumstances. Typical values are shown in Table 13.6, from which it can be seen that sweat is considerably less concentrated than blood plasma. Hence, sweating causes an increase in plasma electrolyte concentrations. Only in events of long duration (more than three hours) is it considered essential to replace lost sodium during the event, to guard against hyponatraemia (low plasma sodium concentration) (Gisolfi and Duchman, 1992). However, sports drinks conventionally contain added sodium, chloride and other electrolytes, at levels similar to those found in sweat.

The key function of sodium is its role in assisting the absorption of glucose and water from the small intestine, both by complex formation (Schultz and

Table 13.6 Electrolytes in sweat, plasma and muscle (mmol/l).

	Sweat	Plasma	Muscle
Sodium	40–60	140	9
Chloride	30–50	101	9
Potassium	4–5	4	162
Magnesium	1.5–5	1.5	31

(Costill and Miller, 1980).

Curran, 1970) and by a phenomenon known as 'solution drag' (Fordtran, 1975). Sodium also aids post-activity recovery by:

1 raising plasma osmolality – this reduces urine output, so enabling the body to retain more ingested water;
2 reducing the thirst-quenching capability of water, thus enabling more to be ingested.

13.3.2.4 Vitamins

Thiamine is proportionately linked with carbohydrate metabolism, and the 'antioxidant' vitamins A, C and E can aid the elimination of free radicals formed within muscle at an increased rate during strenuous activity (Davies *et al.*, 1982). Vitamins can thus be of value during post-activity recovery.

13.3.3 The absorption of drinks

Once swallowed, a drink passes via the oesophagus into the stomach. From there, it is released to the small intestine, and only then can its components be absorbed and utilised.

13.3.3.1 Gastric emptying

This is the transfer of the stomach contents to the intestine, which is a controlled process influenced by several factors:

1 The volume of fluid in the stomach – the greater its contents, the faster it will be emptied. Carbon dioxide liberated from carbonated drinks gives apparent additional fluid and, hence, faster emptying.
2 The concentration of carbohydrate in the fluid – any carbohydrate content will slow emptying, but the effect does not become noticeable until a concentration of about 40 g/l is reached (Vist and Maughan, 1995). It should be noted that the slower release of a more concentrated solution can still deliver carbohydrate to the intestine at a faster rate than is achieved by the more rapid emptying of a weaker solution.
3 The osmolality of the fluid – although conflicting evidence has been published, it appears on balance that increasing osmolality delays emptying, though the effect may only be slight at the concentration of typical sports drinks.
4 Strenuous activity – emptying is slowed when exercise levels exceed a rate of about 70% $VO_{2\,max}$ (Fordtran and Saltin, 1967).

Optimised gastric emptying would thus seem to be achieved by the regular drinking of large volumes of highly carbonated water or dilute carbohydrate solution – not a comfortable regime during strenuous activity.

13.3.3.2 Absorption from the intestine

Plain water crosses the walls of the intestine by osmotic action alone. However, if sodium and sugars are also present, then the active transport mechanisms described earlier operate, and water absorption can be enhanced. The effect is dependant on concentrations, and the maximum rate of water uptake occurs when the concentrations give a slightly hypotonic solution (200–250 mOsm/kg, cf. 287 mOsm/kg for isotonicity) (Wapnir and Lifshitz, 1985). Conversely, when the lumen contents are significantly hypertonic, water is secreted from plasma into the intestine by osmotic action; this is a dehydrating effect.

The type of sugar seems to have relatively little effect on absorption rate, with the exception of fructose. This is not actively absorbed, nor does it benefit from the presence of sodium.

13.3.3.3 Consumption pattern
Many regimes for ingesting water, electrolytes and carbohydrates in association with strenuous activity have been published (see, for example, Gisolfi and Duchman, 1992; Olsson and Saltin, 1971). In general, carbohydrate and electrolyte supplementation are not considered necessary during the activity, until the duration exceeds about an hour but, in longer events and to aid post-activity recovery, such supplementation plays a key role.

13.3.4 Formulation
13.3.4.1 Preliminary stage
Before detailed formulation commences it is necessary to define the target end product broadly, particularly with respect to:
1 carbohydrate system;
2 electrolyte contents;
3 osmolality.

Carbohydrate system
First, a quantitative decision must be made, balancing energy content against the inhibition of gastric emptying at higher concentrations. Typically, sports drinks contain, in total, 60–80 g of carbohydrate per litre.

Next, it must be decided which carbohydrate or carbohydrates will be used. Key factors are molecular weight (and, hence, contribution to osmolality) and electrolyte content.

Commercially available carbohydrates, all contributing 4 kcal/g dry basis, include:
- *Glucose*: low molecular weight (MW) of 180, hence a major impact on osmolality. Low electrolyte content.
- *Fructose*: MW 180, as for glucose, but not rapidly absorbed from intestine, nor quickly available for muscle utilisation.
- *Sucrose*: MW 342, so lower osmolality contribution per gram than the monosaccharides. Negligible electrolytes. Inverts to glucose and fructose in acidic solution, effectively halving the MW and giving a rise in osmolality.
- *Glucose syrup*: a complex mixture of sugars produced by the hydrolysis of starch. Available in various grades of dextrose equivalent (DE), typically

varying from 42 to 95. As DE increases, the average MW reduces and sweetness increases (average MW at 42 DE is 412, and at 95 DE is 186). Usually a moderate electrolyte content.
- *High fructose corn syrup*: made by the partial enzyme conversion of glucose to fructose in a high DE glucose syrup. The grade widely available in Europe contains, on a dry basis, 42% fructose and 52% glucose, with remaining sugars present as higher saccharides. MW similar to 95 DE glucose syrup. Sweeter than glucose syrup, and generally a low electrolyte content. The presence of fructose is not advantageous, but low levels of this syrup give no more fructose than partly inverted sucrose.
- *Maltodextrins*: produced by the partial hydrolysis of starch. They vary from 15–30 DE and have a very high average MW (typically 1100 for 15 DE). High electrolyte content, virtually no sweetness.

By blending from the above, a very wide range of osmolality contributions can be achieved for any given carbohydrate content.

Electrolyte content

Although only sodium plays an active role in the absorption of water and carbohydrate, sports drinks are typically also fortified with potassium, magnesium, calcium and chloride. Levels usually approximate to those found in sweat (see Table 13.6).

The electrolyte contributions of the production water and carbohydrate must first be deducted, together with those from any known additives (such as sodium benzoate or citrate, acesulfame-K or potassium sorbate). Salt additions can then be calculated to achieve the target values. Suitable additives include:
- sodium: chloride, citrate;
- potassium: chloride, citrate, phosphates;
- calcium: chloride, lactate;
- magnesium: chloride, sulphate.

Osmolality

The majority of sports drinks are formulated to be isotonic – that is, to have an osmotic pressure matching that of blood serum. This is considered to optimise absorption from the intestine although, as stated earlier, water uptake is optimal with slightly hypotonic solutions.

Blood serum osmolality varies slightly within the individual, and from person to person, but is generally taken to be 287 mOsm/kg.

The osmotic pressure of a solution is a colligative property – that is, it is proportional to the number of solute particles present in the solution, rather than to their weight. It is usually determined by measuring another colligative property, the depression of freezing point relative to water.

For all non-ionic substances, a molar solution will depress freezing point by the same amount (1.86°C) but, for ionic substances, the depression of a molar solution will be 1.86 multiplied by the number of ionic components generated per molecule.

Sports drinks are complex solutions of non-ionic and ionic substances; the latter will be dissociated to varying extents, according to their nature and the other solutes present. Their osmolality cannot, therefore, be precisely calculated; theoretical estimates must be checked by measurement, and their compositions must be fine-tuned to achieve the target value.

13.3.4.2 Palatability

It is crucial that the consumer finds the product pleasant and comfortable to drink, both during and after heavy physical activity. Sipping small amounts in a laboratory during a day of sedentary work cannot give a valid appraisal, and field testing with athletes is essential. Particular aspects needing attention are carbonation, sweetness, saltiness, acidity and flavour intensity.

13.3.4.3 Prototype development

Having outlined the product composition, it is useful to calculate the approximate osmolality from published tables – for example, in the *CRC Handbook of Chemistry and Physics* (CRC Press Inc.), *The Merck Index* (Merck & Co. Inc.) and *The British Pharmacopoeia* (HMSO). Ingredients for which osmolality values are not available can be allowed for by using values for substances of similar molecular weight and ionic character.

The approximation will show whether major adjustments are necessary prior to bench work – the value at this stage should be about 80% of the target level. A reduction in osmolality can be achieved by substituting low MW sugars with a maltodextrin, or by reducing the total carbohydrate content. Similarly, an increase can be achieved by the opposite moves.

A first prototype can then be put together, with the decided carbohydrate system and with salts added to approximate to the target electrolyte levels. Sweetness must be adjusted, probably by adding non-nutritive sweetener(s), an acidity level must be selected, and flavour, colour and preservative must be added at appropriate levels. If the drink is to be carbonated, a level must be chosen (a low level is preferable).

All the adjustments will contribute to the osmolality, which should be measured to assess what further fine-tuning will be required. It is important to measure the osmolality of the drink, as presented to the small intestine, by which time carbonation will have been virtually eliminated.

13.3.4.4 Final formulation

It is usually necessary to undertake a series of adjustments to achieve the target composition. If, for example, the osmolality has to be increased by replacing maltodextrin with monosaccharides, then additional salts will be required to maintain electrolyte levels. These, in turn, will increase the osmolality and require a further reduction in maltodextrin. Table 13.7 indicates how the components of a typical isotonic sports drink contribute to its osmolality.

Table 13.7 Osmolality of a typical isotonic sports drink.

	% of total osmolality
Carbohydrate system	70
Salts	10
Juice/flavour/acid/sweeteners	15
Carbonation	5
	100

13.3.4.5 Powdered sports drinks

It is straightforward to prepare sports drinks in powder form, for dissolving at point of use. Isotonicity will depend on the accuracy with which the instructions for use are followed, and electrolyte content will vary with the composition of the water used.

13.4 Niche drinks

This section examines various beverage types outside the mainstream of soft drinks markets.

13.4.1 Alcoholic-type drinks

These are drinks with a low alcohol content, designed to mimic stronger versions (Note: the currently popular and highly successful so-called 'alcopops' are, because of their high alcohol content, excluded from consideration as soft drinks). The first drink of this type in the UK was probably ginger beer. This was introduced in the early 19th century, with a strength of about 5–8% ABV (alcohol by volume), and was made by the fermentation of sugar solution containing ginger root. In 1885, it was regulated by the Customs and Inland Revenue Act to an alcohol content of less than 2% proof spirit (1.14% ABV). To meet the limit, fermentation had to be arrested when, hopefully, the alcohol content was still below 2% proof spirit.

Though still a popular soft drink, ginger beer can now be produced from compounds manufactured by the flavour industry, most of which are made by direct extraction rather than fermentation. However, there are a number of very popular brands of ginger beer that are produced by fermentation of ginger root with added carbohydrates. These drinks must be sold with an alcohol content of less than 0.5% ABV.

In the United Kingdom, The Food Labelling Regulations, 1996, as amended (Statutory Instrument 1996, No 1499) make clear that any product with an alcohol content above 1.2% ABV must declare the fact on the label, but do not make clear whether products below that level may still be considered a soft drink.

These same regulations allow the description 'alcohol free' for products with less than 0.05% ABV.

The issue is further complicated by the UK licensing laws. The Licensing Acts of 1964 and 2003 both define an intoxicating beverage as any product with an alcohol content above 0.5% ABV. However, the 1964 Act states that any product with an alcohol content above 0.5% ABV may not be sold without the appropriate licence. It thus appears that a product with an alcohol content above 0.5% ABV, but less than 1.2% ABV, may still be sold as a soft drink, but only on licensed premises. Consideration should also be given to the UK Consumer and Unfair Trading Regulations, 2008 (Statutory Instrument 2008, No 1277) which state that failure to indicate that a product contains alcohol, even if the level is below 0.5% ABV, might be seen as a material omission in providing the consumer with information that might affect their purchase of the product. Companies seeking to develop products that contain low levels of alcohol should seek formal legal advice, particularly since regulations will be likely to differ from country to country.

The usual alcoholic-type soft drink currently on the UK market is lemonade shandy, which simulates the traditional pub blend of 50/50 lemonade and beer (usually bitter). This was introduced as a soft drink in the early 1960s and, in order to be sold in unlicensed premises, had to be below 2% proof spirit. The Food Labelling Regulations, 1984, added a minimum alcohol content of 1.5% proof spirit, but both limits were replaced by a new maximum of 0.5% ABV imposed by the Licensing (Low Alcohol Drinks) Act, 1990.

Lemonade shandy is made with 'shandy ale', a bitter beer brewed to 6.5% ABV to minimise transport costs. For colouring the product, it is important to use an ammonia caramel, as the sulphite ammonia caramels used for conventional soft drinks will react with tannins in the beer and precipitate out.

Even at this low alcohol level, microbiological spoilage is a hazard. Suggested manufacturing processes are sorbic acid/sorbate preservation, with either sterile filtration or flash-pasteurisation of the syrup, or in-pack pasteurisation.

Ginger beer, shandy, lager and lime, cider shandy, rum and cola, and also apple drinks with a cider content of 5–10%, have also been produced in low-alcohol versions, but are now much less popular.

Product developers working on products containing low levels of alcohol should familiarise themselves with local regulations, which may differ from country to country.

13.4.2 Energy drinks

These products aim primarily to provide a boost to mental energy, or 'buzz'. They are marketed as 'pick-me-ups', and command a high profit margin. An impressive list of ingredients is essential for credibility, and typical components include caffeine, taurine, glucuronolactone, inositol, maltodextrin, vitamins and herbal extracts such as guarana, ginseng and schizandra.

Claims made for these products must be carefully worded. If they are specific to the product, they must be capable of substantiation and, if they refer to benefits conferred by the ingredients, then there must be sound scientific evidence available in support. Exactly what evidence is required is not laid down in law, so it is left to the manufacturer to decide if a claim can be justified if challenged.

In UK regulatory terms, 'energy' is always taken to mean nutritional energy. Arguably, therefore, to justify the description 'energy drink', there should be a higher calorie content in the drink than in a standard one. Low-energy drinks are defined, and must contain at least 25% fewer calories than the corresponding standard drink. A logical extension of this would be that energy drinks not presently defined should contain at least 25% more calories than a standard drink.

The additional calories to achieve this level are often added as maltodextrins; their low sweetness does not make the drink cloying.

13.4.3 Functional drinks or nutraceuticals (see also chapter 12)

All food and drink has a function – providing nutrition – and should also give pleasure. Increased awareness of the additional health benefits conferred by some ingredients has allowed these properties to be exploited in the form of functional drinks, also known as nutraceuticals. These are drinks that claim to have health-giving properties, and they originated in Japan in 1988. In the UK, they must avoid any medicinal claims in order to be classified as foodstuffs. Such claims are specifically excluded by the UK Food Labelling Regulations, 1996, which prohibit 'a claim that a food has the property of preventing, treating or curing a human disease or any reference to such a property'. As for energy drinks, any claim that is made must be capable of substantiation.

Functional drinks continue to be very successful in Japan, and the major ingredients exploited there are fibre, calcium, iron and oligosaccharides. Label text refers to beneficial effects on the digestive system, bones, teeth and other parts of the body, as appropriate. In Europe, these drinks are only beginning to be developed. Examples that are available in the UK include:

- *ACE drinks*, containing beta-carotene (Vitamin A precursor) and Vitamins C and E. These materials are antioxidants, and there is evidence that eliminating free radicals in the body will protect against cancer and cardiovascular diseases, particularly in older people. Suggested levels to achieve beneficial effects are, per day, 6–20 mg beta-carotene, 100–150 mg Vitamin C, and 60–100 mg Vitamin E.
- *Fibre drinks*, containing both soluble and insoluble fibres (non-starch polysaccharides). These drinks add bulk to the diet and increase stool weight (a low stool weight is associated with an increased risk of bowel cancer and gallstones). Fibre also binds cholesterol (a risk factor for heart disease), thus reducing its adsorption from the intestine. To support a claim, the quantity of

drink consumed in a day must provide at least 3 g of fibre (i.e. one-sixth of the recommended daily allowance of 18 g).

13.4.4 Powder drinks

Powder drinks form an insignificant part of the UK market, but have substantial sales in the USA and some other parts of the world. Their formulation closely mirrors that of liquid soft drinks, with the exception of water content. The components that do differ are:
- flavouring and clouding emulsions which, after conventional preparation, are spray-dried;
- fruit materials, which are spray- or freeze-dried.

13.4.4.1 Powder drink manufacture

Production must be carried out in an area of controlled (low) humidity. To prevent separation after blending, all components should be as similar as possible in particle size and bulk density. The ingredients must be sieved before addition, and it is often preferable to prepare a pre-blend of some ingredients, in order to facilitate the addition of small mass items and mixing in. Mixing is carried out in ribbon blenders, and can be on either a batch or a continuous basis. Packaging is in sachets or jars.

Long-term stability is better than for liquid drinks, and powder drinks are ideal for vitamin fortification, because of their slow decay rate in the absence of water and without ingress of air. Anti-caking agents may be sometimes incorporated.

Effervescence can be achieved by the incorporation of sodium bicarbonate, but care needs to be taken to avoid excessive levels of sodium and saltiness.

Normal dose rates are about 100–120 g of powder to make one litre of sugar sweetened drink, and as little as 10–20 g for low-calorie drinks and unsweetened drinks, where the consumer adds sugar to taste.

13.5 Dispensed soft drinks and juices

13.5.1 Introduction

Dispensing systems are used in licensed and catering outlets as a convenient, rapid and economical way of serving ready-chilled drinks without the need to handle bottles, cans or other unit packages.

There are two basic systems – pre-mix and post-mix:
- Pre-mix: the outlet receives bulk stainless steel containers (tanks) of carbonated soft drink (approx 18.75 US gallons). The drink is dispensed by CO_2 top pressure via a chiller.
- Post-mix: the outlet receives a concentrated syrup in non-returnable packaging, usually a 10 litre 'bag-in-box' (BiB). The syrup is diluted with chilled carbonated water at the point of dispensing.

13.5.2 Pre-mix and post-mix compared

The two systems offer different benefits:
- Pre-mix: control of product quality remains with the manufacturer, and all the water used is treated.
  - requires cheap and simple equipment at the outlet;
  - gives typically 94 × 20 cl servings per tank;
  - distribution economies are relatively small;
  - containers are returnable, and washing for re-use needs care to ensure that valves are cleaned and do not leak.
- Post-mix: relies on the quality of water at the outlet.
  - product composition depends on the settings of the dispensing equipment;
  - requires a carbonator as well as a chiller at the outlet;
  - gives typically 320 × 20 cl servings per 10 l syrup bag;
  - distribution economies are substantial;
  - uses non-returnable packaging and, for 'bag-in-box' systems packaging, stocks take very little storage space;
  - requires a simplified manufacturing process at the factory;
  - aseptic packaging of syrups is an option.

Pre-mix systems are best suited for manufacturers operating in small distribution areas, and for outlets with relatively low throughputs.

In the UK, where currently two manufacturers dominate the industry, virtually all dispensing is by the post-mix system.

13.5.3 Equipment

A typical post-mix unit is shown schematically in Figure 13.1. Water is supplied via a pressure reducer and non-return valves to a carbonator. If the water quality requires it, particulates can be removed by fitting a cartridge filter and

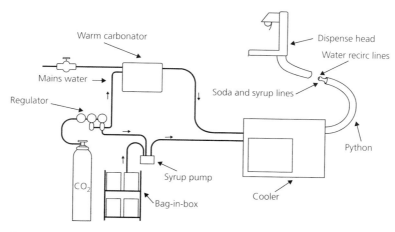

Figure 13.1 Post-mix unit.

chlorine can be removed with a carbon filter. The carbonator is pressurised to a regulated level with carbon dioxide from a gas cylinder.

After carbonation at ambient temperature, the carbonated water passes through a cooling coil in a cooler and on to the dispense head, where the dispensing valves are housed. The cooler is a tank of water with refrigerated walls, on which a layer of ice (the 'ice bank') accumulates. This ice allows a reserve of cooling capacity to be built up during less busy periods.

Syrup is pumped from a bag-in-box (BiB) container or syrup reservoir by a pressure-activated pump, through another cooling coil in the cooler, and on to one of the dispense valves. When the valve is opened to dispense product, pressure in the syrup line drops, and the pump is activated until the valve shuts. The ratio of syrup to carbonated water is adjusted by flow restrictors in the feed pipes to the valve.

The cooler is connected to the dispense head by a 'python'. This is an insulated bundle of tubes, one feed tube for each syrup, and a loop line of chilled carbonated water, continuously circulating from the cooler, around the dispense head and back to the cooler via the carbonator. The circulation is powered by a lightweight plastic pump when water is not being drawn off, and it is carried out to ensure that cold drink can be delivered instantly, even when the dispense head is at a considerable distance from the cooler.

Carbonated water is teed-off to each dispense valve and, when the valve is opened, it feeds annularly round a central syrup flow (Figure 13.2). Mixing occurs as the liquids leave the valve.

An alternative system is to operate with cold carbonation, in which case one, or sometimes two, carbonators are located within the cooler water bath. This system is able to operate at a much lower carbonation pressure (typically 2.7 bar compared, with 5.4 bar for an ambient carbonator).

Cold carbonation gives a more compact installation, and is generally the preferred choice for large installations.

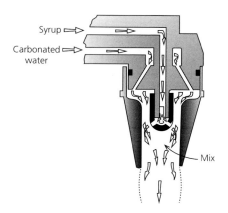

Figure 13.2 Post-mix dispensing valve.

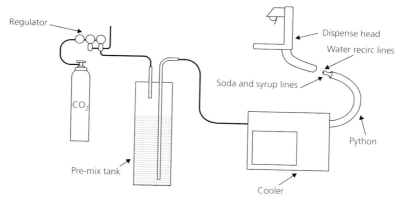

Figure 13.3 Pre-mix unit.

Figure 13.3 shows the much simpler equipment needed for pre-mix dispensing. Top pressure is applied from a carbon dioxide cylinder to a pre-mix tank, factory-filled with carbonated drink. When the dispense valve is opened, the pressure delivers product from the base of the tank, via the dip tube, through the cooler to the valve. Product in the python feed to the valve is kept cold by the pumped circulation of chilled water from the cooler bath.

13.5.4 Outlets
There are several important considerations for outlets:
1 Water quality (relevant for post-mix only): the dispenser must be plumbed in to the mains water supply, rather than be fed from a storage tank. It is essential to guard against carbonated water being forced back into the water main, should water pressure drop. This is usually achieved by fitting two non-return valves in series.
2 Sales volume: the rate of sale of drinks influences the sizing of the chiller, and must exceed a critical level to justify making an installation.
3 Distance between chiller and dispense head: the rate of heat gain in the python is obviously proportional to its length, so a larger chilling capacity is required when distances are long.
4 Large new outlets: dispensing arrangements should be planned from an early stage, to allow concealed pipe runs and proper planning of locations.

13.5.5 Hygiene
A rigorous hygiene programme must be operated at the outlet, with daily cleaning of the dispensing valves and drip trays, and of the disconnects, when syrup containers are changed. Periodic sanitation of syrup lines is required. This must be done more often when fruit drinks or juices are dispensed.

Hygiene training must be given by the soft drinks company, backed up with simple instruction charts fixed near the equipment.

13.5.6 Post-mix syrup formulation

With certain exceptions the same formulation can be used for a dispensed drink as is used when it is bottled or canned. The key differences are:

1 *Pulp*. No matter how fine it is, any pulp in a carbonated post-mix syrup will inevitably lead to a blockage in the system, for it will rise to the top in the container and then enter the valve in a concentrated form. Accordingly, fruit drinks must be reformulated for dispensing by replacing conventional concentrated juices and comminutes with either clarified or low-pulp concentrated juice. A fortification of the flavour system may be necessary, to compensate for the loss of character associated with clarification.
2 *Preservation*. The dispensed product will have a shelf life of perhaps 20 minutes, at most, so the syrup needs only to contain sufficient preservative to maintain its own microbiological integrity.
3 *Dispensing accuracy*. Typically, a dispenser will be capable of producing drinks with an accuracy of ± 5% for both water and syrup supplies. When calculating ingredient additions, the possibility that these tolerances will operate in opposing directions should be taken into consideration, as well as the syrup tolerance.

Example:
Formulate a post-mix syrup with a ratio of 1 : 5.4 for a product containing the maximum amount of sucralose to meet a legal limit of 300 mg/l.

Nominal syrup ratio is 1 volume syrup plus 5.4 volumes water (15.625% syrup). Maximum concentration occurs when syrup quantity is +5% of target and water quantity is −5% of target − that is, maximum drink composition is 1.05 volumes syrup, plus 5.13 volumes water (16.99% syrup). Thus, to ensure a maximum sucralose content of 300 mg/l in the drink, the syrup must contain a maximum of $(300 \times 100)/16.99$ mg/l of sucralose (i.e. = 1765 mg/l).

Assuming that the syrup is also manufactured to an accuracy of ±5%, then the target sucralose concentration must be $(1765 \times 100)/105$ mg/l of sucralose (i.e. 1680 mg/l).

At a target ratio of 1 plus 5.4, 1680 mg sucralose per litre in the syrup would give 262 mg/l in the drink.

4 *Carbonation*. This will be effectively identical for all heads on any dispenser, and is conventionally set at 3.5 volumes. This must be kept in mind when formulating dispensed drinks, which are usually carbonated to a lower level (e.g. orange drink, normally about 2.5 volumes).
5 *Viscosity*. Syrup viscosity influences its flow rate through a dispensing valve so, if a formulation change significantly alters viscosity, then all dispensers will have to be reset.

13.5.7 Post-mix syrup packaging

Modern systems use BiB packaging for syrups. A double bag is used: the inner bag is blended polymers (for example LDPE and EVA) and the outer protective bag is a laminate, with an inner layer of metallised polyester sandwiched between LDPE. The bags are sealed with a coupling valve, designed to connect only to the drink manufacturer's dispensing equipment and, after filling, are dropped into cardboard outer boxes.

Earlier systems used syrup packed into 5 litre HDPE jars, which served to fill reservoirs coupled to the dispenser, giving ample opportunity for contamination, dilution, or for use of syrups from other manufacturers ('pirating').

13.5.7.1 Post-mix syrup production

Syrup manufacture follows the same procedure as bottling or canning syrups, after which it is packed off.

13.5.7.2 Product quality at the outlet

There are four key criteria:

1 Concentration: measured either by a 'Brixing cup', which checks volumetrically the volumes of syrup and water delivered by a dispensing valve, or by hand refractometer. In the latter case, the refractometric solids of the actual syrup being used should first be measured. so that allowance can be made for the syrup inversion that will have occurred since manufacture.
2 Taste and flavour: to check the cleanliness of the system and filters.
3 Carbonation: measured by collecting dispensed product in a container with a sealable lid incorporating a pressure gauge.
4 Temperature: normally targeted at 4.5°C, measured in a glass pre-chilled by filling and emptying before taking the test drink.

13.5.7.3 Uncarbonated systems

1 Bowl dispensers: these are very simple devices for dispensing diluted squashes. A Perspex bowl fitted with a chilled base and a paddle stirrer is filled with squash and water, and product is dispensed from a tap. Usually the squash is purpose-made for catering outlets and is double normal strength (i.e. for dilution at 1 + 9, rather than 1 + 4).
2 Juice dispensers: these operate on a post-mix basis, using as a 'syrup' aseptically filled concentrated juice (diluted to a suitable viscosity). Sometimes, a low level of sulphur dioxide is added to protect juice in the dispense valve which, for citrus juices, must be modified to cope with high pulp content.

One or more valves on a carbonate dispenser can be modified to handle juice, replacing carbonated water with chilled water from a separate cooling coil in the chiller.

Alternatively, a purpose-built dispenser can be used. One counter-top system stores a bag-in-box (BiB) container of concentrate in a refrigerated compartment above a water chiller. Concentrate is transferred directly to the dispense valve by a peristaltic pump acting on a tube that is replaced with every BiB. This ensures that the only permanent part of the equipment contacting the juice is the valve.

13.6 Ingredient specifications

In this section, the purpose and preparation of ingredient specifications are discussed.

13.6.1 Why have specifications?

A specification serves to define the ingredient being purchased and should form part of the contract between buyer and seller. For example, it is not sufficient to order 1 tonne of citric acid: the order should be for '1 tonne of citric acid, anhydrous, to our specification reference…'. In the UK, a properly drawn up and agreed specification can also form part of a due diligence defence against a prosecution under the Food Safety Act, 1990.

13.6.2 What a specification should include

There are three key aspects that should be included in all specifications:
1. It should describe the item – including, if appropriate, a quantified definition of its functionality.
2. It should specify the required purity or composition of the item, and the maximum acceptable levels of contaminants (organic, inorganic and microbiological).
3. It should identify how the item will be packaged and labelled, how it should be stored, and its minimum (and maximum) shelf life or retest date.

For fruit juices, this is confirmation of authenticity and freedom from adulterants.

13.6.3 Preparation of a specification

A good starting point is always the supplier's own specification, which can be expanded to include requirements specific to the required end use (for example, particle size for powdered drink ingredients).

Additional information sources include:
1. The various European Union Directives specifying criteria of purity:
   - 95/31/EC Sweeteners for use in Foodstuffs
   - 95/45/EC Colours for use in Foodstuffs
   - 96/77/EC Food Additives other than Colours and Sweeteners for use in Foodstuffs
2. The various Pharmacopoeias:
   - British
   - European
   - US

3 The Food Chemicals Codex, 4th edition.
4 The Merck Index.

The specification must be discussed and agreed individually with all potential suppliers, who must be provided with a copy of the final document.

Although the preparation of the specification appears to be a straightforward process, there are pitfalls that must be considered:

1 When the supplier is not the manufacturer, but is an agent buying on the open market, is the supplier's expertise sufficient to ensure that all sources are capable of meeting the specification?
2 When the supplier manufactures to a standard grade and will not deviate from it, is the required deviation essential and realistic? If so, an alternative supplier must be found.
3 Seasonal variations influence fruit material availability. It is generally possible to source fruit materials to quite tight specifications, but this can incur a heavy cost penalty for blending or selection. Setting as wide a tolerance as possible saves money and adds flexibility in poor seasons.
4 Specified parameters must be unambiguous, and test methods must be defined – for example, is 'Brix' to be measured by density or by refractometer? Is it a corrected value or not? Subjective parameters such as aroma and flavour are particularly difficult, and the goodwill and cooperation of the supplier is important. 'Matches previously accepted delivery' can lead to a gradual drift in quality, and 'Matches original sample' relates to an ageing and deteriorating specimen. Many disputes involving ingredients arise because supplier and user employ different test methods.
5 It must be stipulated that no changes to the material or its manufacturing process may be made without prior notification.

13.6.4 Supplier performance

A specification agreed with a supplier is not a guarantee that all deliveries will comply fully with it and arrive in full and on time. A Certificate of Analysis adds confidence and, if the supplier is accredited to the ISO 9000 series, the British Retail Consortium (BRC) Technical Standard or a national equivalent it shows that the quality systems have been independently audited and found to be acceptable. However, there is no substitute for regular performance reviews backed up, as appropriate, by audits of the manufacturing facilities and quality systems (including incoming raw materials). This is a very onerous and time-consuming exercise, and is well beyond the resources of most soft drink manufacturers.

A Food Safety Act infringement caused by a substandard ingredient can be countered with a defence of 'due diligence'. This requires the defendant to establish that 'all reasonable checks' were made to ensure the ingredient was satisfactory and capable of performing as required. Case history is limited, but the operative word for small companies must be 'reasonable'.

13.7 Complaints and enquiries

Tactful and sympathetic handling of communications with consumers will pay dividends. Complaints can often be defused with minimum loss of goodwill, and promptly answered enquiries will generate it. It has been said the performance of a company is best judged by its complaint-handling performance

13.7.1 Complaints

No matter how much care a manufacturer takes, it is inevitable that occasionally consumers will believe they have cause for complaint about a product. It is often said that fewer than one in ten people who are dissatisfied with a soft drink bother to make a complaint, but that most of them will tell their friends and family.

Complaints reach the manufacturer either directly from the consumer, via a retailer or from a government agency (in the UK either a Trading Standards Officer (TSO) or an Environmental Health Officer (EHO)).Occasionally consumer complaints will arise in the form of a solicitors' letter Most complaints fall into one of the following categories:

1 Sub-standard composition:
   - ingredients omitted, added twice or the wrong ingredient added;
   - syrup over- or under-dosed.
2 Sub-standard formulation:
   - label claim not met (e.g. a vitamin claim or 'no added preservatives').
3 Deteriorated product:
   - improper storage (e.g. too hot or in direct sunlight);
   - aged product;
   - loss of carbonation (PET bottles or faulty closures);
   - microbiological spoilage (fermentation, mould growth or bacterial spoilage);
   - metal pick-up (old or damaged cans).
4 Contaminated product:
   - accidental (glass in product, inadequate washing of returnable bottles) microbial contaminants or chemical residues;
   - deliberate (extortion or grudge).
5 Damage, illness or injury caused by product:
   - staining of carpets or textiles after spillage or leakage;
   - glass injuries from exploding or dropped bottles;
   - internal injuries from glass in product;
   - illness due to consumption (contamination, revulsion due to off-flavours, high metal content; pathogens unlikely since pH generally too low).
6 Personal feelings:
   - dislike of the product;
   - objection to an ingredient used.

The nature of the complaint dictates the appropriate response, which could range from a conciliatory letter ('I am sorry you do not like our product...') to a full public recall of product, and cessation of production until a fault is resolved (for example, a major incident of fermentation in a glass bottled product). In all cases, the cycle of events must be to understand the problem, to identify the cause, to decide the response and then to take appropriate action.

Experience shows that for any complaint raised by an enforcement authority, a response (even of a holding nature pending investigation) within one working day should be regarded as essential.

Increasingly nowadays, a consumers will telephone or email a complaint, often very angry, and always expecting an immediate resolution of the issue. A calm, polite and measured response is called for, without waffle or technical jargon.

It is an excellent maxim always to consider the situation from the complainant's point of view. If the product is clearly at fault, then an apology should be offered; this often serves to disarm a complainant expecting to do battle. Note, however, that admitting liability should be avoided, as it restricts negotiations if legal proceedings ensue, and may invalidate liability insurance.

All complaints should be dealt with as rapidly as possible, and a telephoned response to a letter not only saves time, but gives an opportunity to impress the complainant that the matter is being taken seriously.

Threats to 'take it to the newspapers/police/public health' can be made when a complainant is very upset, or feels that compensation should be offered and wants to talk-up the amount. In such cases, attempts at dissuasion are likely to be counterproductive. It is generally best to agree that the complainant is at liberty to take this action, and to suggest the most appropriate body (EHO or TSO), but also to point out that it will inevitably delay investigations.

A typical complaint should be processed through some or all of the following stages:

1 Receipt of complaint:
   - note date of arrival;
   - Send acknowledgement if investigation is likely to be lengthy.
2 Follow up if full information not provided:
   - nature of complaint;
   - product and pack;
   - production code and time;
   - where and when purchased.
3 Initial assessment and categorisation:
   - nature of fault;
   - degree of seriousness;
   - extent of fault (specific to one package, to a quantity of badly stored product or to part or all of a production batch).
4 Is there a need to obtain the complaint sample?
   - is there a parallel EHO or TSO investigation?
5 Check with production site:
   - were there any abnormalities noted at the time?
   - was it the start or end of a batch?

6 Assess product if available:
   - 'tastes nasty' could be caused by microbial spoilage, adverse storage, compositional faults or contamination.
7 More detailed investigation if appropriate:
   - Chemical and/or microbiological analysis of drink
   - Physical and/or chemical examination of foreign bodies
8 Brief legal advisor or insurers if appropriate.
9 Formulate internal action to be taken, if any:
   - modification of product formula;
   - changes to factory procedures;
   - product recall.
10 Prepare and issue response to complainant.

The final response to the complainant should include a simple explanation of the cause of the fault, what is being done to rectify it, what steps (if any) are being taken to prevent a recurrence, and an apology.

13.7.2 Enquiries

Consumers are increasingly more curious about the composition of the soft drinks they buy, and are also more concerned about health and dietary considerations. Many soft drink companies now operate telephone consumer response services, and some also have pages on the internet. These make enquiries much easier for the consumer, and undoubtedly encourage communication with the manufacturers, which has benefits for both parties. The consumer has an easy route to obtain product information and to enhance his relationship with a favourite brand of drink, and the manufacturer can build on the consumer's loyalty and establish a database of known consumers for subsequent consumer research or marketing activity.

Responding to enquiries can be speeded up and handled by non-technical staff if an information database is established. Typical product enquiries for which answers should be available are:
- How many calories are in…?
- Which of your products are suitable for:
  - vegetarians?
  - diabetics?
  - Muslims?
- Why do you use
  - preservatives?
  - sweeteners?
  - 'additives'?

More general enquiries tend to originate from the education system, and it is interesting to speculate how many GCSEs the average consumer response operative has earned on behalf of enquiring young consumers.

The predictable 'How do you make soft drinks?' is readily answered with some publicity material. In the UK, there is some excellent published material available

from the British Soft Drinks Association, but responding to more advanced queries can be time-consuming. Examples of more complex questions are:
- 'My A-level project is about Vitamin C in foods: how much Vitamin C is in your soft drinks, and what is the best analytical method for checking?'
- 'For my final year degree project, I am researching the contribution of soft drinks to the diet. Please advise nutrition information and consumption patterns for all your drinks, covering *per capita* consumption for the following age ranges…'

13.8 Health issues

13.8.1 Soft drinks and dental damage

Soft drinks have long been blamed for causing damage to teeth, especially among children. In this section, the validity of this is discussed in the context of the widespread use of fluoridated toothpaste, mechanisms of damage are reviewed, and ways of minimising damage are considered.

13.8.1.1 Causes of dental damage

Figure 13.4 shows the structure of a tooth, which relies on the integrity of the hard enamel external layer to protect the vulnerable interior.

There are three routes to tooth damage, all of which involve the breaching of the enamel and all of which have existed since mankind evolved:
- *Trauma*. Accidental or deliberate damage to the tooth structure.
- *Tooth wear*. This occurs in three separate ways:
  1 By abrasion – gnawing on bones or chewing food contaminated by grit (e.g. crudely milled grain). In developed countries, a more frequent cause is

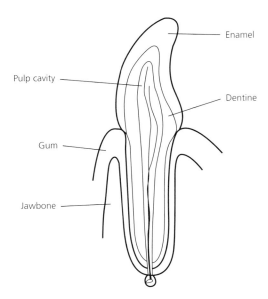

Figure 13.4 Tooth structure.

aggressive tooth brushing, especially when the enamel has recently been softened by acidic foodstuffs.
2 By attrition – tooth-to-tooth contact, caused by poor alignment of the upper and lower rows of teeth, and by habitual grinding of the teeth.
3 By erosion – chemical damage caused by acids. Chewing, sucking or drinking acidic foods, liquids or medications brings the teeth into direct contact with acids that will dissolve calcium and phosphorus from the enamel layer. Additionally, gastric reflux (the regurgitation of acidic stomach contents) can cause erosion on the back surface of the front teeth. This is specially frequent in infants, and in people with eating disorders such as bulimia nervosa.

- *Caries.* Oral bacteria, in association with glycoproteins from saliva, form a sticky coating on the surface of the teeth known as plaque. When carbohydrates are eaten, they provide substrate for the plaque bacteria to ferment into organic acids (mainly lactic acid). Sugars are directly metabolised, and starch must first be digested to sugars by salivary amylase. The acids are retained in contact with the tooth surfaces by the plaque and, as in erosion, can attack the enamel by demineralisation. Once the mouth has cleared, fermentation ceases, and saliva acts to raise the pH at the tooth surface, allowing remineralisation to take place.

Unlike erosion, caries damage is localised to areas where plaque most readily accumulates, namely between the teeth, at the gum margins, and in fissures and irregularities on the tooth surfaces.

13.8.1.2 Can soft drinks cause dental damage?

The answer is clearly 'yes'. Except for low-calorie drinks, they provide a nutrient source of mono- and disaccharides for acid formation and, like all acidic foodstuffs (including low-calorie drinks), they have the potential to cause erosion. However, there are two mitigating factors serving to reduce greatly the damage that soft drinks might at first be thought to cause. These are:

1 Their very brief residence time in the mouth. Unlike sucked sweets or chewed starchy or sugary foods, there is no debris left behind after swallowing, reducing available substrate for the oral bacteria.
2 The saliva flow stimulated by their acidity neutralises acidity in the mouth and plaque, and contains calcium and phosphate for remineralisation.

The effect on plaque pH as a consequence of consuming various foods demonstrates these effects clearly (Schachtele and Jensen, 1981). Demineralisation can occur at pH 5.7 or below, so measurements of the time plaque pH remains below 5.7 after consumption of different foodstuffs allows ranking in order of potential tooth damage:

Least damaging:	cheddar cheese
	skimmed milk
	soft drinks
	potato crisps
Most damaging:	cake

13.8.1.3 Fluorides

The introduction of fluoridated toothpaste during the 1970s caused a much greater than expected improvement in the dental health of the nation, which far outweighs dietary-initiated tooth decay in those that use it. There are now very few brands of toothpaste on sale in the UK that do not contain fluoride.

Fluoridated drinking water is currently available to only 12% of the UK population, a figure unlikely to grow, due to strong opposition from pressure groups opposed to mass medication. Fluoride protects teeth in four ways (Shellis and Duckworth, 1994):

1 It reduces mineral solubility.
2 It inhibits mineral dissolution.
3 It promotes remineralisation of the enamel.
4 It inhibits acid production by plaque bacteria.

Figure 13.5 shows how the incidence of tooth decay amongst 12-year-old children has declined over recent years, despite a significant increase in their consumption of soft drinks.

13.8.1.4 Making soft drinks more tooth-friendly

The beneficial protective effects of fluoride are obviously only available to those who use it, and recent studies have shown that there are substantial variations, both geographically and by social class, in rates of toothpaste use (Davies and Hawley, 1995).

Dentists believe that there is a strong case to formulate soft drinks to minimise the damage they could potentially cause to teeth. They have recently put forward proposals to reduce their erosivity and cariogenic properties, and these are:

1 a reduction in sugar content, to combat caries;
2 an increase in pH to 5.4 or above, to reduce erosion.

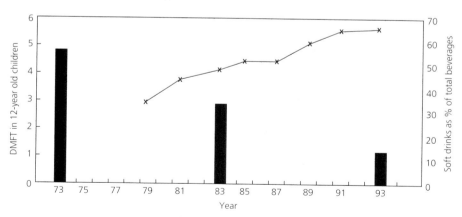

Figure 13.5 Tooth decay and soft drink consumption 1973–93.
Bar chart: decayed, missing and filled teeth (DMFT) in 12-year-old children (O'Brien, 1994).
Line graph: soft drinks as a percentage of the total beverages consumed by 10- to 12-year-old children (data from Nestlé National Drinks Surveys).

Practical steps against caries
Sugar-free drinks are widely available, and are targeted at all age ranges, rather than just at slimmers (a reduction in sugar content would have little effect; it is its total absence that is necessary).

Practical steps against erosion
The proposed pH target is not sensibly achievable. pH 5.4 is generally recognised as the cut-off point for erosion but, unfortunately, this is far higher than the pH barrier of 4.5, above which pathogenic organisms can survive and grow. Any soft drink with such a high pH would need to be retorted – patently not a practicable proposition, especially for carbonated drinks. Fruit juices and fresh fruit all have pHs below 5.4, and cannot be modified. Typically, soft drinks have a pH in the range 2.4 to 3.5, and it would be possible to move most of them towards the higher end of this range, although this would still be well below the target value.

It has been shown that the titratable acidity of a drink is a better indicator of its erosivity than its pH (Grenby *et al.*, 1989). However, significantly reducing acidity would have a much greater effect on the character of a drink than would buffering it to a modestly higher pH.

Given that their normal consumption is unlikely to damage teeth, the best approach to reducing the possibility of erosion by soft drinks and juices would be to promote guidelines for consuming them in the best way, namely:
- Swallow drinks rather than sipping them over prolonged periods.
- Do not swill them around the mouth.
- Do not drink them last thing at night.

13.8.2 Effect of colourings and preservatives
Synthetic food colourings and benzoic acid have often been associated with largely anecdotal evidence of causing hyperactivity in children. Food colourings and preservatives that are permitted food additives in the United Kingdom and the European Union are among the substances most tested for adverse effects. Safety assessments are based on reviews of all toxicological data, including testing on animals and human subjects. From the available data, the maximum level of an additive that has no demonstrable toxic effect is determined. This is called the no-observed-adverse-effect-level (NOAEL), and it is used as the basis for establishing the 'acceptable daily intake' (ADI) figure for each colouring, preservative or other additive. The ADI is assumed as the amount of the substance that may be safely consumed in the diet daily, over the lifetime span, without adverse effect on health.

Set against this toxicological data is a large volume of mostly anecdotal information and claims of hyperactivity in children being caused by food colourings. It is likely that many parents and guardians can relate to the effects alleged to be caused by the colourings. These alleged effects relate almost exclusively to synthetic colourings, and not the natural colourings that are now widely used on a '*quantum sufficit*' (q.s.) basis.

The claims of causation of hyperactivity relate mainly to the following colours, which are sometimes referred to as the ' Southampton six' as a result of work carried out at the University of Southampton. These colourings remain on the permitted list, but future restrictions could be applied.
- E 102 Tartrazine
- E104 Quinoline yellow
- E110 Sunset yellow
- E122 Carmoisine
- E124 Ponceau 4R
- E129 Allura red

Similar anecdotal accusations have been made about the presence of benzoic acid and benzoates, despite the significant technological benefit of antimicrobial activity.

Consumers are made aware of the presence of these colourings by their appearance in lists of ingredients. Despite these substances having formal approval by most regulatory authorities, they have effectively been blacklisted by many institutional and commercial bodies, and are now almost completely excluded from use in soft drinks in the UK and EU. They remain in wide use in many other countries.

Media concern arises from time to time, usually in response to public concern over a specific incident, and the issues of the alleged adverse effect of colourings are again placed before the public.

13.8.3 Obesity

Obesity is now widespread in most developed societies, and is of increasing concern to public health authorities worldwide. The most widely used definition of obesity is now that of Body Mass Index (BMI), which is obtained by dividing body mass by the height squared. An index of more than 30 kg/cm^2 is now considered to indicate that an individual is obese. The condition is a primary potential cause of serious illnesses, including type 2 diabetes, cardiovascular diseases and various cancers.

Carbohydrate intake, and particularly sugars, is considered to be a significant contributory factor. Since many soft drinks and most fruit juices contain around 10–12 % sugars, the industry has often been cited as a major contributor to obesity. In the USA, it has been claimed that carbonated soft drinks are the largest source of calories in the average diet, and there is pressure in several countries to introduce sugar tax on products such as soft drinks. In the United Kingdom, just over 60% of soft drinks are either low calorie or contain no added sugar.

In a balanced approach however, it is important to recognise that there is no scientific evidence that any one sector of the food and beverage industry is the sole contributor to obesity. The soft drinks industry has taken steps to reduce sugar content in many products, both by direct reduction and by the use of intense sweeteners. In some instances, reduction of portion sizes (e.g. 330 ml reduced to 250 ml) is considered a useful approach.

The need to improve consumer education is also very important. This is particularly important for fruit juices, where the naturally occurring sugars present are sometimes confused with added sugars. The other merits of fruit juice consumption, for example in the UK public health campaign, where the benefits of portions of juice are considered as 'one of your five a day' (portions of fruit and vegetables), needs to be emphasised.

13.9 Alternative processing methods

Many soft drinks, and all fruit juices, other than those designated as very short shelf life 'freshly squeezed' products, need pasteurisation to eliminate or reduce spoilage organisms and enhance shelf life. The usual way to achieve this is by pasteurisation, using appropriate equipment employing steam or hot water as the heating medium. Even the use of very short periods of pasteurisation (e.g. 15–20 seconds at around 90°C) is considered to have a slight effect on reduction of apparent freshness of taste. Alternative processing techniques are available and, of these, the most promising are the use of microwaves as a heat source, and high pressure.

13.9.1 Microwave pasteurisation technology
13.9.1.1 Introduction
Microwave volumetric heating (MVH) is associated with significantly more efficient microbiological kill and, when combined with rapid cooling, it minimises the damage to nutrients and other functional components such as flavour, colour, texture and micronutrients. These significant advantages, together with other processing and supply chain benefits, position microwave volumetric heating as a transformational tool for the beverage industry.

13.9.1.2 Conventional thermal processing – the limitations
Conventional thermal pasteurisation processes rely on the transfer of heat from a range of heat sources by conduction and convection. This process has two limitations:
- *It is a relatively slow process*. Transfer of heat energy is passed through adjacent molecules in a sequential process with those closest to the heat source being heated up first. In a large volume, the time to get the whole product up to the target temperature can be significant ('come up' time of minutes to hours). This process can sometimes be speeded up by either creating turbulence or by increasing the size and temperature of the heating surface. In most beverage processing operations the use of plate heat exchangers can significantly speed this up.
- *Contact with a hot surface.* To transfer heat into a product, a thermal gradient must be created and therefore the source of heat must be hotter than the target temperature. When the product comes in contact with this surface or

interface, the layer in contact with the interface will be heated beyond the target pasteurising temperature. This can cause fouling, the creation of flavour taints and damage to the functional components.

Many thermal processes used in the manufacture of drink products are subject to guidance which dictates the minimum time and temperature combination to be used, in order to minimise the microbiological risk to the consumer and to obtain the desired shelf life.

13.9.1.3 Microwaves and microwave volumetric heating

Microwaves are electromagnetic radiation with a frequency between radio waves and infrared light – the microwave part of the spectrum. They penetrate most foods almost instantly, causing certain molecules, particularly water, to rotate, creating heat in a process that is known as dielectric heating. The efficiency with which different foods can be heated using microwaves is, therefore, related to the molecules they contain, the shape of these molecules and their resultant dielectric constants. This is well understood and, generally, most foods containing water can be very efficiently heated using microwaves. As a result, the microwave oven has become a standard item in most modern domestic and catering kitchens, and is accepted as a safe, convenient and efficient method of rapidly reheating previously cooked foods and cooking vegetables.

The use of microwaves on an industrial scale has largely been frustrated by uneven heating. Microwaves produced in a classic system are very diffuse, uneven, and only penetrate to depths of 3–6 mm. Thereafter, conduction and convection are still required to dissipate the heat throughout the product. Generally, microwave energy is delivered much faster than conventional methods of heat transfer can dissipate, and overheating of the surface layers occurs.

The last ten years has seen the development of several solutions. Industrial-scale microwave systems designed for the pasteurisation or sterilisation of ready meals in individual trays use additional pressure to achieve a much more even distribution of heat. This has involved using specialist equipment, or the use of a pressure-relieving valve on each individual tray. Several of these systems are now in commercial production around the world, and are being used to produce food products with better flavours and textures. Developmental, bespoke microwave equipment has also been built to process liquids and semi-solid materials. Products such as sweet potato puree have benefited from this process, as they are difficult to cook by conventional means.

Technologies have recently focused entirely on heating liquids, semisolids and suspensions. Developed from first principles, microwave volumetric heating (MVH) is a new method of delivering microwaves instantly and evenly, through the entire volume of a 42 mm tube transparent to microwaves. This allows products flowing through the tube to be instantly and evenly heated on an industrial scale. The unique delivery system focuses the microwave energy on the tube, which enables the microwave energy to achieve a much greater penetration depth than in conventional microwave systems.

Refinements to the system have also overcome the other principal problems that can occur, such as laminar flow and the different absorption rates of different foods. As a result, MVH eliminates the historical issue of hot and cold spots by design. This can be consistently and reliably demonstrated by the high levels of microbiological kill that are achieved with this system. An additional unique feature is that it employs a series of small microwave generators or magnetrons. Not only does this mean that the energy is delivered in relatively small amounts, but also that the system is modular, allowing virtually any size of machine to be built very cost-effectively. Currently, systems processing 100–5000 kg per hour have been developed.

13.9.1.4 Microwave volumetric heating – the practical benefits

Originally designed for processing difficult high protein liquids such as blood, MVH offers several practical benefits:

- *No hot surfaces.* The microwave transparent tube remains at the same temperature as the product inside, eliminating burn on, fouling and flavour taints. The reduction of fouling may also have downstream benefits for cleaning times and frequencies, chemical use and the volume of high BOD/COD waste water.
- *Precise temperature control.* The processing temperature can be very accurately controlled (within 0.1°C for homogeneous liquids), and none of the product is heated beyond the target temperature.
- *Increased yields.* Yield increases of up to 10% have been achieved, due to significantly reduced evaporative losses and reductions in waste through continuous processing.
- *Modular and scalable.* The technology is truly scalable, and allows flexible delivery of energy by varying the number and positioning of the magnetrons.
- *'Plug and play'.* No ancillary equipment, such as boilers and their associated buildings and infrastructure, are required.
- *Small building footprint.*
- *Scale-up is easy.* Products from small development machines are the same as the large industrial machines. Consequently, product development is cheap quick and easy.
- *Renewable energy and reduced carbon footprint.* Carbon footprint reductions of up to 80% are possible, particularly where bulk pasteurisation vessels can be replaced.

13.9.1.5 Microwave volumetric heating – the technical benefits

The technical benefits of MVH are now creating considerable interest. Some of the technical effects of heating with microwaves have been documented historically, but the equipment now produced allows many of these effects to be much more accurately controlled enabling measured and studied.

MVH inactivates microorganisms more efficiently

Because the thermal energy is delivered instantly and evenly throughout the whole body of the flowing liquid, MVH is a highly efficient means of inactivating microorganisms. Challenge tests with common food pathogens and spoilage organisms (or their surrogates), suspended in milk and phosphate buffered saline (PBS), indicate that the microorganisms are killed at lower temperatures. The MVH inactivation curves of a range of vegetative cells is typically displaced to the left of that obtained by conventional thermal inactivation methods (Figure 13.6), and is characterised by a very rapid decrease in cell numbers over a short temperature range. This is also seen as lower D and Z-values. Similarly, spores are very efficiently inactivated, with 6-log reduction typically being achieved between 110°C and 120°C with a holding time of 1–2 minutes.

Researchers working with microwaves over the years have often documented this effect, and there has been speculation about a 'non-thermal effect' of the microwaves on microorganisms. It is now generally accepted that, if there is such an effect, it is very difficult to prove, and the principal mechanism of bacterial inactivation when microwaves are employed is due to the thermal effects (FSA report).

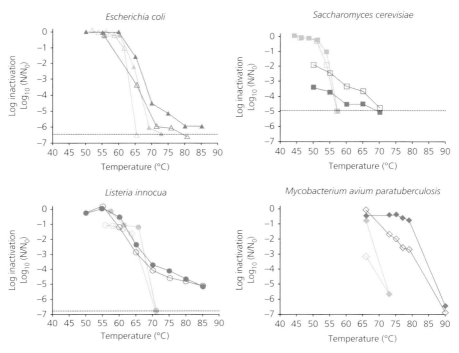

Figure 13.6 Inactivation of vegetative cells.
Key: Solid symbols = cells suspended in whole milk, Open symbols = cells suspended in PBS. Grey line = conventional thermal heating, Green line = MVH heating. Solid horizontal line = limit of detection.

The two immediate, and highly significant, advantages that the rapid MVH inactivation of microorganisms gives to producers is, firstly, an extension (by approximately 50%) to chilled shelf life and, secondly, a greater safety margin and protection against pathogens and potential zoonotic organisms. Working within different sectors, an extension to shelf life has now been realised across a broad range of products, including freshly squeezed fruit juices, dairy-based sweet and savoury sauces, and meat emulsions, such as haggis and black pudding. In a ten-month study funded by Innovate UK, the chilled shelf life of milk and cream was extended by one and two weeks, respectively, without loss of nutrients, functionality and organoleptic qualities. Furthermore, the end of shelf life in these dairy products was determined by a decrease in organoleptic quality, rather than microbiological safety. Following thorough validation, MVH offers the tantalising possibility of producing safe food at lower cooking temperatures, with all the nutritional and environmental benefits that that may bring.

MVH maintains micronutrients and functional components
The impact of MVH at different temperatures (and cooling methods) on the nutrients and functional components present in a range of fruit juices and milk has been validated at both a commercial scale and through trials.

Fruit and vegetable juices processed with MVH, and rapidly cooled in-line, were compared to raw juice over a prolonged period (up to 12 weeks, stored at 4°C). Organoleptic assessment by an untrained panel (made up of at least 20 participants) found that it was very hard to distinguish between MVH processed juices and raw material (stored at −20°C and thawed 24 hours beforehand) (see Figure 13.7). Similarly, taste panels (including triangle taste testing) showed that milk processed by MVH at temperatures up to 85°C was identical to conventionally pasteurised milk. The use of heat recovery/pre-conditioning did not affect these results.

The same is true of temperature-sensitive vitamins such as Vitamin C (Figure 13.8) and anti-oxidants (Figure 13.9). The Vitamin C concentration is reduced by 8–14% on processing, and the rate of decline on storage at 4°C is the same. The level of antioxidants is unchanged after processing, and the rate of decline of storage at 4°C is the same.

13.9.1.6 Conclusion
The true potential of using the new generation of microwave technologies has yet to be fully realised, but it is clear that MVH is a powerful tool that can deliver a wide range of products with improved microbiological safety and extended shelf life, without compromising quality. When combined with versatility, lean manufacturing, yield improvement, the utilisation and reduction of waste, reduced energy consumption, the switch to renewables and supply chain logistics, new doors start to open in food and drink processing. The new generation of advanced microwave technologies are on centre stage to transform the food and drink industry.

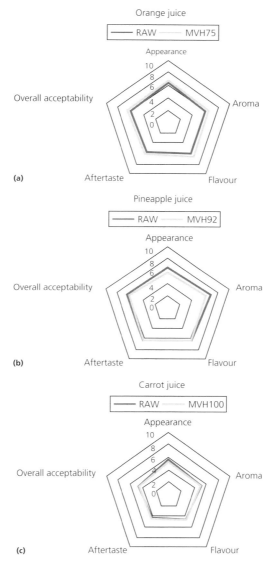

Figure 13.7 Organoleptic assessment of fruit and vegetable juices. **a)** Orange juice processed at 75°C, week four. **b)** Pineapple juice processed at 92°C, week four. **c)** Carrot juice processed at 100°C, week one.

13.9.2 High-pressure processing

High-Pressure processing (HPP) is a cold pasteurisation technique by which products, already sealed in their final packages, are introduced into a vessel and subjected to a high level of isostatic pressure (300–600 MPa/43,500-87,000 psi), transmitted by water.

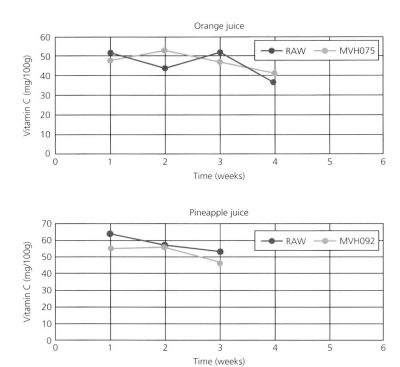

Figure 13.8 Vitamin C in orange and pineapple juice.

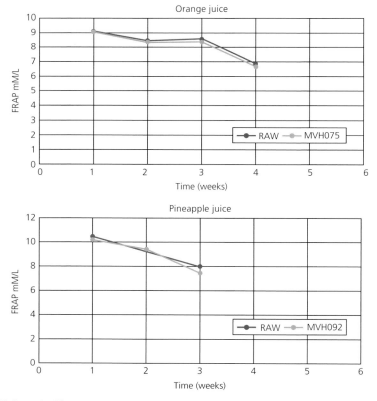

Figure 13.9 Antioxidants (FRAP) in orange and pineapple juice.

Pressures above 400 Mpa, at temperatures of 4–10°C, inactivate the cells of yeasts, moulds and bacteria, and extend the shelf life of the products. The technique is particularly relevant to fruit juice processing, as it allows the production of packaged products with the taste and nutrition of freshly squeezed juice.

The process involves packaging freshly pressed juices and transferring the closed containers to a high-pressure vessel. Pressure is then applied and, after its release, the containers have to be removed from the vessel, labelled, and collated into the final secondary packaging unit.

The batch nature of the process is its principal disadvantage as it limits production quantities and adds cost. HPP juices need to be stored ideally at below 10°C to maintain their fresh taste and appearance.

The shelf life of products processed in this way is not comparable with conventionally packaged items from aseptic operations, but it does compare very favourably with unprocessed juices and short shelf life juices subjected to so-called light pasteurisation.

13.9.3 Irradiation

Food irradiation is a processing technique that exposes food to electron beams, X-rays or gamma rays. The process produces a similar effect to pasteurisation, cooking or other forms of heat treatment, but with less effect on appearance and texture. Irradiated food has been exposed to radioactivity, but does not become radioactive itself.

Food absorbs energy when it is exposed to ionising radiation. The amount of energy absorbed is called 'absorbed dose', and is measured in units called grays (Gy) or kilograys (kGy), where 1 kGy = 1,000 Gy. The energy absorbed by the food causes the formation of short-lived molecules known as free radicals, which kill bacteria and other organisms that cause food poisoning and spoilage. They can also delay fruit ripening and help to stop vegetables, such as potatoes and onions, from sprouting.

Decades of research worldwide have shown that irradiation of food is a safe and effective way to kill microorganisms in foods and extend its shelf life. Food irradiation has been examined thoroughly by joint committees of the World Health Organisation (WHO), the United Nations Food and Agriculture Organisation (FAO), by the European Community Scientific Committee for Food, the United States Food and Drug Administration and by a House of Lords committee. In 2011, the European Food Safety Authority reviewed the evidence, and reasserted the opinion that food irradiation is safe.

The European Commission has agreed a list of irradiated foods that can be freely traded across the European Union (EU). The list is not complete and, at present it, has only one food group: dried aromatic herbs, spices and vegetable seasonings. Until this list is complete, EU member states may continue to allow the irradiation of additional categories of foods.

In the UK, there are seven categories of food which may be irradiated. For each category of food, the 'maximum overall average dose' that can be used is specified in units of kilograys (kGy):
- fruit, 2 kGy
- vegetables, 1 kGy
- cereals, 1 kGy
- bulbs and tubers, 0.2 kGy
- dried aromatic herbs, spices and vegetable seasonings, 10 kGy
- fish and shellfish, 3 kGy
- poultry, 7 kGy.

These categories of food can also be irradiated and used as ingredients in other food products.

However, if irradiated fruit were to be used to produce juices, the juice, when packaged, would still require conventional pasteurisation. Irradiation of the packaged juice would not then be permitted, although many containers for fruit juices, such as the bags used in producing bag-in-box products, are sterilised by irradiation before filling.

References

Costill, D.L. and Miller, J.M. (1980). Nutrition for endurance sport. *International Journal of SportsMedicine* **1**, 2–14.

Davies, K.J.A., Quintanilha, A.T., Brooks, G.A. and Packer, L. (1982). Free radicals and tissue damageproduced by exercise. *Biochemical and Biophysical Research Communications* **107**, 1198–1205.

Davies, R.M. and Hawley, G.M. (1995). Reasons for inequalities in the dental health of children. *Journal of the Institute of Health Education* **3**, 88–89.

Englyst, H.N. and Cummings, J.H. (1988). An improved method for the measurement of dietary fibre as non-starch polysaccharides in plant foods. *Journal of the Association of Official Analytical Chemists* **71**, 808–14.

Fordtran, J.S. (1975). Stimulation of active and passive sodium absorption by sugars in the human jejunum. *Journal of Clinical Investigation* **55**, 728–37.

Fordtran, J.S. and Saltin, B. (1967). Gastric emptying and intestinal absorption during prolonged severe exercise. *Journal of Applied Physiology* **23**, 331–35.

Gisolfi, C.V. and Duchman, S.M. (1992). Guidelines for optimal replacement beverages for different athletic events. *Medicine and Science in Sports and Exercise* **24**, 679–87.

Grenby, T.H., Phillips, A., Desai, T. and Mistry, M. (1989). Laboratory studies of the dental properties of soft drinks. *British Journal of Nutrition* **62**, 451–64.

McCance and Widdowson (2015). *The Composition of Foods*, 7th Edition. The Royal Society of Chemistry, Cambridge, UK.

O'Brien, M. (1994). *Children's Dental Health in the United Kingdom 1993*. OPCS, an imprint of HMSO, London.

Olsson, K.E. and Saltin, B. (1971). Diet and fluids in training and competition. *Scandinavian Journal of Rehabilitation Medicine* **3**, 31–38.

Saltin, B. and Costill, D.L. (1988). Fluid and electrolyte balance during prolonged exercise. In: Horton, E.S. and Terjung, R.L. (eds). *Exercise, Nutrition and Metabolism*, pp. 150–58. Macmillan, New York.

Schachtele, C.F. and Jensen, M.E. (1981). Human plaque pH studies: estimating the acidogenic potential of foods. *Cereal Foods World* **26**, 14–18.

Schultz, S.G. and Curran, P.F. (1970). Coupled transport of sodium and organic solutes. *Physiological Review* **50**, 637–718.

Shellis, R.P. and Duckworth, R.M. (1994). Studies on the cariostatic mechanisms of fluoride. *International Dental Journal* **44**, 263–73.

Vist, G.E. and Maughan, J.R. (1995). The effect of osmolality and carbohydrate content on the rate of gastric emptying of liquids in man. *Journal of Physiology, London* **486**, 523–31.

Wapnir, R.A. and Lifshitz, F. (1985) Osmolality and solute concentration – their relationship with oral rehydration solution effectiveness: an experimental assessment. *Pediatric Research* **19**, 894–98.

Index

Note: Page numbers in *italics* refer to Figures; those in **bold** to Tables.

acceptable daily intake, 93, **95**, 118–122, 386
acesulfame K, **95**, 134, **135**, 246, 247, *247*, 248, 360, 361, 367
   analysis of, 247
acidulants, analysis of, 72, **99**, **128**, 135, 138
ADI *see* acceptable daily intake
adulteration of fruit juices, 10, 133
adverse reactions to foods and food ingredients, 123
alcohol *see also* ethanol, levels in fruit juice volatiles
   limits in soft drinks, 369
alcoholic type soft drinks, 370
algal polysaccharides, 97, 237
American Association for Laboratory Accreditation, 236
analysis of natural sweeteners
   by HPLC, 241
   by refractometry, 232, 240
analytical methods for soft drinks ingredients
   AOAC, 233, 234, 236, 246, 250, 252, 253, 256, 257, 259–261, 270, 273, 275, 277
   British Standards Institute BSI, 236, 273
   European Standards Organisation CEN, 234, 273
   International Fruit Juice Union IFU, 233–234, 240–242, 244, 250–253, 263, 264, 268, 269, 275, 277, 278, 280
anthocyanins, analysis of, 267, 269
anti-oxidants, 20, 29, 194, 392
apple processing yields, 37
ascorbic acid effects on colours, 102
aseptic filling
   closures, 144, 179
   pack types, 2, 31, 48, 136, 141, 143, 144, *201*, 291, 373
   plant schematic, 62, 144, 179

processes, 62, 173
   sterilants, 199 *see also* hydrogen peroxide; peracetic acid
aspartame
   analysis of, 134, 247, 248
   as source of phenylalanine, 4, 134, 246

bacteria
   Acetobacter, 295
   Alicyclobacillus, effects of, 75, 295, 296
   Gluconobacter characteristics, 295
   heat resistant spore forming organisms, 297
   Lactobacillus, 295
barrier plastics, 13, 142
baume, degrees, 132, **132**
benzene in carbon dioxide, 150
benzoic acid and benzoates
   analysis of, 119, 387
   benzene production from, 300, 305
beverage consumption
   charts, 23–24, *23*, *24*
   drivers, 28–29
   share by category, *24*
   share by markets and regions, **25**, 25–26
   trends, 16–23
beverage ingredients
   impurity limits
   standardisation
beverage processing, 388
BFS *see* blow-fill-seal filling
BHA *see* butylated hydroxy anisole
BHT *see* butylated hydroxy toluene
biofilm reduction, 82
biological oxygen demand, 67, 73, 84, **85**, 86, 390
bixin and norbixin, 271
blow-fill-seal filling, 198, 199, 201

Chemistry and Technology of Soft Drinks and Fruit Juices, Third Edition. Edited by Philip R. Ashurst.
© 2016 John Wiley & Sons, Ltd. Published 2016 by John Wiley & Sons, Ltd.

BOD *see* biological oxygen demand
bottled water, market leaders, 21
bottles
   Codd, 91
   glass, 127, 144, 146, 154, 180, 196, 202, 203, 209–211, 218
   Hamilton, 90
   metal, 218
   polyethylene terephthalate, 13, 139, 142, 146, 152, 154, 161, 166, 169, 172, 180, 181, 193, 195–197, 202, 204, 206, 207, 210, 211, 218
   returnable, 85, 181, 188, 190, 213, 226, 380
   selection, 144
Brix acid ratio, 60
Brix, degrees, 11, 59, 129, 132
Brix standard for reconstituted juices, 9, 55
bromate formation, 238
butylated hydroxy anisole, 95, 98, 121
butylated hydroxy toluene, 98, 121

caffeine
   analysis of, 257–258
   effects of, 311
cans
   corrosion of, 72, 198
   dimension nomenclature, 215, *216*
   ends, 198, 215
   manufacturing, 196, 213, 214
   plastic, 142, 218, 226
   shaped, 214
carbohydrates, as ingredients, 126 *see also* individual materials
carbonated beverages, background, 146
carbonation
   chart, *153*, 153–156
   levels in products, 5
   measurement of, 154, *154*, 240
   plant, 150
   volumes Bunsen, 153
carbon dioxide
   characteristics, **149**
   contaminants and impurities, 293
   level required to inhibit microbial growth, 292
   loss from containers, 154, 166, 202
   phase diagram, 147, *147*
   production sources, 132
   quality standards, 149
cariogenic properties of beverages, 385

carotene beta, 13, 109, **114**, 137, 261, 262, 267, 270, *271*, 352, 359, 371
carotenoids
   analysis of, 270, 271
   synthetic, 262, 264
chemical analysis
   by gas chromatography-mass spectrometry (GC-MS), 232
   by high performance liquid chromatography (HPLC), 232
   by high performance liquid chromatography-mass spectrometry (HPLC-MS), 232
   by liquid chromatography (LC), 233
China, market competition, 22
CIP *see* clean-in-place systems
citrus comminutes, 2, *130*, 358, 359
citrus fruit varieties, 38
claims permitted for products, 7
clean-in-place systems
   typical cycle, 169, *170*, *171*, 187
closures
   crown corks, 213
   foil seals, 199, 201
   liners, 90, 145, 212
   metal roll-on (RO) and roll-on pilferproof (ROPP), 140, 203
   plastic, 198, 199, 203, 210–212, *212*, 221, 222
   selection criteria, 198
   sports types, 193, 204, 212, *212*
   vacuum seal, 210–211
clouding agents, 4, **128**
codex alimentarius, 92, 104
coffee, 1, 15, *16*, 20, 21, *24*, 25, 28, 30, 94, 126, 193, 258, 312, 313, 329, 334, 335, 340
cola drinks, 2, 100, 102, 135
cold pasteurised fruit juices, 291
colour contamination with non-approved dyes, 265
colourings
   adverse effects on children, 386 *see also* Southampton 6
   analysis of, in soft drinks, 261–263
   as ingredients, 386–387
   legislation, 3, 113
   natural, 137, 386
   synthetic, 386
colours in soft drinks, assessment of, 236
colours synthetic, characterisation of, **115**, 258, 261, 262, 265, 267

combibloc packs, 142 *see also* laminated packaging
complaint and enquiry management, 380–383
compound ingredients, 105, 137
condensation on containers, 13
control measures for minimising cleanliness, good hygiene practice, 300
Coriolis principle, 156
corrosion of cans, 72, 198
Crown corks *see under* closures
Cryptosporidium, 75, 299
cyclamates, analysis of, 134

Dalton's law of partial pressures, 155
deaeration, 157–158, 171
deterioration of colour and flavour, 2, 95, 113, 121, 126, 141
diet formulations, 240
dilutable soft drinks, 2, 128, **128**, 136–138
dimethyl dicarbonate, analysis of, 135
dispensed soft drinks and juices
    equipment, *373*, 373–375, *374*, *375*
    formulation, 376
    hygiene, 375–376
    outlets, 375
    packaging, 377
    pre-and post-mix systems, 376–378
distribution, 9, 18, 41–43, 108, *109*, 110, 117, 124, 150, 177, 181, 190, 195, 196, 198, 199, 203–205, 207, 218, 219, 222, 223, 373, 389
drying methods for extracts, 322

EDTA *see* ethylene diamine tetra-acetic acid
effluent, clean up and re-use
    contaminants, 85
EIGA *see* European Industrial Gases Association, specification for CO_2
energy drinks, 7, 8, 16, 19–20, 25, 30, 132, 213, 232, 246, 259, 334, 335, 360, 371
environment
    bioplastics, 227, 228
    container weight reduction, 227
    Industry Council for Packaging and Environment, 227
    packging considerations, 225
    production risks, 182, **182**
enzyme linked assay, 243, 253
enzymes
    amylases, 45, 260, 384
    cellulases, 39, 45
    juice clarification, 39, 42, 48–50
    naturally occurring, 43, 44, 72
    pectolytic, 10, 44, 48
EQCS *see* European quality control system for fruit juices
ethanol, levels in fruit juice volatiles, 53–54
ethylene diamine tetra-acetic acid, 121–122, 242, 272
EU Directives
    Colours and Sweeteners, 239, 378
    Drinking water, 68, 238
    Flavourings, 104
    Food Additives, 91–92, 111, 239, 245, 378
    Fruit juices and fruit nectars, 54–55
EUREGAP protocols, 301
European Industrial Gases Association, specification for CO_2, 149, **149**
European quality control system for fruit juices, 276
extract characteristics, specifications, 331–333
extraction processes
    counter current, 47, 52, 106, 107
    decoction, 311, 314, 315
    percolation, 315–316
    supercritical fluid, 123, 318–319
    typical operation, 169
    ultrasonic, 320
extract types
    cost, 329–330, *330*
    dry, 316
    infusions, 314–315
    liquid, 314, 316, 317, 331
    organic, 329, 330
    particle size, 324
    pH, 328
    soft, 314, 316–317, *318*, 319, 323, 329
    solvents, 103, 258, 314, 315, 318, 328
    temperature, 317, 324, 326
    time, 324–325, 326, *326*
    tinctures, 314, 317, 331
extrusion blow moulding, 199, 201, 207

factory layout flows
    materials, 183, *184*
    people, 183, *185*
    process, *182*, 182–183
    waste, 183, *185*
FEMA GRAS listings for flavourings, 105
fibre
    analysis of, 259–260
    description, 359
filling and packaging, 139–140
filling equipment, types of, 180

filling, general principles, 160–171
filling problems, 158, 159
flavonoids, 12–13, 311–312
flavourings
   beverage applications, 106
   emulsions, 107
   essences, 106–107
   as ingredients, 120–122
   legislation, 104–105 see also EU
      Directives, Flavourings
   preparations, 105
   substances, 105
fleshy fruits, 32–33, 39–43, 46
flexible pouch packs, 142, 221–222
floc formation in products, 75, 96–97, 237
Food Safety Act UK, 54, 68, 378, 379
food safety considerations, 122–123
form-fill-seal operations, 5, 13, 142, 180–181, 198, 201–202
form-fill-seal plastic cup containers, 5, 142
French AFNOR for assessing fruit juice quality, 276
fructose syrups, 133
fruit and juice processing for products, 31–32
fruit juice
   adulteration and methods for detecting, 10–12
   aroma volatiles, 51–52
   authenticity, 57 see also under adulteration
   definitions, 8–9
   extraction, 174
   flavour components, 49–50
   permitted additions, 238–239
   quality parameters, 58
   raw material suppliers, 277–278
   storage, 48–49
   validated analytical methods, 233, 246, 250
fruit juice content in soft drinks, methods of assessing, 280–282
fruit juice extraction, 48–50
fruit juice legislation and regulations
   AIJN, 56
   Codex, 54
   EU, 54–56
   Germany, 57 see also RSK system for assessing juice quality
   minimum brix requirements, 55, **55**
   UK, 54
fruit juices and nectars, processing and packaging, 19
fruit juices concentrated, as ingredients, 128–129

fruit juices processing technology, 9–10, 39–47
fruit nectars, 8–9, 54–55
fruit processing plant
   citrus, 36, 38–39
   juice concentration, 49–50
fruit types classification and processing, 32–39
functional drinks
   nutraceuticals, 371–372
future beverage trends, 123–124, 172–173

galenicals, 314
gas laws
   Charles's Law, 153
   Henry's Law, 153
gas permeability data for containers, 209
Geosmin see mycotoxins
Giardia, 75
glucose syrups
   dextrose equivalent (DE), 131, 366–367
   modified, 132–133
glycosides in stone fruits, 12, 40, 134, 327, 338, 339, 345, 350
good manufacturing practice, 62, 105, 186, 191, 237, 294, 322

HACCP see Hazard Analysis Critical Control Points
Hazard Analysis Critical Control Points, 68, 150, 186, 236, 301, 306
haze generation in products, 45, 293–295, 326, 332
HDPE see high density polyethylene containers; polyethylene containers
health issues relating to soft drinks
   colourings and preservatives, 136, 386–387
   dental damage, 383–384
   fluorides, 70, 71, 74, 187, 237, 385
   obesity, 6, 8, 17, 28, 356, 387–388
   Southampton six colours, 262, 387
herbal drinks
   analysis references, 310, 312, 313, 333
   early examples, 310, 312, 313, 333
herbal extract use in beverages
   carbonated and dilutable drinks, 334
   energy and sports drinks, 335
   juices and flavoured drinks, 9, 54, 55, 60, 271, 297, 386
   mineral water based products, 334
herbs in common use in extracts, 113, 311, 314–316, 321, 323–329, 331, 334, 337–350, 352

high density polyethylene containers, 13, 140, 194, 199, 207, 211, 332, 377
high fructose glucose syrup, 132, 133
high pressure processing, synergistic effects with essential oils, 291, 393–395
high shear stirrers, 138
homogenisation, 106, 111, 130, 139, 390
hot filling, 142–144, 177, 194, 197, 198, 206, 210, 211, 293, 303
HPLC analysis, 268–270, 274
HPP *see* high pressure processing, synergistic effects with essential oils
hydrogen peroxide
   as sterilant, 199

identification and interpretation schemes for microorganisms
   molecular identification, 302
   non-molecular methods, 302
   sample isolation, 301–302
   yeast characteristics, 302, **303**
INCPEN *see* Industry Council for Packaging and Environment
Industry Council for Packaging and Environment, 227
ingredient control for extracts, 260–261, 311–313
ingredient specifications for soft drinks, supplier performance, 379
International Society of Beverage Technologists, 152
invert sugar, 131
iron and manganese removal from water, 72, 76
irradiation, 395–396
ISBT *see* International Society of Beverage Technologists
isofumigaclavine *see* mycotoxins
ISO standard 9000, 235, 379
ISO standard 17025, 235–236
isotonic products, 261

JECFA *see* Joint Expert Committee on Food Additives
Joint Expert Committee on Food Additives, 22, 111, 118, 119, 122, 149

kerosene taint in products, 300 *see also* moulds, 1,3-pentadiene producers

labelling regulations EU, essential information, 124
label selection, 3, 10, 54, 57, 88, 105, 123, 136, 146, 190, 194, 224, 225, 271, 281, 291, 316, 325, 327, 331, 336, 359, 371, 380, 395, xvii
laboratory accreditation
   AIJN Expert group, 56, 57, 275, 276, 279
   CLAS, 235
   FAPAS, 235
   ISO 9000 standard, 235, 379
   LABCRED, 235
   UKAS, 236
laboratory equipment, basic quality assurance, 232, 235, 236, 264, 277
laboratory equipment for juice production, 36, 38, 64
Lactobacilli, 75, 295
laminated packaging, 48, 218
laminated paperboard containers
   briks, 1, 196
   composition, 219
   production and filling, 1, 48, 142, 181, 190, 218, 222
limits for impurities
   arsenic, 70, 93, 237
   copper, 71, 80, 82, 93, 237, 300
   lead, 92
limonin, 12, 13

manufacturing operations, soft drinks, 85, 137–139
markets for beverages
   adult soft drinks, 29, 327
   bottled water, 15–18, 20, 25, 298
   carbonated soft drinks, 2, 15, 17–18, 20, 23, 122, 146, 149, 256, 292, 372, 387
   energy drinks, 7, 8, 16, 19–20, 25, 30, 132, 213, 232, 246, 259, 334, 335, 360, 371
   fruit juices and nectars, 54–56, 126, 142–145, 234, 238
   milk and milk flavoured drinks, 15, *16*, 22–23, 25
   natural products, 28, 29, 111, 192, 318
   tea and coffee, 20, 329
   top countries 2009, 27
   top countries 2013, 27
   wine, 15, 22, 89, 99, 117, 280, 282, 312, 321, 329
medlar fruits, 36

methaemoglobinaemia, 98
microbial growth factors, 86, 292, 300
microbiological safety problems
   E.coli and enterococci, 299
   Salmonella, 299
microwave processing technology
   effect on functional ingredients, 28, 120–123, 297, 334
   effect on microorganisms, 49, 56, 70, 75, 116–117, 293
miscellaneous Food Additive regulations UK, 245
missiling of caps, 211
moulds
   associated with poor hygiene, 297
   heat resistant spoilage types, 199, 292, 293, 297, 303, 304, 306
   identification, 298
   1,3-pentadiene producers, 120, 141
   preservative resistant, 295, 300, 303
multiple packs, 52, 223
mycotoxins
   toxins produced, 75, 298, 338, 340

nectars, 8, 9, 15, 54–56, 63, 126, 142–145, 234, 238, 245, 291
neohesperedin dihydrochalcone, 134
new product development, 8, 19, 21, 28, 32, 193, 195, 218, 306, 323
NFC *see* not from concentrate fruit juice
NHDC, *see* neohesperedin dihydrochalcone
nitrogen injection, 202
non-carbonated drinks, manufacturing and problems, 2, 97, 140
non-thermal techniques for soft drink processing, 299, 391
not from concentrate fruit juice, 9, 11, 19, 279, 333
nutrition
   declaration, 4
   five-a-day, 58, 334, 388
   fruit juices, 12
   information, 360–363, 383
   soft drinks, 2, 7–8, 94, 123, 128, 274, 291, 356–358, 371
nutritional components of beverages, 3, 4, 357–360

off-flavours in citrus juices, 62
orange juice, concentrated, 2, 127, 129

osmolality, measurement of, 261, 367–368
oxygen in products, 207, 219, 257, 272, 292–296
oxysporone *see* mycotoxins
ozonolysis, 237–238

PAA *see* peracetic acid
packaging
   consumer interactions, 193
   decoration, 189–190, 193–195, 207, 214, 224–225
   general considerations, 193–195
   glass, 13, 127, 139–140, 144–146, 154, 161, 178, 180, 193–194, 194, 196–198, 202–203, 205, 209, 211–212, 218, 226, 323
   labels, 189–190, 194, 224–225
   plastic, 13–14, 142, 144–145, 180, 188, 194, 196–199, 201, 203, 205, 208–212, 218–219, 221, 224–228, 314, 323, 332
   pre-decorated, 194–195, 198
   role of, 195
   shrink sleeving, 194, 198, 207, 211, 225
packaging machinery, typical layout, 188, *188*
packaging materials, selection of, 195, 198, 223–224
pack or cheese fruit press, 40
pack sleeving, 224
pallet sizes and handling, 223
partial pressures *see under* Dalton's law
pasteurisation
   flash, 138–139, 141, 143, 177–178, 179, 297, 370
   in-pack, 2, 139, 142–144, 177, 179, 181, 370
pathogens in products, 290–291, 391–392
   see also microbiological safety problems
pathogens in water, 67, 73, 75, 84, 87
patulin *see* mycotoxins
PEN *see* polyethylene naphthalate containers
pentadiene, 1,3-*see under* moulds
peracetic acid
   as sterilant, 198, *200*
per capita consumption of juices, 16, 18, 19, 21
PET *see* polyethylene terephthalate containers
pigments, natural, 264, 267, 270
plastics, properties for containers, **208**, 209
polyethylene containers, 13, 140, 194, 199, 207, 211, 377

Index

polyethylene naphthalate containers, 14
polyethylene terephthalate containers
   barrier material inclusion, 205
   bottle manufacture, 139–140, 196
   crystallinity, 204–206
   gas permeability, 208–209
   hot filling of, 206, 211
   mould spores contamination risk, 293
   production, 195–196, 203, 205
   stretching, 204–205
   use of preforms, 188
polypropylene containers, 202, 207
polystyrene use in containers, 13, 202
polyvinyl chloride containers, 13, 207
post mix, 5, 372, 373, *373*, *374*, 377
powder drinks, 372
PP *see* polypropylene containers
pre-mix, 5, 372, 373, 375, *375*
preservatives
   effects on microorganisms, 116–117
   maximum permitted levels, 115–116, 118, 120
process control, 181
processing
   cold filling, 181
   hot filling, 178–179
product blending systems
   batch operations, 176
   continuous, 176
   flip-flop, 176
product fortification
   amino acids, 231
   caffeine, 231
   herbal extracts, 231
   vitamins, 231, 272, 359, 372
protein drinks, 29
PVC *see* polyvinyl chloride containers

quantitative ingredient declaration regulations, 4, 281
QUID *see* quantitative ingredient declaration regulations
quinine
   analysis of, 258
   minimum requirement in tonic water, 128, 257, 366

raw material for herbal extracts, availability, 321–323
RECOUP *see* recycling of used plastics
recycling of used plastics, 194, 227
refractometric solids, 59, 241, 377 *see also* Brix
regulatory issues relating to herbal extracts, 335–336
   Council of Europe classification (Blue Book), 105, 336
roquefortine C *see* mycotoxins
RSK system for assessing juice quality, 176

saccharin
   analysis of, 134
   to include preservatives and colourings, 246–248
saponins, 120–121
secondary packaging, 198, 203, 219, 222–224, 395
sensory evaluation of products, 236
shrink film use on packs, 203, 222
SNIF-NMR techniques for fruit juice authenticity testing, 277–278
soda fountain, 146
soft drinks
   concentrated, 2–3, 5–6
   definition of, 1–2
   ready to drink, 2, 4–5, 20, 28, 30, 111, 315
Soft Drinks Regulations UK 1964, 3, 127, 131, 239, 280 *see under* UK regulations
sorbic acid and sorbates, analysis of, 119–120
sorbitol, 13, 242–244, *243*, *244*
Southampton 6 colours, label warning, 262
   see also health issues
spoilage problems in beverages
   carbon dioxide tolerant and resistant microorganisms, 290, 297
   case studies, 303–304
   general spoilage organisms, 294–298
   pathogens, 290, 291
   risk organisms related to pH, 292, **292**
   sampling for, 301
   sources of, 293–294
   unusual spoilage organisms, 290, 291
sports drinks
   absorption, 365–366
   electrolytes, **364**, 364–365
   energy, 334–335
   formulation, 366–369
   osmolality, 367–368
   powder variants, 369
   vitamins, 365
   water loss replacement, 364
squashes and cordials, 2, 3, 6, 15, 251
stabilisers, 102, 109, 120, 137, 258, 359

stability of extracts, 331–332
stable isotope ratios, 133, 276 *see also* SNIF-NMR techniques for fruit juice authenticity testing
stack burn, 139
Stevia, analysis of, 28–29, 134, 239, 249
sucralose, analysis of, 134, 239, 247–248
sucrose, 5, 7, 11, 59, 109–110, 129, 131–134, 239–242, 248, 260, 292, 298, 357
sugar issues in carbonated drinks, 6, 8, 17–18
sulphur dioxide, analysis of, 117–119, 251–252
superfruits, 29, 232, 333–334, 351
sweeteners, typical levels in fruit and vegetable juices, 239, **240**
syrup preparation for carbonated products, 156, *157*
syrup proportioning, 5, 100, 132–133, 156, *157*, 239, 245

taints in water, 75
tamper evidence, 190, 195, 210, 211
tamper evident closures, 210–211
taurine, analysis of, 259
tea
  anti-oxidants, 20, 311
  flavoured and herbal, 25, 193, 311–312
Tetra Paks, 140, 144 *see also* laminated packaging
thermogenic drinks, 124
tocopherols, 13, 121, 274
Tonic Water, Indian, 128
tristimulus colour measurement, 264
tritium in water, 73
Tryptoquivalins *see* mycotoxins

UK regulations
  Miscellaneous Food Additives 1995, 245
  Soft Drinks 1964, 3, 127, 131, 280
  Sweeteners 1995, 239, 245
ultrafiltration in juice processing, 50, 77
US Food Safety Modernization Act, 235
US Nutrition labelling and education act, 281

vitamin B class, analysis of, 274
vitamin C, analysis of, 12, 58, 61, 127, 242, 249, 272–275
vitamin P, 12
vitamins
  analysis of in soft drinks using immunological procedures, 275

  fat soluble, analysis of, 272–274, 359
  in soft drinks, 272–275, 359
volumes of carbon dioxide, 5, 116

Waste Resources and Action Programme, 227
water
  chemical descriptions, 73–75
  chemical treatments, 77–79
  constituents, **71**, 74–75
  contaminants, 68, 84–85
  customer standards, 68–69
  disinfection techniques, 80
  effect of impurities, 77, 80, 97–98, 148
  filtration, 76–77
  hydrological cycle, 66
  process requirements for beverages, 94–98
  sources, 66–68, 72–75
  testing, 83–84
  treatment by ion exchange, 12, 75–76
  use ratio in manufacturing, 65, 84
water activity of products, 291, **303**
water contaminants
  pesticides, 63, 67, 73, 74, 228, 237, 321, 329
  polyaromatic hydrocarbons (PAHs), 74, 237
Water Natural Mineral and Spring, 69
water quality
  assessment, 237
  environmental impact, 75
  grey water use, 85
  parameters and values, **70–71**, 70–72
  standards, 68–69, **70–71**
whey protein, 8
white milk consumption, 23
whole fruit as ingredient, 6, 36, 46, 128
WRAP *see* Waste Resources and Action Programme

yeasts
  classification characteristics, 294
  effects of spoilage, 290, 294–295
  examples of species in factory environment, 294, **295**
  heat resistance, 292
  nutrient requirements, 290, 291
  synonyms of most important, 295, **296**
  taxonomy, 294–295

Zygosaccharomyces bailii, 290, **295**, **303**